T0208992

Springer Undergraduate Mathematics Series

More information about this series at http://www.springer.com/series/3423

David F. Griffiths · John W. Dold
David J. Silvester

Essential Partial Differential Equations

Analytical and Computational Aspects

 Springer

David F. Griffiths
University of Dundee
Fife
UK

John W. Dold
School of Mathematics
The University of Manchester
Manchester
UK

David J. Silvester
School of Mathematics
The University of Manchester
Manchester
UK

ISSN 1615-2085   ISSN 2197-4144   (electronic)
Springer Undergraduate Mathematics Series
ISBN 978-3-319-22568-5   ISBN 978-3-319-22569-2   (eBook)
DOI 10.1007/978-3-319-22569-2

Library of Congress Control Number: 2015948777

Mathematics Subject Classification (2010): 35-01, 65L10, 65L12, 65L20, 65M06, 65M12, 65N06, 65N12

Springer Cham Heidelberg New York Dordrecht London

Printed on acid-free paper

Springer International Publishing AG Switzerland is part of Springer Science+Business Media
(www.springer.com)

In mathematics you don't understand things.
You just get used to them.

John von Neumann

"Begin at the beginning", the King said, very
gravely, "and go on till you come to the end:
then stop".

Lewis Carroll, *Alice in Wonderland*

Preface

This textbook is a self-contained introduction to the mathematical aspects of partial differential equations. The book is aimed at undergraduate students studying for a mathematics degree. Selected chapters could also be of interest to engineers and scientists looking to develop an understanding of mathematical modelling and numerical analysis.

The material is organised into 13 chapters, with roughly equal emphasis placed on *analytical* and *numerical* solution techniques. The first four chapters provide a foundation for the study of partial differential equations. These chapters cover physical derivation, classification, and well-posedness. Classical solution techniques are discussed in Chaps. 8 and 9. Computational approximation aspects are developed in Chaps. 6 and 10–12. A clear indication is given in each of these chapters of where the basic material (suitable perhaps for a first course) ends and where we begin to probe more challenging areas that are of both a practical and theoretical interest. The final chapter defines a suite of projects, involving both theory and computation, that are intended to extend and test understanding of the material in earlier chapters.

Other than the final chapter, the book does not include programming exercises. We believe that this strategy is in keeping with the aims and objectives of the SUMS series. The availability of software environments like MATLAB (www.mathworks.com), Maple (www.maplesoft.com) and Mathematica (www.wolfram.com) means that there is little incentive for students to write low-level computer code. Nevertheless, we would encourage readers who are ambitious to try to reproduce the computational results in the book using whatever computational tools that they have available.

Most chapters conclude with an extensive set of exercises (almost 300 in all). These vary in difficulty so, as a guide, the more straightforward are indicated by* while those at the more challenging end of the spectrum are indicated by*. Full solutions to all the exercises as well as the MATLAB functions that were used to generate the figures will be available to authorised instructors through the book's website (www.springer.com). Others will be able to gain access to the solutions to odd-numbered exercises though the same web site.

Distinctive features of the book include the following.

1. The level of rigour is carefully limited—it is appropriate for second-year mathematics undergraduates studying in the UK (perhaps third or fourth year in the USA). The ordering of topics is logical and new concepts are illustrated by worked examples throughout.
2. Analytical and numerical methods of solution are closely linked, setting both on an equal footing. We (the authors) take a contemporary view of scientific computing and believe in mixing rigorous mathematical analysis with informal computational examples.
3. The text is written in a lively and coherent style. Almost all of the content of the book is motivated by numerical experimentation. Working in the "computational laboratory" is what ultimately drives our research and makes our scientific lives such fun.
4. The book opens the door to a wide range of further areas of study in both applied mathematics and numerical analysis.

The material contained in the first nine chapters relies only on first-year calculus and could be taught as a conventional "introduction to partial differential equations" module in the second year of study. Advanced undergraduate level courses in mathematics, computing or engineering departments could be based on any combination of the early chapters. The material in the final four chapters is more specialised and, would almost certainly be taught separately as an advanced option (fourth-year or MSc) entitled "numerical methods for partial differential equations". Our personal view is that numerical approximation aspects are central to the understanding of properties of partial differential equations, and our hope is that the entire contents of the book might be taught in an integrated fashion. This would most probably be a double-semester (44 hours) second- (or third-) year module in a UK university. Having completed such an integrated (core) course, students would be perfectly prepared for a specialist applied mathematics option, say in continuum mechanics or electromagnetism, or for advanced numerical analysis options, say on finite element approximation techniques.

We should like to extend our thanks to Catherine Powell, Alison Durham, Des Higham and our colleagues at Manchester and Dundee, not to mention the many students who have trialled the material over many years, for their careful reading and frank opinions of earlier drafts of the book. It is also a pleasure to thank Joerg Sixt and his team at Springer UK.

May 2015   David F. Griffiths
                                                                              John W. Dold
                                                                            David J. Silvester

Contents

Chapter 1
Setting the Scene

Abstract This chapter introduces the notion of a partial differential equation. Some fundamentally important PDEs are identified and some classical solutions are discussed. The chapter motivates the analytical and numerical solution techniques that are developed in the remainder of the book.

Partial differential equations (PDEs) underpin all of applied mathematics and enable us to model practical problems like forecasting the weather, designing efficient aeroplanes and faster racing cars, and assessing the potential returns from investments in financial stocks and shares.

There are two very different questions that will be considered in this book. The first question is practical: can we find a solution to a given PDE problem—either analytically (that is writing down an explicit formula for the solution in terms of known quantities like position and time), or else numerically (that is, using a computer to approximate the continuous solution in a discrete sense). It turns out that analytic solutions can only be obtained in special cases, whereas computers enable the possibility of calculating a numerical solution in any case where the solution makes sense from a practical perspective.

The second question is more subtle, seemingly magical: can we infer generic properties of a solution *without* actually solving the PDE problem? We will show that this is indeed possible. The essential idea is to characterise PDEs into different types so that the solutions have similar properties. A suitable classification is touched upon in this first chapter: it will be developed in the rest of the book.

We will need to establish some notation to begin with. Consider a function, say $u(x, y, z, \ldots, t)$ of several independent variables, perhaps representing the temperature of the air at a given point in space and at a given point in time. The *partial derivative* of u with respect to x is defined to be the rate at which u changes when x varies, with all the other independent variables held fixed. It will be written as

$$\frac{\partial u}{\partial x} \quad \text{or} \quad u_x \quad \text{or} \quad \partial_x u.$$

© Springer International Publishing Switzerland 2015
D.F. Griffiths et al., *Essential Partial Differential Equations*, Springer
Undergraduate Mathematics Series, DOI 10.1007/978-3-319-22569-2_1

Other first-order derivatives, u_y, u_z, u_t, etc., are written analogously. Second derivatives are written in the form

$$\frac{\partial^2 u}{\partial t \partial x} \quad \text{or} \quad u_{tx} \quad \text{or} \quad \partial_t \partial_x u$$

and represent rates of change of first derivatives. Note that in practical applications, u will usually be at least twice continuously differentiable, so the order of differentiation can be commuted and we have $u_{tx} = u_{xt}$.

We now define a PDE having solution $u(x, y, z, \ldots, t)$ to be a relationship of the form

$$F(u, x, y, \ldots, t, u_x, u_y, \ldots, u_t, u_{xx}, u_{yx}, \ldots, u_{tx}, \ldots) = 0, \qquad (1.1)$$

where F is some given function. The *order* of the PDE is the highest degree of differentiation that appears in the expression (1.1).

1.1 Some Classical PDEs . . .

The most straightforward case is when there are just two independent variables, x and t say, and when the PDE is *first order* (that is, no second or higher derivatives of u are present). A simple example is the following:

| **one-way wave** | $u_t + cu_x = 0$ | **(pde.1)** |
| **equation** | | |

which is also known as the *advection equation*. Here $u(x, t)$ might represent the height of a travelling wave at a point x on a straight line at time t, with c a given constant which represents the speed of propagation of the wave. If the speed of propagation is not constant but instead depends on the height of the wave then we have a more complicated first-order PDE:

| **inviscid Burgers'** | $u_t + uu_x = 0.$ | **(pde.2)** |
| **equation** | | |

The physical intuition that motivates using **(pde.2)** as a mathematical model of a real-life flow problem is developed in Chap. 3. A technique for determining analytic solutions of PDEs like **(pde.1)** and **(pde.2)** is described in Chap. 9. The numerical solution of **(pde.1)** is discussed in Chap. 12.

Second-order PDEs are extremely important—they frequently arise in modelling physical phenomena through Newton's second law of motion. The three classical examples of second-order PDEs in two variables are stated below. First, we have

Laplace's equation
$$u_{xx} + u_{yy} = 0. \tag{pde.3}$$

In this case $u(x, y)$ might represent the electrostatic potential at a point (x, y) in a square plate. Look out for (**pde.3**) across the mathematical landscape—it pops up everywhere; from complex variable theory to geometric mappings to (idealized) fluid mechanics.[1] Second, we have the

heat equation
$$u_t - \kappa u_{xx} = 0. \tag{pde.4}$$

Here $u(x, t)$ might represent the temperature in a wire at point x and time t with $\kappa > 0$ a given constant which represents the thermal conductivity. We will expand on the physics that is built into (**pde.4**) in Chap. 3. Third, we have the

wave equation
$$u_{tt} - c^2 u_{xx} = 0. \tag{pde.5}$$

This PDE is sometimes referred to as the *two-way wave equation* to distinguish it from (**pde.1**). The reason for the nomenclature will become apparent in the next section when we construct a solution $u(x, t)$ that satisfies (**pde.1**).

These classical second-order PDEs also have analogues in higher dimensional space: that is when the position is a vector $\vec{x} \in \mathbb{R}^d$ ($d = 2, 3$) rather than a scalar $x \in \mathbb{R}$:

heat equation (in $\mathbb{R}^3$)
$$u_t - \kappa(u_{xx} + u_{yy} + u_{zz}) = 0, \tag{pde.6}$$

wave equation (in $\mathbb{R}^2$)
$$u_{tt} - c^2(u_{xx} + u_{yy}) = 0. \tag{pde.7}$$

In (**pde.6**) the solution $u(x, y, z, t)$ might represent the temperature at time t in a solid block of material at position (x, y, z). Similarly, in (**pde.7**) the sought-after function $u(x, y, t)$ might represent the height at time t of a wave at a point (x, y) on a two-dimensional ocean surface. Time independent (steady-state or equilibrium) solutions of (**pde.6**) and (**pde.7**) satisfy Laplace's equation in $\mathbb{R}^3$ and $\mathbb{R}^2$, respectively.

The following second-order PDE has achieved a certain notoriety in recent years through its use as a model for pricing options in the financial marketplace:

---

[1] The left-hand side of (**pde.3**) is called the *Laplacian* of u and is often denoted by $\nabla^2 u \equiv u_{xx} + u_{yy}$. Some texts use the alternative notation $\Delta u \equiv u_{xx} + u_{yy}$.

**Black–Scholes
equation**
$$u_t + \tfrac{1}{2}\sigma^2 x^2 u_{xx} + rxu_x - ru = 0. \qquad \textbf{(pde.8)}$$

In this setting $u(x, t)$ typically represents the option price at time t and is a function of the stock price x (which also depends on t), with the interest rate r and the volatility σ assumed to be known. The model (**pde.8**) is an example of an *advection–diffusion* equation (because it includes first and second derivatives of x). What is remarkable about the Black–Scholes equation is that the underlying stock price fluctuates unpredictably over time, yet the optimal option price can be readily computed by solving a deterministic PDE.

Third-order PDEs seldom feature as models of physical systems but there is one important exception:

**Korteweg-de Vries
equation**
$$u_t + 6uu_x + u_{xxx} = 0. \qquad \textbf{(pde.9)}$$

It is commonly referred to as the KdV equation and is used primarily as a model for wave propagation. It has analytic solutions $u(x, t)$, called *solitons*, that are very persistent (see Exercise 1.5). Solving (**pde.9**) has led to the design of extremely efficient long distance communication networks.

1.2 ... and Some Classical Solutions

To solve any of the PDEs in the previous section we shall have to integrate functions in multiple dimensions—this is not going to be easy! The alternative strategy is to cheat. Specifically, we can postulate a hypothetical function u, and then differentiate it enough times to see if it does indeed satisfy the PDE being considered. Some specific examples will be used to illustrate this process.

Example 1.1 Find PDEs that have the solution $u(x, t) = x^2 t^2 + A(t) + B(x)$, where $A(t)$ and $B(x)$ are arbitrary functions.

By differentiating $u(x, t)$ we discover the PDEs

$$u_t = 2x^2 t + A'(t), \qquad\qquad u_{tx} = 4xt,$$
$$u_{txt} = 4x, \qquad\qquad u_{txtx} = 4$$

all of which have the same solution u. Moreover, a given function u will satisfy any number of PDEs—just keep differentiating! There is also an endless list of other relationships, such as $u_{tx} = tu_{txt}$ and $u_{txt} = xu_{txtx}$.

Note that two specific differentiations are needed to kill the two functions of integration $A(t)$ and $B(x)$. This has significant ramifications. Clearly there are infinitely

Fig. 1.1 A solution
$u = F(x - ct)$ of the
one-way wave equation with
$F(x) = e^{-(x+1)^2}$ and wave
speed $c = 1$. It is plotted at
intervals of half a time unit
for $0 \le t \le 4$

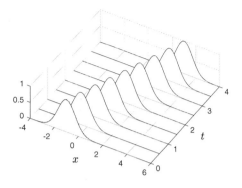

many functions which satisfy the following PDE, arbitrarily chosen from the list
above,

**test
equation**
$$u_{tx} - 4xt = 0. \qquad \textbf{(pde.10)}$$

(Two possibilities are $u_1 = x^2 t^2$ and $u_2 = x^2 t^2 + e^t + x^3$.) To fix the solution to
(**pde.10**)—that is to determine A and B precisely—we need *two* additional pieces
of information. We will return to this issue in the next chapter. ◊

Example 1.2 Find a solution to (**pde.1**), that is, $u_t + c u_x = 0$.

Let $u(x, t) = F(x - ct)$ with $c \ne 0$ (constant). The first partial derivatives are

$$u_t = -c F'(x - ct)$$
$$u_x = F'(x - ct)$$

and combine to give $u_t + c u_x = 0$.

The special form of this solution explains why the wave equation (**pde.1**) is only
"one-way". With $u(x, t) = F(x - ct)$ the initial profile defined by the function $F(x)$
simply translates to the right with speed c (see Fig. 1.1). In contrast the solution of
the two-way wave equation (**pde.5**) can be expressed as the sum of two waves that
travel in opposite directions (see Exercise 1.1). ◊

Example 1.3 Find a PDE satisfied by the special function $u(x, t) = t^{-1/2} e^{-x^2/4t}$.

Calculating partial derivatives:

$$u_t = -\frac{1}{2} x t^{-3/2} e^{-x^2/4t} + \frac{1}{4} x^2 t^{-5/2} e^{-x^2/4t}$$
$$= \frac{1}{4t^2} (-2t + x^2) u(x, t)$$
$$u_x = -\frac{1}{2} x t^{-3/2} e^{-x^2/4t} = -\frac{1}{2} \frac{x}{t} u(x, t)$$

Fig. 1.2 The fundamental
solution of the heat equation
for $0.25 \le t \le 10$

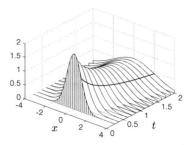

$$u_{xx} = -\frac{1}{2t}u(x,t) - \frac{1}{2}\frac{x}{t}u_x$$

$$= -\frac{1}{2t}u(x,t) - \frac{x^2}{4t^2}\frac{x}{t}u(x,t)$$

so that $u_t - u_{xx} = 0$, see (**pde.4**).   ◇

The solution u shown in Fig. 1.2 has a "Gaussian" profile[2] in space for each time t.
We deduce from the PDE that at locations where $u_{xx} < 0$ (so u is convex in x)
the solution decreases in time at fixed x whereas, when $u_{xx} > 0$ (so u is concave)
the solution increases in time. The points of inflection, where u_{xx} changes sign,
separate regions where the solution is increasing in time from those where it is
decreasing. In this example, the points of inflection are located at $x = \pm\sqrt{2t}$ and the
curve corresponding to the positive root $(x = \sqrt{2t})$ is shown in Fig. 1.2 as a thick
transverse curve.

The solution u in this example is very special. When it is translated through a
distance s as in the following expression,

$$u(x,t) = \frac{1}{\sqrt{4\pi t}} \int_{-\infty}^{\infty} e^{-(x-s)^2/4t} g(s)\mathrm{d}s, \tag{1.2}$$

then it is called a *fundamental* solution of the *heat equation* (see Exercise 1.6) for
$t > 0$. It can be shown that $u(x,t)$ in (1.2) satisfies the initial condition $u(x,0) = g(x)$
for all $x \in \mathbb{R}$ in the limit $t \to 0$.

Example 1.4 Find a PDE satisfied by the special function $u(x,y) = \frac{1}{2}\ln(x^2 + y^2)$.

Calculating partial derivatives:

$$u_x = x(x^2 + y^2)^{-1}$$
$$u_{xx} = (x^2 + y^2)^{-1} - 2x^2(x^2 + y^2)^{-2}$$
$$u_{yy} = (x^2 + y^2)^{-1} - 2y^2(x^2 + y^2)^{-2}$$

---

[2] Also known as a "normal distribution" in probability theory, where it is extremely important—but
that is a topic for another textbook!

Fig. 1.3 A solution
$u = \ln(1/r)$ of Laplace's
equation in $\mathbb{R}^2$, $0.1 \leq r \leq 1$

so that $u_{xx} + u_{yy} = 0$, see (**pde.3**). ◇

Functions that satisfy *Laplace's equation* are known as *harmonic* functions. Thus,
with $r^2 = x^2 + y^2$,

$$u(x, y) = \tfrac{1}{2} \ln r^2 = \ln r,$$

is an example of a harmonic function in two dimensions. We show the negative of this
function $(\ln(1/r) = -\ln r)$ in Fig. 1.3. An important feature of harmonic functions
is easily deduced from the PDE: when the surface $z = u(x, y)$ is convex in the
x–direction $(u_{xx} < 0)$ then it must be concave in the y–direction $(u_{yy} > 0)$ and vice
versa. This shape behaviour is evident in Fig. 1.3.

Note that $u = \tfrac{1}{2} \ln(x^2 + y^2 + z^2)$ does *not* satisfy Laplace's equation in three
dimensions so we have to look for a different special function in this case.

Example 1.5 Show that $u(x, y, z) = (x^2 + y^2 + z^2)^{-1/2}$ is a harmonic function in
$\mathbb{R}^3$.

Calculating partial derivatives:

$$u_x = -x(x^2 + y^2 + z^2)^{-3/2}$$
$$u_{xx} = -(x^2 + y^2 + z^2)^{-3/2} + 3x^2(x^2 + y^2 + z^2)^{-5/2}$$
$$u_{yy} = -(x^2 + y^2 + z^2)^{-3/2} + 3y^2(x^2 + y^2 + z^2)^{-5/2}$$
$$u_{zz} = -(x^2 + y^2 + z^2)^{-3/2} + 3z^2(x^2 + y^2 + z^2)^{-5/2}$$

we find that $u_{xx} + u_{yy} + u_{zz} = 0$. Therefore $u(x, y, z) = 1/r$, with $r^2 = x^2 + y^2 + z^2$,
is an example of a harmonic function in $\mathbb{R}^3$. ◇

A general method for determining analytic solutions to the three classical PDEs
(**pde.3**), (**pde.4**), (**pde.5**) is developed in Chap. 8. Numerical methods for generating computational solutions to all of the PDE models in Sect. 1.1 are described
in Chaps. 10–12. More complicated physical models involve "systems" of coupled PDEs. Examples include Maxwell's equations governing electromagnetism, the
Navier–Stokes equations governing incompressible fluid flow, and Einstein's equations which model the evolution of the universe and the formation of black holes.
These models, though extremely important, are beyond the scope of this textbook.

Exercises

1.1 Given that A and B are arbitrary functions and c is a constant, determine whether or not the given function u is a solution of the given PDE in each of the following cases:

(a) $u(x, y) = A(y); u_y = 0.$
(b) $u(x, y) = A(y); u_{xy} = 0.$
(c) $u(x, t) = A(x)B(t); u_{xy} = 0.$
(d) $u(x, t) = A(x)B(t); uu_{xt} - u_x u_t = 0.$
(e) $u(x, y, t) = A(x, y); u_t = 0.$
(f) $u(x, t) = A(x+ct) + B(x-ct); u_{tt} + c^2 u_{xx} = 0.$

1.2 Find PDEs that are satisfied by each of the following functions:

(a) $u(t, x) = e^t \cos x$
(b) $u(x, y) = x^2 + y^2$
(c) $u(t, x) = x^2 t$
(d) $u(t, x) = x^2 t^2$
(e) $u(x, y) = e^{-x^2}$
(f) $u(x, y) = \ln(x^2 + y^2)$

In each case try to find more than one suitable PDE.

1.3 In each of the following cases (a)–(c), find second-order PDEs that are satisfied by the given function. For cases (d)–(f) find a first-order PDE involving u_x and u_t.

(a) $u(x, t) = A(x + ct) + B(x - ct)$, where c is a constant.
(b) $u(x, t) = A(x) + B(t).$
(c) $u(x, t) = A(x)/B(t).$
(d) $u(x, t) = A(xt).$
(e) $u(x, t) = A(x^2 t).$
(f) $u(x, t) = A(x^2/t).$

1.4 Show that

$$u = f(2x + y^2) + g(2x - y^2)$$

satisfies the PDE

$$y^2 u_{xx} + \frac{1}{y}u_y - u_{yy} = 0$$

for arbitrary functions f and g.

1.5 Show that $u(x, t) = \frac{1}{2}c \operatorname{sech}^2 \frac{1}{2}\sqrt{c}(x - ct - x_0)$ is a solution of the KdV equation (**pde.9**). This is an example of a *soliton*, a solitary wave that travels at a speed c proportional to its height.

1.6 By differentiating under the integral, show that the function u defined by (1.2) satisfies the heat equation (**pde.4**) for any initial function $g(x)$.

1.7 Show that if ϕ satisfies the heat equation then the change of dependent variable

$$u(x,t) = -2\frac{\partial}{\partial x}\log\phi\,(x,t)$$

(commonly referred to as the Cole–Hopf transformation) satisfies the nonlinear PDE $u_t + u u_x = u_{xx}$ (also known as the viscous Burgers' equation to distinguish it from (**pde.2**)). Hence use Example 1.3 to determine a solution u to this nonlinear PDE.

1.8 Suppose that a and b are positive numbers and that $\phi(x,t) = e^{-a(x-at)} + e^{b(x+bt)}$. Show that ϕ is a solution of the heat equation and use the Cole–Hopf transformation to determine the corresponding solution $u(x,t)$ of the viscous Burgers' equation $u_t + u u_x = u_{xx}$. Show that

(a) $u(x,t) \to 2a$ as $x \to -\infty$ and $u(x,t) \to -2b$ as $x \to \infty$.
(b) $u(x,t)$ is constant along the lines $x - (b-a)t = $ constant.

Describe the nature of the solution in the cases $a < b$, $a = b$ and $a > b$.

1.9 Determine a solution of Burgers' equation via the Cole–Hopf transformation based on $\phi(x,t) = 1 + e^{-2a(x-2at)} + e^{-a(x-x_0-at)}$. Graph the solution on the interval $-20 < x < 40$ at times $t = 0, 5, 10, 15$ when $a = 1$ and $x_0 = 10$.

1.10 Show that $u(x,y) = \tan^{-1}(y/x)$ is a harmonic function in the first quadrant $\{(x,y); x > 0, y > 0\}$ of $\mathbb{R}^2$ and has the properties

$$u(x,0) = 0 \quad \text{and} \quad \lim_{x \to 0+} u(x,y) = \tfrac{1}{2}\pi \ (\text{for } y > 0).$$

Thus u has different constant values on the positive coordinate axes and is undefined at the origin.

Chapter 2
Boundary and Initial Data

Abstract This chapter introduces the notions of boundary and initial value problems. Some operator notation is developed in order to represent boundary and initial value problems in a compact manner. Familiarity with this notation is essential for understanding the presentation in later chapters. An initial classification of partial differential equations is then developed.

Our starting point here is a simple ordinary differential equation (ODE): find $u(t)$ such that $u' = 2t$. Integrating gives the solution $u = t^2 + C$ where C is the "constant" of integration. To compute C and thus get a unique solution we need to know $u(t)$ at some specific time $t = T$; for example, if $u(0) = 1$ then $C = 1$ and $u = t^2 + 1$.

To build on this, suppose that $u(x, t)$ satisfies the simple PDE

$$u_t = 2t. \tag{2.1}$$

Writing this in the form $\partial_t(u - t^2) = 0$, it is seen that $u(x, t) - t^2$ does not vary with t (but it may vary with x), so the PDE is readily integrated to give the solution

$$u(x, t) = t^2 + A(x), \tag{2.2}$$

where $A(x)$ is an arbitrary "function" of integration. We may regard the PDE (2.1) as an (uncountably) infinite set of ODEs, one for each value of x. The arbitrary function of integration is then seen to be a consequence of requiring a different constant of integration at each value of x. For a unique solution, we must specify an additional initial condition; for example,

$$u(x, 0) = g(x), \tag{2.3}$$

where $g(x)$ is a given function. Putting $t = 0$ in (2.2) and using (2.3) gives a solution that is uniquely defined for all time:

$$u(x, t) = t^2 + g(x). \tag{2.4}$$

© Springer International Publishing Switzerland 2015
D.F. Griffiths et al., *Essential Partial Differential Equations*, Springer
Undergraduate Mathematics Series, DOI 10.1007/978-3-319-22569-2_2

This combination of PDE (2.1) and initial condition (2.3) is referred to as an *initial value problem* (IVP).

Applying this logic to the solution of the test equation (**pde.10**), we deduce that the function $u(x, t)$ in Example 1.1 will uniquely solve the PDE if we augment (**pde.10**) by an additional initial condition, say,

$$u(x, 0) = g(x) \tag{2.5}$$

together with an additional "boundary condition", say,

$$u(0, t) = a(t). \tag{2.6}$$

Insisting on continuity of u at the origin requires $u(0, 0) = g(0) = a(0)$ and we then obtain the unique solution

$$u(x, t) = x^2 t^2 + g(x) + a(t) - g(0).$$

The combination of PDE (**pde.10**) with initial condition (2.5) and boundary condition (2.6) is referred to as an *initial–boundary value problem* (IBVP) or simply as a *boundary value problem* (BVP). Further insight into these issues may be found in the exercises at the end of the chapter.

Let us move on to consider the heat equation (**pde.4**). For a unique solution, we will need an initial condition and two boundary conditions. For example, it will be shown in Chap. 7 that the following BVP has a uniquely defined solution $u(x, t)$:

$$\left.\begin{aligned} u_t - \kappa u_{xx} &= 0 && \text{in } (0, 1) \times (0, T] \\ u(x, 0) &= g(x) && \text{for all } x \in [0, 1] \\ u(0, t) &= a_0(t); \quad u(1, t) = a_1(t) && t > 0, \end{aligned}\right\}$$

$$\tag{2.7}$$

where $g(x)$, $a_0(t)$ and $a_1(t)$ are given functions. Two boundary conditions are needed because of the second derivative with respect to x. Those used in (2.7), where the value of u is specified, are known as Dirichlet boundary conditions. It is not necessary for the values at corners to be uniquely defined. For instance, it is not necessary for $\lim_{x \to 0} g(x)$ to equal $\lim_{t \to 0} a_0(t)$. This might model the situation where one end of an initially "hot" bar of material is plunged into an ice bath. A solution of the heat equation with this property is given in Exercise 2.3.

Alternative, equally viable, BVPs would be obtained by replacing one or both boundary conditions in (2.7) by conditions on the x derivative, for example,

$$u(0, t) = a_0(t); \quad u_x(1, t) = a_1(t) \quad \text{for all } t > 0, \tag{2.8}$$

where the boundary condition at $x = 1$ is known as a Neumann condition, or

$$u(0, t) = a_0(t); \quad \alpha u(1, t) + \beta u_x(1, t) = a_1(t) \quad \text{for all } t > 0, \tag{2.9}$$

where the boundary condition at $x = 1$ is known as a Robin condition. Note that choosing $\alpha = 1$ and $\beta = 0$ reduces this to a Dirichlet condition and the combination $\alpha = 0$ and $\beta = 1$ leads to a Neumann boundary condition.

Note however that the following combination of boundary condition does not make sense:

$$u(0, t) = a_0(t); \quad u_{xx}(1, t) = a_1(t) \quad \text{for all } t > 0,$$

since the value of $u_{xx}(1, t)$ provided by the boundary condition would generally conflict with the value obtained from the PDE as $x \to 1$. This leads to the general rule of thumb that the order of derivatives appearing in boundary conditions must be lower than the highest order derivative terms appearing in the PDE.

Turning to the wave equation (**pde.5**), we need *two* initial conditions because of the second derivative with respect to t, for example,

$$\left.\begin{aligned}
u_{tt} - c^2 u_{xx} &= 0 \quad \text{in } (0, 1) \times (0, T] \\
u(x, 0) = f_1(x); \ u_t(x, 0) &= f_2(x) \quad \text{for all } x \in [0, 1] \\
u(0, t) = g_0(t); \ u(1, t) &= g_1(t) \quad \text{for all } t > 0.
\end{aligned}\right\} \tag{2.10}$$

Replacing the Dirichlet boundary conditions by either Neumann or Robin boundary conditions would also lead to legitimate BVPs. In this case, for $u(x, t)$ to be a continuous function, the initial and boundary conditions need to be equal where they meet so that $f_1(0) = g_0(0)$ and $f_1(1) = g_1(0)$.

In our final example in this section we consider Laplace's equation (**pde.3**) on a domain Ω in two dimensions which has a boundary that we will denote by $\partial \Omega$ (Fig. 2.1). In order to obtain a unique solution it is necessary to specify a condition at every point on this boundary.

Since we have a second-order PDE the possible types of boundary condition are Dirichlet, Neumann and Robin and these take the form:

- For a Dirichlet boundary condition, the value of u is specified

$$u(\vec{x}) = g_D(\vec{x}) \quad \text{for all } \vec{x} \in \partial \Omega. \tag{2.11a}$$

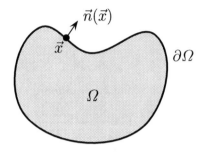

Fig. 2.1 A domain Ω in $\mathbb{R}^2$ with boundary $\partial\Omega$. Also shown is the outward normal vector $\vec{n}(\vec{x})$ at a point $\vec{x} \in \partial\Omega$

- For a Neumann boundary condition, the value of the (outward) normal derivative of u is specified, that is

$$\frac{\partial u}{\partial n}(\vec{x}) = g_N(\vec{x}) \quad \text{for all } \vec{x} \in \partial\Omega. \tag{2.11b}$$

The outward normal derivative of $u(x, t)$ is the rate of change of u with distance moved in the normal direction with any other independent variables, such as tangential displacement and time, being held fixed. For a unit normal vector $\vec{n}$ it is the component of the gradient ∇u in the normal direction, which can be written as $\partial_n u = u_n = \vec{n} \cdot \nabla u$.

- For a Robin boundary condition, a linear combination of the value of u and its (outward) normal derivative is specified, that is

$$\alpha u(\vec{x}) + \beta \frac{\partial u}{\partial n}(\vec{x}) = g_R(\vec{x}) \quad \text{for all } \vec{x} \in \partial\Omega, \tag{2.11c}$$

where α and β are usually constant, but in some situations could depend on $\vec{x}$ or even u.

- Finally, we could also mix the three types by partitioning $\partial\Omega$ into nonoverlapping pieces so that $\partial\Omega = \partial\Omega_D \cup \partial\Omega_N \cup \partial\Omega_R$ and then specify a boundary condition of Dirichlet, Neumann and Robin type on $\partial\Omega_D$, $\partial\Omega_N$ and $\partial\Omega_R$, respectively.

Note that whenever Laplace's equation is solved in $\Omega \subset \mathbb{R}^2$ the boundary $\partial\Omega$ is one-dimensional. When solving the equation in $\mathbb{R}^3$ the boundary is two-dimensional, so that there are two tangential derivatives and one normal derivative at every point on the boundary.

2.1 Operator Notation

Before embarking on a more detailed study of PDEs we introduce some notation that will enable us to write PDEs, boundary conditions and BVPs in a compact fashion much as linear algebraic equations are commonly expressed using matrices and vectors.

A calligraphic font $(\mathcal{L}, \mathcal{M}, \mathcal{B}, \ldots)$ will be used for symbols that denote differential operators in the *space variables* only. For example, defining

$$\mathcal{L}u(x,t) = -\kappa u_{xx}(x,t), \quad (x,t) \in (0,1) \times (0,T)$$

would allow us to write the heat equation (**pde.4**) as

$$u_t + \mathcal{L}u = 0.$$

Similarly, defining the boundary condition operator

$$\mathcal{B}u(x,t) = \begin{cases} u(0,t) & \text{for } t > 0, x = 0 \\ u_x(1,t) & \text{for } t > 0, x = 1, \end{cases} \tag{2.12}$$

allows the conditions (2.8) to be expressed as

$$\mathcal{B}u = f(x,t), \quad t > 0, \ x = 0, 1,$$

where $f(0,t) = a_0(t)$ and $f(1,t) = a_1(t)$. Then, by defining

$$\mathscr{L}u(x,t) = \begin{cases} u_t(x,t) + \mathcal{L}u(x,t) & \text{for } (x,t) \in (0,1) \times (0,T) \\ \mathcal{B}u(x,t) & \text{for } (x,t) \in \{0,1\} \times (0,T) \\ u(x,0) & \text{for } t = 0, x \in [0,1] \end{cases} \tag{2.13}$$

and

$$\mathscr{F}(x,t) = \begin{cases} 0 & \text{for } (x,t) \in (0,1) \times (0,T) \\ f(x,t) & \text{for } (x,t) \in \{0,1\} \times (0,T) \\ g(x) & \text{for } t = 0, x \in [0,1] \end{cases} \tag{2.14}$$

the BVP (2.7) could be written in compact form, so that

$$\mathscr{L}u = \mathscr{F}. \tag{2.15}$$

In the sequel, symbols in script font $(\mathscr{L}, \mathscr{M}, \mathscr{F}, \ldots)$ will be reserved for statements of BVPs.

Example 2.1
Consider the BVP defined by Laplace's equation in the unit square $0 \le x, y \le 1$ with

(a) Dirichlet boundary conditions on the vertical edges: $u(0,y) = \cos \pi y$ and $u(1,y) = y - 1$ for $0 < y < 1$.

(b) A Neumann condition[1] on the lower edge: $-u_y(x, 0) = 1$ for $0 < x < 1$.
(c) A Robin condition on the upper edge: $u_y(x, 1) + u(x, 1) = 0$ for $y = 1$,
     $0 < x < 1$.

Define suitable forms for $\mathscr{L}$ and $\mathscr{F}$ so that it can be expressed as $\mathscr{L}u = \mathscr{F}$.

Here we define the BVP terms directly without first defining spatial differential operators $\mathcal{L}$ and $\mathcal{B}$:

$$\mathscr{L}u(x, y) = \begin{cases} u_{xx}(x, y) + u_{yy}(x, y) & \text{for } 0 < x, y < 1 \\ u(0, y) & \text{for } x = 0, 0 < y < 1 \\ u(1, y) & \text{for } x = 1, 0 < y < 1 \\ -u_y(x, 0) & \text{for } y = 0, 0 < x < 1 \\ u_y(x, 1) + u(x, 1) & \text{for } y = 1, 0 < x < 1 \end{cases}$$

and

$$\mathscr{F}(x, y) = \begin{cases} 0 & \text{for } 0 < x, y < 1 \\ \cos \pi y & \text{for } x = 0, 0 < y < 1 \\ y - 1 & \text{for } x = 1, 0 < y < 1 \\ 1 & \text{for } y = 0, 0 < x < 1 \\ 0 & \text{for } y = 1, 0 < x < 1. \end{cases}$$

It can be observed that boundary values have not been specified at the corners of the domain as these do not affect the solution in the interior. An example of a problem with a discontinuity is given in Exercise 1.10.   ◇

2.2 Classification of Boundary Value Problems

We will categorise BVPs (that is, PDEs and associated initial or boundary conditions) into those that are linear and those that are nonlinear in the next two sections. A formal definition of a well-posed boundary value problem is the subject of the final section.

2.2.1 Linear Problems

Linear BVPs have a number of useful properties, some of which will be investigated in this section. Our first goal is to identify which problems are linear.

---

[1] The outward normal direction on $y = 0$ is in the direction of $\vec{n} = (0, -1)$.

Definition 2.2 (*Linearity*) An operator $\mathcal{L}$ is *linear* if for any two functions u and v and any $\alpha \in \mathbb{R}$ the following two properties are satisfied:

(a) $\mathcal{L}(u + v) = \mathcal{L}(u) + \mathcal{L}(v)$;
(b) $\mathcal{L}(\alpha u) = \alpha \mathcal{L}(u)$.

An operator that does not satisfy these conditions is said to be *nonlinear*. We will explore various kinds of nonlinearity in the next section.

Example 2.3 Show that the BVP defined by (2.13)–(2.15) is linear.

The spatial operator $\mathcal{L}u(x, t) = -\kappa u_{xx}(x, t)$ is linear since

$$\mathcal{L}(u + v) = -\kappa (u + v)_{xx}$$
$$= -\kappa u_{xx} - \kappa v_{xx} = \mathcal{L}u + \mathcal{L}v$$
$$\mathcal{L}(\alpha u) = -\kappa (\alpha u)_{xx}$$
$$= -\alpha \kappa u_{xx} = \alpha \mathcal{L}u.$$

Similarly, the boundary condition operator $\mathcal{B}$ satisfies

$$\mathcal{B}(u + v) = \begin{cases} u + v \\ (u + v)_x \end{cases} = \begin{cases} u + v & \text{for } t > 0, x = 0 \\ u_x + v_x & \text{for } t > 0, x = 1, \end{cases}$$
$$= \mathcal{B}u + \mathcal{B}v$$

and

$$\mathcal{B}(\alpha u) = \begin{cases} \alpha u \\ (\alpha u)_x \end{cases} = \begin{cases} \alpha u & \text{for } t > 0, x = 0 \\ \alpha u_x & \text{for } t > 0, x = 1, \end{cases}$$
$$= \alpha \mathcal{B}u$$

so it is also linear. Note that the same approach shows that conventional Dirichlet/Neumann/Robin boundary conditions are always linear. This means that linearity of a BVP normally depends only on the linearity of the PDE component of the problem.[2]   ◊

Also, from the definition of $\mathcal{L}$ in (2.13), we see that

$$\mathcal{L}(u + v) = \begin{cases} (u \mid v)_t \mid \mathcal{L}(u + v) \\ \mathcal{B}(u + v) \\ u + v \end{cases} = \begin{cases} (u_t + \mathcal{L}u) + (v_t + \mathcal{L}v) \\ \mathcal{B}u + \mathcal{B}v \\ u + v \end{cases}$$

---

[2]Nonlinear boundary conditions such as $u_n = e^u$ on $\partial \Omega$ are certainly possible, but will not be considered here.

so that $\mathscr{L}(u + v) = \mathscr{L}u + \mathscr{L}v$. (The proof that $\mathscr{L}(\alpha u) = \alpha \mathscr{L}u$ is left as an exercise.) We conclude that the given BVP for the heat equation (**pde.4**) is linear.

Other examples of linear BVPs are associated with the classical second-order PDEs: the wave equation (**pde.5**), Laplace's equation (**pde.3**), as well as the Black–Scholes equation (**pde.8**). As shown in the next definition and the subsequent theorems, a linear BVP leads to a "principle of superposition" which allows us to combine solutions together.

Definition 2.4 (*Homogeneous BVP*) Suppose that $\mathscr{L}$ is a *linear* operator associated with the BVP $\mathscr{L}u = \mathscr{F}$, then the *homogeneous* BVP is the corresponding problem $\mathscr{L}u = 0$. To generate a homogeneous BVP, any terms that are independent of u must be removed from the PDE and all boundary and initial data must be set to zero.

The statements of the following theorems may be familiar from studies of ordinary differential equations and linear algebra.

Theorem 2.5
Suppose that u_1 and u_2 are any two solutions of a homogeneous boundary value problem $\mathscr{L}u = 0$, then any linear combination $v = \alpha u_1 + \beta u_2$, with constants α, β, is also a solution.

Proof
$$\mathscr{L}v = \mathscr{L}(\alpha u_1 + \beta u_2) = \alpha \underbrace{\mathscr{L}u_1}_{0} + \beta \underbrace{\mathscr{L}u_2}_{0} = 0. \qquad \square$$

Theorem 2.6 *Suppose that u_* is a "particular" solution of the linear boundary value problem $\mathscr{L}u = \mathscr{F}$, and that v is a solution of the associated homogeneous problem, then $w = u_* + v$ is also a solution of the BVP $\mathscr{L}u = \mathscr{F}$.*

Proof
$$\mathscr{L}(w) = \mathscr{L}(u_* + v) = \underbrace{\mathscr{L}(u_*)}_{\mathscr{F}} + \underbrace{\mathscr{L}(v)}_{0} = \mathscr{F}. \qquad \square$$

The superposition principle will prove to be invaluable in Chap. 8, where we will construct analytic solutions to BVPs like (2.7).

Theorem 2.7 *(Uniqueness) A linear boundary value problem $\mathscr{L}u = \mathscr{F}$ will have a unique solution if, and only if, $v = 0$ is the only solution of the homogeneous problem $\mathscr{L}v = 0$.*

Proof We suppose that there are two solutions, u_1 and u_2. Hence $\mathscr{L}u_1 = \mathscr{F}$ and $\mathscr{L}u_2 = \mathscr{F}$. Subtracting these equations from each other and using the linearity of $\mathscr{L}$ gives
$$\mathscr{L}v = 0, \qquad v = u_1 - u_2.$$

Thus, if $v = 0$ is the only solution of the homogeneous problem, we must have $u_1 = u_2$. $\qquad \square$

When the homogeneous problem has a nontrivial solution v, then αv, for any constant α, is also a solution. Hence $u_2 = u_1 + \alpha v$ and there are therefore an infinite number of solutions corresponding to different choices of α. This is typical of linear problems: they have either no solutions, one solution or an infinite number of solutions.

Example 2.8 Consider the solution of the wave equation $u_{xx} - u_{yy} = 0$ and Laplace's equation $u_{xx} + u_{yy} = 0$ in the rectangle $0 < x < 1,\ 0 < y < 1/2$ with the boundary conditions given in Fig. 2.2 (with n a positive integer).

The wave equation has a particular solution $u_* = \sin n\pi x \cos n\pi y$ that also satisfies the boundary conditions. However, it is readily shown that

$$v(x, y) = \sin(2m\pi x)\sin(2m\pi y)$$

solves the corresponding *homogeneous* BVP for any integer m and so the wave equation has the nonunique solutions

$$u(x, y) = u_*(x, y) + \alpha v(x, y)$$

for each constant α and each integer m. In contrast, it is readily checked that

$$u(x, y) = \sin n\pi x\, \frac{\sinh n\pi(1 - y)}{\sinh n\pi}$$

solves the BVP for Laplace's equation. In fact, this is the only solution. It will be shown in Chap. 7 that a solution of Laplaces equation always has its maximum and minimum values on the boundary. This ensures that the only solution of the corresponding homogeneous BVP is zero.   ◊

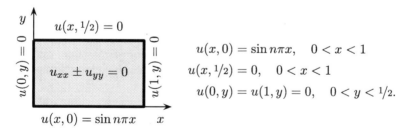

$$u(x, 0) = \sin n\pi x, \quad 0 < x < 1$$
$$u(x, 1/2) = 0, \quad 0 < x < 1$$
$$u(0, y) = u(1, y) = 0, \quad 0 < y < 1/2.$$

Fig. 2.2 Domain and boundary conditions for Example 2.8

2.2.2 Nonlinear Problems

Not all of the PDEs that are listed in Sect. 1.1 are linear and the theorems given in the previous section do not apply to these.

Example 2.9   The inviscid Burgers' equation (**pde.2**), which we write as $u_t + \mathcal{L}(u) = 0$ with $\mathcal{L}(u) = uu_x$ satisfies

$$\mathcal{L}(\alpha u) = (\alpha u)(\alpha u)_x$$
$$= \alpha^2 uu_x \neq \alpha \mathcal{L}(u)$$

and violates the second condition in Definition 2.2. This shows that this PDE is *nonlinear*. The KdV equation (**pde.9**) can be shown to be nonlinear using the same argument.   ◊

It is sometimes useful to classify the degree of nonlinearity of a PDE or associated BVP. A standard classification is as follows:

Linear: The PDE should satisfy Definition 2.2. In effect, all coefficients of u and any of its derivatives must depend only on the independent variables $t, x, y, \ldots$.

Semi-linear: The coefficients of the *highest derivatives* of u do not depend on u or any derivatives of u.

Quasi-linear: The coefficients of the highest derivatives of u depend only on lower derivatives of u.

Otherwise the PDE is *fully nonlinear*. Some examples are listed below.

$$u_{xxx} - 4u_{xxyy} + u_{yyzz} = f(x, y, z) \qquad : \text{``linear''}$$
$$u_x^2 u_{tt} - \tfrac{1}{2}u_{xxxxx} = 1 - u^2 \qquad : \text{``semi-linear''}$$
$$u_{tt}u_{xxx} + u_x u_{ttt} = f(u, x, t) \qquad : \text{``quasi-linear''}$$
$$\exp(u_{xtt}) - u_{xt}u_{xxx} + u^2 = 0 \qquad : \text{``fully nonlinear''}.$$

Additional nonlinearity "classification" exercises are given in Exercise 2.8.

We will see in Chap. 9 that the character of nonlinear *first-order* PDEs is completely governed by the nature of the nonlinearity, so it is important to classify the nonlinearity correctly. This can be readily achieved if the first-order PDE is written in the "additive form",

$$au_t + bu_x + cu = f, \qquad (2.16)$$

where a and b are functions of t, x, u, u_t and u_x; c is a function of t, x and u, and f is a function of t and x. The classification is then immediate:

Linear: If a, b and c depend only on t and x, and not on u or any of its derivatives, then the PDE is *linear*.

Semi-linear: If a and b do not depend on u or any of its derivatives, but c depends on u, then the PDE is *semi-linear*.

Quasi-linear: If a and/or b depend on u but not on any derivatives of u, then the PDE is *quasi-linear.*

Otherwise the PDE is *fully nonlinear*. Some examples are listed below.

$$2\cos(xt)u_t - xe^t u_x - 9u = e^t \sin x \qquad : \text{``linear''}$$

$$x\cos(t)u_t + tu_x + u^2 \cdot u = \frac{x}{t}u\sin(u) \qquad : \text{``semi-linear''}$$

$$uu_t + u^2 u_x + u = e^x \qquad : \text{``quasi-linear''}$$

$$u_t + \tfrac{1}{2}u_x^2 - u = \cos(xt) \qquad : \text{``fully nonlinear''}.$$

2.2.3 Well-Posed Problems

So far so good. Well-conceived boundary value problems typically have unique solutions. However, an unfortunate complication is that such BVPs can still be inordinately sensitive to the problem data (for example, the boundary data). Thus, even though such problems might have practical applications, they are much too demanding for a textbook at this level. Accordingly our aim is to filter out such BVPs and focus on those that are relatively well behaved.

An overview of the situation can be obtained by considering a generic *linear* BVP written in the notation introduced in Sect. 2.1, namely,

$$\mathscr{L}u = \mathscr{F}. \tag{2.17}$$

Suppose that we now make a "small" change $\delta\mathscr{F}$ to the data $\mathscr{F}$ and we denote the subsequent change to the solution by δu. Thus,

$$\mathscr{L}(u + \delta u) = \mathscr{F} + \delta\mathscr{F}. \tag{2.18}$$

Then, since $\mathscr{L}$ is a linear operator, (2.17) may be subtracted from (2.18) to give

$$\mathscr{L}(\delta u) = \delta\mathscr{F}, \tag{2.19}$$

so we see that δu satisfies the same BVP as u with $\mathscr{F}$ replaced by $\delta\mathscr{F}$. We now get to a definition which is the crux of the issue.

Definition 2.10 (*Well-posed BVP*) A boundary value problem which has a *unique* solution that varies *continuously* with the initial and boundary data is said to be *well posed*. A problem that is not well posed is said to be *ill posed*.

In the context of (2.19) this means that δu should be "small" whenever $\delta\mathscr{F}$ is "small" in the sense that there are norms[3] $\|\cdot\|_a$, $\|\cdot\|_b$ and a constant C that does not depend on u, $\mathscr{F}$ or $\delta\mathscr{F}$ so that

$$\|\delta u\|_a \leq C\|\delta\mathscr{F}\|_b \tag{2.20}$$

which must hold for all admissible choices of $\delta\mathscr{F}$. By choosing $\delta\mathscr{F} = -\mathscr{F}$ we deduce that $\delta u = -u$ and (2.20) then implies that

$$\|u\|_a \leq C\|\mathscr{F}\|_b. \tag{2.21}$$

This reflects the general approach for linear problems: beginning with the homogeneous problem, the change in the data from zero to $\mathscr{F}$ causes the solution to change from zero to u.

The good news here is that the BVPs defined at the start of the chapter, namely, (2.7), (2.10) and Laplace's equation with either Dirichlet, Neumann or mixed conditions are well posed in the sense of satisfying this definition.[4] The bad news is that boundary value problems which have a unique solution are not automatically well posed. A pathological example is given next.

Example 2.11   Consider the BVP obtained by setting $\kappa = -1$ in (**pde.4**), so that

backward heat equation	$u_t + u_{xx} = 0,$	(**pde.11**)

and subjecting it to the initial data $u(x, 0) = 0$.

This BVP has the unique solution, $u(x, t) = 0$. However, if we make *tiny* changes in the initial data to say $u(x, 0) = 10^{-99}\cos(nx)$, then the unique solution changes to

$$u(x, t) = 10^{-99}e^{n^2 t}\cos(nx).$$

The ratio of solution to data (initial value) is $u(x, t)/u(0, x) = \exp(n^2 t)$. This can be made as large as we wish, even for very small values of t, by taking a large enough value of n. This happens in spite of the fact that the size of the change in initial data would probably be subatomic in any practical example—it reflects the fact that "anti-diffusive" behaviour violates the second law of thermodynamics and is something like having time running backwards!   ◊

Our second example extends the PDE from Example 2.8 and has a rearranged boundary condition.

---

[3]The examples of ill-posed problems that we shall give are clear cut without the need to specify precisely which norms are used.

[4]This proof is deferred to Chap. 7.

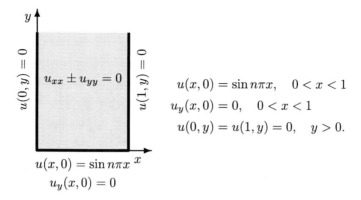

Fig. 2.3 Domain and boundary conditions for Example 2.12

Example 2.12 Consider the solution of Laplace's equation $u_{xx} + u_{yy} = 0$ and the wave equation $u_{xx} - u_{yy} = 0$ in the semi-infinite strip with the boundary conditions given in Fig. 2.3 (with n a positive integer).

Now the wave equation has the unique solution

$$u(x, y) = \sin n\pi x \cos n\pi y$$

in which its magnitude (amplitude) is the same as that of the data.[5] Laplace's equation, however, has the solution

$$u(x, y) = \sin n\pi x \cosh n\pi y$$

which, for any $y > 0$ can be made as large as we wish by taking a suitably large value of n. Hence this type of BVP for Laplace's equation is ill posed.

A well-posed problem could be recovered by replacing one of the conditions applied at $y = 0$ by a condition such as $u(x, y) \to 0$ as $y \to \infty$ for all $x \in (0, 1)$.

Exercises

2.1 For each of the cases (a)–(f) in Exercise 1.3, can you determine the functions A and/or B using the initial condition $u(x, 0) = f(x)$, where f is some given function? Give expressions for A and/or B wherever they can be determined.

2.2 For each of the cases (a)–(f) in Exercise 1.3, can you determine the functions A and/or B using the alternative "initial condition" $u(x, 1) = g(x)$, where g is some given function?

---

[5] Note that we cannot conclude that this problem is well posed since we would need to consider all possible choices of the data in order to make that claim.

2.3 Let the function g be defined by

$$g(x) = \begin{cases} 0, & x < 0 \\ 1, & x > 0 \end{cases}$$

and suppose that $u(x, t)$ is defined by (1.2). (See also Exercise 1.6.) Use the result that

$$\int_0^\infty e^{-s^2}\, ds = \tfrac{1}{2}\sqrt{\pi}$$

to show that $u(0, t) = \tfrac{1}{2}$ for $t > 0$ and $u(x, 0) = 1$ for $x > 0$.

2.4 By following the example in Sect. 2.1, define suitable forms for $\mathscr{L}$ and $\mathscr{F}$ so the BVP (2.10) can be written as $\mathscr{L}u = \mathscr{F}$.

2.5 Show that $\mathscr{L}$ defined by (2.13) satisfies $\mathscr{L}(\alpha u) = \alpha \mathscr{L}u$.

2.6 Show that $\mathscr{L}$ defined in Example 2.1 is a linear operator.

2.7*Consider the backward heat equation (see Example 2.11) for $u(x, t)$, corresponding to having a negative thermal diffusivity coefficient $\kappa = -1$:

$$u_t + u_{xx} = 0.$$

Confirm that, for constant values of A and T a solution, for any $t < T$, is given by

$$u(x, t) = \frac{AT^{1/2}}{(T-t)^{1/2}} \exp\left(-\frac{x^2}{4(T - t)}\right).$$

Use this to show that solutions can exist with, initially, $|u(0, x)| \leq \varepsilon$ for any $\varepsilon > 0$ but which become infinite in value after any given subsequent time. Deduce that the backward heat equation is not well posed for $t > 0$ when subjected to initial conditions at $t = 0$.

2.8 Determine the order and categorise the following PDEs by linearity or degree of nonlinearity.

(a) $u_t - (x^2 + u)u_{xx} = x - t$.
(b) $u^2 u_{tt} - \tfrac{1}{2}u_x^2 + (uu_x)_x = e^u$.
(c) $u_t - u_{xx} = u^3$.
(d) $(u_{xy})^2 - u_{xx} + u_t = 0$.
(e) $u_t + u_x - u_y = 10$.

2.9 Categorise the following second-order PDEs by linearity or degree of nonlinearity.

(a) $u_t + u_{tx} - u_{xx} + u_x^2 = \sin u$.

(b) $u_x + u_{xx} + u_y + u_{yy} = \sin(xy)$.

(c) $u_x + u_{xx} - u_y - u_{yy} = \cos(xyu)$.

(d) $u_{tt} + xu_{xx} + u_t = f(x, t)$.

(e) $u_t + uu_{xx} + u^2 u_{tt} - u_{tx} = 0$.

Chapter 3
The Origin of PDEs

Abstract This chapter is a self-contained introduction to mathematical modelling. Some classical partial differential equations are derived from Newton's laws of motion or else are developed by considering the conservation of quantities like mass and thermal energy. The resulting PDEs can be readily extended to give mathematical models of plate bending, the dispersal of pollutants and the flow of traffic on congested motorways.

In general, the "derivation" of a PDE is an exercise in combining physical conservation principles with modelling assumptions. Typically, the more assumptions that are made, the simpler the resulting PDE model. The validity of these assumptions determines the success (or otherwise) of the mathematical model in predicting properties of the real world.[1] Our aim here is to sketch the origin of some of the PDEs that feature in later chapters of the book.

3.1 Newton's Laws

Newton's second law of motion: *mass × acceleration = force* is arguably the most important relation in the whole of applied mathematics. The following example gives a classic illustration of the key role that Newton's second law plays in the derivation of simple mathematical models.

3.1.1 The Wave Equation for a String

We consider the motion of a thin, inextensible string of density ρ per unit length, stretched to a tension T between two fixed points. The adjectives describing the situation have quite precise meanings—by "thin" we mean that the thickness of the string may be ignored, "inextensible" means that it does not stretch under tension

---

[1]Einstein summarised the situation succinctly: *"As far as the laws of mathematics refer to reality, they are not certain, as far as they are certain, they do not refer to reality."*

© Springer International Publishing Switzerland 2015

D.F. Griffiths et al., *Essential Partial Differential Equations*, Springer Undergraduate Mathematics Series, DOI 10.1007/978-3-319-22569-2_3

Fig. 3.1 Sketch of a portion
of a string of arc length δs.
The tension $T_\pm$ at its ends
acts in the direction of the
tangents which make angles
$\alpha_\pm$ with the horizontal

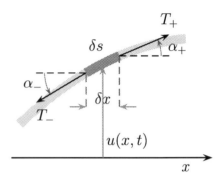

and a "string" has no resistance to bending. The equation of motion is derived by
considering a small length of the string (of arc length δs) whose centre is a height
$u(x, t)$ above the x-axis. Fig. 3.1 shows a sketch of the situation (with a greatly
exaggerated vertical scale).

The two key modelling assumptions are:

(a) The vertical motion is small, so that the angles $\alpha_\pm$ are small and we may make
the approximations $\sin \alpha_\pm \approx \alpha_\pm$, $\cos \alpha_\pm \approx 1$ and $\tan \alpha_\pm \approx \alpha_\pm$.
(b) Horizontal motion is negligible compared to the vertical motion. The horizontal
components of the forces acting on the portion of the string must therefore
balance: $T_+ \cos \alpha_+ = T_- \cos \alpha_-$ so, in view of assumption (a), $T_+ = T_-$. Thus
the tension T is the same at all points along the string.

The mass of the small portion of string is $\rho \delta s$ and its vertical acceleration is $u_{tt}(x, t)$.
By Newton's second law mass$\times$acceleration is balanced by the vertical component
of force, and so

$$\rho \, \delta s \, u_{tt}(x, t) = T_+ \sin \alpha_+ - T_- \sin \alpha_-. \tag{3.1}$$

We have already seen that $T_+ = T_- = T$ and, by assumption (a), $\delta x \approx \delta s$. The
angles $\alpha_\pm$ are the angles that the tangents make with the horizontal and so

$$\tan \alpha_\pm = u_x(x \pm \tfrac{1}{2}\delta x, t).$$

Since $\tan \alpha_\pm \approx \alpha_\pm$ we find that $\alpha_\pm = u_x(x \pm \tfrac{1}{2}\delta x, t)$ so (3.1) leads to

$$\rho \, \delta x \, u_{tt}(x, t) = T \, (\alpha_+ - \alpha_-) = T \left(u_x(x + \tfrac{1}{2}\delta x, t) - u_x(x - \tfrac{1}{2}\delta x, t) \right)$$
$$= T \delta x \, u_{xx}(x, t) + \mathcal{O}(\delta x^2)$$

which, on division by δx and letting $\delta x \to 0$, gives the wave equation (**pde.5**), that
is, $u_{tt} = c^2 u_{xx}$ with a wave speed $c = \sqrt{T/\rho}$.

3.2 Conservation Laws

Suppose that we are interested in a property of a material (the heat content or concentration of a dye) that varies continuously with space (x, y, z) and time t. The principle behind conservation laws is to focus on the changes and causes of change in that property in a volume, V say, (it may be helpful to imagine a ball) of material that is enclosed by a surface, S say. There are three key ingredients—the quantity Q of the property per unit volume, the rate F at which it is being produced or destroyed (that is, the net production) per unit volume and the net "flux" (or flow) of that property into/out of the volume through its surface. If $\vec{q}$ denotes the flow at any point in V then the flux outwards at a point on the surface is $\vec{q} \cdot \vec{n}$, where $\vec{n}$ is the outward normal vector at that point. A *conservation law* is the statement that

The rate of change of Q is equal to the difference between net production and net outward flux.

This is expressed mathematically as

$$\frac{d}{dt} \int_V Q \, dV = \int_V F \, dV - \int_S \vec{q} \cdot \vec{n} \, dS. \tag{3.2}$$

The surface integral may be converted to a volume integral by means of the divergence theorem (see Theorem C.1). Moreover, the time derivative and the first integral may be interchanged provided that the volume V does not vary with t and provided that both Q and Q_t are continuous. Making these simplifications gives

$$\int_V \left(\frac{\partial Q}{\partial t} + \vec{\nabla} \cdot \vec{q} - F \right) dV = 0,$$

where $\vec{\nabla} \cdot \vec{q}$ denotes the divergence of $\vec{q}$ (often written as div $\vec{q}$). Since this relationship holds over every volume V, the PDE

$$\frac{\partial Q}{\partial t} + \vec{\nabla} \cdot \vec{q} = F \tag{3.3}$$

must hold at every point in the domain of interest. The quantities Q, $\vec{q}$ and F featuring in (3.3) are functions of the dependent variable (usually denoted by u), the coordinates (x, y, z, t) and any constants (or functions) that describe the properties of the specific medium, such as conductivity and density. Some specific examples are discussed next.

3.2.1 The Heat Equation

The thermal energy or "heat" per unit mass in a substance is $E = \int_{T_0}^{T} C \, dT$, where T is temperature, T_0 is a base or reference temperature, and $C(T)$ is the coefficient of specific heat or the "specific heat capacity". The heat per unit volume is ρE, where ρ is the density of the material so, with $Q = \rho E$, we have

$$Q = \rho \int_{T_0}^{T} C\,dT.$$

If the density and specific heat capacity are constant, this simplifies to give

$$Q = \rho C(T - T_0).$$

The flux of heat by conduction is, by Fourier's law, $\vec{q} = -\lambda \nabla T$, where $\lambda(T)$ is the coefficient of thermal conductivity. This is a vector directed from regions of high temperature to those of low temperature (see Fig. 3.2). If we assume that there is no other way in which heat can flow (for example that there is no radiation of heat and that there is no movement of the substance itself—as in a solid material at rest) and if we let $F(T, \vec{r}, t)$ represent the net effect of production and destruction of heat per unit volume at a point with position vector $\vec{r} = (x, y, z)$, then, from (3.3), we have the energy conservation law

$$\rho C T_t + \vec{\nabla} \cdot (-\lambda \vec{\nabla} T) = F$$

and if λ is also a constant, this becomes a nonhomogeneous heat equation

$$T_t - \kappa \nabla^2 T = f \tag{3.4}$$

see (**pde.6**), with thermal diffusivity $\kappa = \lambda/\rho C$ and a source term $f = F/\rho C$. In general, (3.4) holds in three spatial dimensions but, if the geometry of the domain, the boundary conditions and the source term f are all independent of one of the coordinates, z, say, then (3.4) reduces to a PDE problem in two space dimensions.

Analogous equations model all diffusive processes by which a property flows from where it is more highly concentrated towards lower concentrations in proportion to the gradient of the concentration at any point. Examples include the dispersion of pollutants, the diffusion of chemicals, some aspects of turbulence and the formation of cracks in solids.

As regards boundary conditions for (3.4), the temperature may be specified on part of the boundary (a Dirichlet BC), whereas another part may be insulated so that there is no flux of heat. This means that the normal derivative of T vanishes leading to a Neumann condition $\partial T/\partial n = 0$. A third type of boundary condition arises from

Fig. 3.2 The flux of heat by conduction is $\vec{q} = -\lambda \vec{\nabla} T$

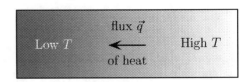

Newton's law of cooling which states that the outward flux of heat is proportional to the difference in temperature between the body and its surroundings. Thus, if $\partial T / \partial n$ denotes the derivative of T in an outward normal direction at the boundary, and the ambient temperature is denoted by T_0, we have the Robin condition

$$\frac{\partial T}{\partial n} = -\mu(T - T_0)$$

where μ is a constant that depends on the materials involved. As noted in the previous chapter, one of these boundary conditions must be imposed on each point of the boundary, together with an initial condition, in order for the solution T to be uniquely defined.

Extension 3.2.1 (The advection–diffusion equation) When the heat conduction takes place in a fluid that is moving with a velocity $\vec{v}$, the flux is modified to become $\vec{q} = \rho C \vec{v} T - \lambda \vec{\nabla} T$, where ρ and C were defined earlier in this section. As a consequence, the energy conservation law (3.3) becomes the *advection–diffusion* equation

$$T_t + \vec{\nabla} \cdot (\vec{v} T) = \kappa \nabla^2 T + f. \tag{3.5}$$

Note that since (3.5) is derived directly from a conservation law, it is referred to as the *conservative form* of the advection–diffusion equation. A non-conservative alternative is developed in Exercise 3.3.[2] $\qquad\qquad \Diamond$

3.2.2 Laplace's Equation and the Poisson Equation

In steady situations where the temperature does not vary with time, the heat equation (3.4) reduces to the *Poisson equation*

$$- \kappa \nabla^2 T = f. \tag{3.6}$$

The Poisson equation pops up everywhere! It is a simple model for a host of physical phenomena ranging from steady diffusive processes to electrostatics and electromagnetism. The importance of keeping the minus sign on the left-hand side will become clear in the next chapter. If, furthermore, we suppose that there are no heat "sources" or "sinks", then $f = 0$ in which case the equation governing the temperature is Laplace's equation: $\nabla^2 T = 0$. This reduces to (**pde.3**) in two dimensions.

---

[2] A PDE does not have to be written in conservative or conservation form for it to be a conservation equation.

Extension 3.2.2 (The biharmonic equation) Starting from the Poisson equation (3.6), if the sources or sinks are such that they themselves satisfy Laplace's equation, $\nabla^2 f = 0$, then

$$\nabla^2 \left(\nabla^2 T \right) = \nabla^2 f = 0.$$

and in this case T satisfies the *biharmonic equation*: $\nabla^4 T = 0$. In two dimensions the biharmonic problem can be written as

$$T_{xxxx} + 2T_{xxyy} + T_{yyyy} = 0, \tag{3.7}$$

since $\nabla^4 = (\nabla^2)^2 = (\partial_x^2 + \partial_y^2)^2$. The biharmonic equation is the most important fourth-order PDE. It plays a fundamental role when modelling the flow of slow-moving viscous fluids and it is the starting point for modelling the distortion of elastic plates when subjected to a heavy load.   ◊

3.2.3 The Wave Equation in Water

It was shown in Sect. 3.1.1 that small disturbances to a taut string are governed by the wave equation. Here we shall show that the same equation may be derived for small amplitude water waves by combining Newton's second law of motion with an appropriate conservation law (namely conservation of mass). Consider a long tank with vertical sides that contains water with an undisturbed depth H above a flat bottom (which is at $y = 0$) and a slightly disturbed surface, which is at a height $y = H + h(x, t)$. The situation is sketched in Fig. 3.3 (left), where the variation in surface height has been greatly exaggerated.

We will make two simplifying assumptions.

- Disturbances are long, or cover a large range of values of x, in comparison to the depth $H + h$ (a "long-wave" or "shallow-water" approximation).
- Disturbances in depth are small compared with the average depth (a "small-amplitude" approximation) so that $|h| \ll H$.

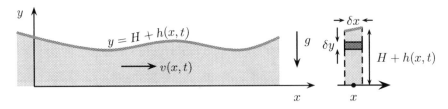

Fig. 3.3 On the *left* is shown a sketch of a tank containing fluid at a depth $H + h(x, t)$ moving with a velocity $v(x, t)$. On the *right* is shown a narrow column of water of width δx and height $H + h(x, t)$. Also *highlighted* is a horizontal slice of this column of height δy

These assumptions lead to the following approximations.

(a) The variation of horizontal component of velocity v with vertical position y is negligible, so that $v \approx v(x, t)$.

(b) The pressure at any point is a combination of the atmospheric pressure P_0 (a constant) and the height of the column of water above that point. The pressure can then be represented by the "hydrostatic" formula $P = P_0 + \rho g(H + h - y)$, where ρ is the density of the water (also constant) and g is the constant of gravitational acceleration.

To proceed we consider a narrow column of water of depth $H + h(x, t)$ and width δx centred on the location x, as depicted in Fig. 3.3 (right). The flux of water in the x-direction is $(H + h)v$, the product of depth and velocity, so the net flux into this strip, per unit time, is "flux in" minus "flux out", that is

$$(H + h)v\big|_{x - \frac{1}{2}\delta x} - (H + h)v\big|_{x + \frac{1}{2}\delta x} \approx -\delta x((H + h)v)_x$$

by Taylor expansion. This can be further simplified to $-\delta x H v_x$ by virtue of the assumption $|h| \ll H$. If water is neither added nor removed from the tank, conservation of (fluid) mass requires that this net influx must be balanced by the rate of change in the volume $(H + h)\delta x$ of the column. Thus, since H does not vary with t,

$$\partial_t((H + h)\delta x) = \partial_t(h\delta x) = -\delta x H v_x,$$

so, on dividing by δx, we obtain

$$h_t + H v_x = 0. \tag{3.8}$$

This is one equation that connects the two dependent variables h and v. A second equation is obtained by applying Newton's second law of motion in the rectangle of size $\delta x \times \delta y$ highlighted in Fig. 3.3. The mass of water in the rectangle is $\rho \delta x \delta y$ and this has momentum (mass×velocity) $\rho v \delta x \delta y$ in the x-direction and therefore acceleration (rate of change of momentum) $(\rho v \delta x \delta y)_t$. The net horizontal force exerted on this rectangle is due to the pressure (force per unit area when dealing with surfaces, here it is the force per unit length of boundary). The net force in the x-direction is then

$$P\delta y\big|_{x - \frac{1}{2}\delta x} - P\delta y\big|_{x + \frac{1}{2}\delta x} \approx -\delta x \delta y P_x.$$

Newton's second law of motion gives $\rho \delta x \delta y v_t = -\delta x \delta y P_x$, which simplifies to

$$\rho v_t = -P_x. \tag{3.9}$$

When (3.9) is combined with the hydrostatic approximation (b), this leads to

$$v_t + gh_x = 0. \tag{3.10}$$

Combining (3.8) with (3.10) it is readily deduced (see Exercise 3.1) that the perturbation h satisfies the wave equation (**pde.5**)

$$h_{tt} - c^2 h_{xx} = 0, \tag{3.11}$$

with wave speed $c = \sqrt{gH}$. This equation models "shallow water waves" (including tsunamis and tidal waves in estuaries) amazingly well. Even in deep sea the wavelength of a disturbance can be much greater than the depth and in such cases the wave speed can reach $\sqrt{gH} \approx 200$ m/s or 400 mph.[3]

The wave equation is pretty important. It is used to model a host of physical phenomena, including: the vibration of drum surfaces, the propagation of light and other electromagnetic waves, the propagation of sound waves and the propagation of disturbances in the earth caused by earthquakes. A more complete review of possibilities can be found in Fowkes and Mahon [5].

3.2.4 Burgers' Equation

Next we examine the flow of water in a gently-sloping river that is assumed to have a constant slope and vertical banks. In this situation the flow velocity v of the water is determined differently from the previous example (Fig. 3.4). The speed of the flow downstream under the force of gravity builds up until resistive forces balance the component of gravitational force acting downstream.

In the simplest model the resistive force R increases in direct proportion to the flow velocity v, so that $R = av$ and the component of gravitational force F downstream increases in proportion to the depth of water h, giving $F = bh$. The flow speed

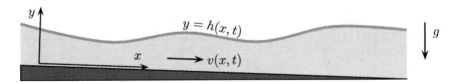

Fig. 3.4 Sketch of a gently sloping river with surface height $h(x, t)$

---

[3]This is roughly how fast a tsunami can cross an ocean. The destructive power of a tsunami is testament to the momentum that it carries.

adjusts itself until these two forces balance, giving $av = bh$, or simply

$$v = Ch$$

for some constant C.

We further assume that rain (or small tributaries) add water and we let seepage into the ground remove water. This will increase the depth of water at any point in the river at the rate $r(h, x, t)$ and the mass conservation law is given by (see Exercise 3.4)

$$h_t + (hv)_x = r. \tag{3.12}$$

Substituting for v using the equation above gives

$$u_t + u u_x = f \tag{3.13}$$

with $u = 2Ch$ and $f = 2Cr$. This is a nonhomogeneous inviscid Burgers' equation (**pde.2**). It is one of the simplest mathematical models used by environment agencies to model flash floods. It also describes numerous other situations in which the speed with which a property moves depends on the property itself.

Extension 3.2.4 (Traffic flow) The flow of traffic might involve movement at a speed v that depends on the density of traffic d in some manner, $v = S(d)$, say. The flux of traffic is dv so that, with inflow from side roads given by $r(d, x, t)$, the conservation law for traffic flow becomes

$$d_t + \big(S(d)d\big)_x = r.$$

If, as one example amongst many possibilities, $S(d) = K/d^2$ with K a constant (a perfectly reasonable model at moderately high traffic density that is below traffic jam conditions) then the PDE flow model would be $d_t + (K/d)_x = r$, or, setting $u = d/K$ and $f = r/K$,

$$u_t - \frac{1}{u^2} u_x = f.$$

Solving this kind of PDE might help ease congestion on the route between Dundee and Manchester! ◊

Exercises

3.1 ☆ Deduce from (3.8) to (3.10) that the height $h(x, t)$ satisfies the wave equation (3.11). Show also that $v_{tt} - gh\, v_{xx} = 0$.

3.2 ☆ How is the governing PDE modified in Sect. 3.2.4 if the resistive force is proportional to the square of the fluid speed: $R = av^2$?

3.3 * If the velocity field $\vec{v}$ in (3.5) is an incompressible flow (so that $\vec{\nabla} \cdot \vec{v} = 0$), deduce that T satisfies the non-conservative form[4] of the advection–diffusion equation

$$T_t + \vec{v} \cdot \vec{\nabla} T = \kappa \nabla^2 T + f$$

in which the term $\vec{v} \cdot \vec{\nabla} T$ represents the derivative of T in the direction of $\vec{v}$ multiplied by the scalar magnitude of v.

3.4 Use an argument similar to that leading to the conservation law (3.8) in order to derive (3.12).

---

[4]This equation is still a conservation equation, having the same validity as (3.5), in spite of not being written in conservative form.

Chapter 4
Classification of PDEs

Abstract This chapter introduces the notion of characteristics. The direction of characteristics is shown to be connected to the imposition of boundary and initial conditions that lead to well-posed problems—those that have a uniquely defined solution that depends continuously on the data. A refined classification of partial differential equations into elliptic, parabolic and hyperbolic types can then be developed.

Some of the more important PDEs were grouped together in Chap. 2 using the criteria of order and linearity. Different types of nonlinearity were categorised and we dipped into the rather deeper waters of well-posed and ill-posed boundary value problems. The classification of PDEs is further developed in this chapter using the notion of *characteristics*. By way of motivation, one might like to identify which boundary conditions for problems involving homogeneous PDEs such as

$$2u_x + 3u_y = 0 \quad \text{and} \quad u_{xx} - 2u_{xy} - 3u_{yy} = 0$$

lead to well-posed problems. Both equations are examples of the general *linear* second-order PDE

$$au_{xx} + 2bu_{xy} + cu_{yy} + pu_x + qu_y + ru = f, \tag{4.1}$$

where the coefficients a, b, c, p, q, r, f are functions of x and y. This general PDE will be the centre of the discussion in this chapter. (Examples with constant coefficients are an important special case of (4.1) that frequently arise in applications.) Defining the operator $\mathcal{L}$ by

$$\mathcal{L} := a\partial_x^2 + 2b\partial_x\partial_y + c\partial_y^2 + p\,\partial_x + q\partial_y + r$$

we can succinctly express (4.1) in the form $\mathcal{L}u = f$. It can be readily verified that $\mathcal{L}(\alpha u + \beta v) = \alpha\mathcal{L}u + \beta\mathcal{L}v$ so $\mathcal{L}$ is a linear operator, even in the case of variable coefficients.

© Springer International Publishing Switzerland 2015
D.F. Griffiths et al., *Essential Partial Differential Equations*, Springer
Undergraduate Mathematics Series, DOI 10.1007/978-3-319-22569-2_4

4.1 Characteristics of First-Order PDEs

We first consider the very special case of (4.1) with $a = b = c = r = 0$, that is

$$pu_x + qu_y = f. \tag{4.2}$$

In many physical applications, see (**pde.1**), one of the independent variables might represent a time-like variable. In stationary applications both variables might be spatial variables. We begin by considering a curve defined by the height of the surface $z = u(x, y)$ in three dimensions above a path $(x(t), y(t))$ in the x-y plane that is parameterised by t. This curve has slope

$$\frac{du}{dt}(x(t), y(t))$$

which, by the chain rule, is given by

$$\frac{du}{dt}(x(t), y(t)) = u_x \frac{dx}{dt} + u_y \frac{dy}{dt}. \tag{4.3}$$

Thus, by choosing the parameterization such that

$$\frac{dx}{dt} = p, \quad \frac{dy}{dt} = q, \tag{4.4}$$

the PDE (4.2) reduces to the ODE

$$\frac{du}{dt} = f. \tag{4.5}$$

The parameter t is not an intrinsic part of the system and can be avoided by writing the three ODEs in (4.4) and (4.5) in the generic form

$$\frac{dx}{p} = \frac{dy}{q} = \frac{du}{f}. \tag{4.6}$$

Paths in the x-y plane described by (4.4) are known as *characteristic curves* or, simply, as characteristics, and equations (4.6) are known as the characteristic equations of (4.2). The relations (4.6) define three equations, of which any two are independent.

For the remainder of this section we shall suppose that p and q are constant coefficients. A more detailed study of the characteristics of variable coefficient and nonlinear first-order PDE problems can be found in Chap. 9.

Example 4.1 Find the general solution of the PDE $pu_x + u_y = u$, where p is constant. Show that the problem is well posed when solved in the infinite strip $\{(x, y) : x \in \mathbb{R}, 0 \leq y \leq Y\}$ and an initial condition $u(x, 0) = g(x), x \in \mathbb{R}$, is applied, where g is a continuous bounded function.

The characteristic equations are

$$\frac{dx}{p} = \frac{dy}{1} = \frac{du}{u},$$

which we may write as

$$\frac{dx}{dy} = p, \quad \frac{du}{dy} = u$$

so that $x = py + k$ and $u = Ae^y$ on any single characteristic. The value of k changes from one characteristic to another and the different values of k distinguish different characteristics. The value of A is also constant along any characteristic but will change, along with k, from one characteristic to another. The appropriate form for the general solution is therefore

$$x = py + k, \quad u = A(k)e^y.$$

Thus the characteristics are simply straight lines along which u varies exponentially. The parameter k may be eliminated to give the general solution directly in terms of x and y:

$$u(x, y) = A(x - py)e^y. \tag{4.7}$$

The unknown function A can be explicitly determined using the initial condition. Setting $y = 0$ in (4.7) we find that $A(x) = g(x)$ which then gives the unique solution

$$u(x, y) = e^y g(x - py).$$

Finally, taking the maximum of both sides with $x \in \mathbb{R}$ and $0 \le y \le Y$ we find

$$|u(x, y)| \le e^Y \max_{x \in \mathbb{R}} |g(x)|$$

from which the well-posedness of the BVP follows.

In the next example we explore how the arbitrary function may be determined by application of a boundary condition on any line through the origin.

Example 4.2 (Half-plane problem) Solve the PDE $pu_x + qu_y = f$ (with p and q constant) in the domain $\alpha x + \beta y > 0$ given that $u = g(x)$ on the line $\ell : \alpha x + \beta y = 0$, where $q > 0$ and $\beta > 0$.

As in the previous example, the characteristic equations (4.4) are readily solved to give

$$x = pt + C_1, \quad y = qt + C_2, \tag{4.8}$$

where C_1 and C_2 are constant along any characteristic. The assumption $q > 0$ means that y increases along a characteristic as t increases—thus the parameter t may be

Fig. 4.1 Characteristics for
Example 4.2 when α, β, p
and q are all positive. The
highlighted characteristic
through the point P intersects
the line ℓ at Q.

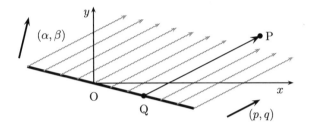

viewed as a time-like variable. This is illustrated in Fig. 4.1 where the arrows on the
characteristics indicate increasing t.

In order to determine the solution at a point P(x, y), we trace the characteristic
through this point backwards in t until it intersects the line ℓ. We shall suppose that
this occurs at the point Q having coordinates $x = \beta s$, $y = -\alpha s$ when[1] $t = 0$. Then,
from (4.8), we find $C_1 = \beta s$, $C_2 = -\alpha s$. Here s is a parameter that plays a role
similar to that of k in the previous example. It varies along the line ℓ and each choice
of s selects a different characteristic:

$$x = pt + \beta s, \quad y = qt - \alpha s. \tag{4.9}$$

Note that the characteristics have inherited their parameterization from that of the
line ℓ. Next, we integrate the ODE (4.5),

$$\int_0^t \frac{du}{dt'} \, dt' = \int_0^t f\left(x\left(s, t'\right), y\left(s, t'\right)\right) dt',$$

to give

$$u(s, t) = u(s, 0) + F(t), \quad F(t) = \int_0^t f\left(x\left(s, t'\right), y\left(s, t'\right)\right) dt'.$$

Here, $A(s) = u(s, 0)$ is constant on any characteristic but varies, with s, from one
characteristic to another. The initial condition gives $u = g(x) = g(\beta s)$ at $t = 0$ from
which we deduce (using (4.9)) that $u(s, 0) = g(\beta s)$ so that

$$u = g(\beta s) + \int_0^t f\left(x\left(s, t'\right), y\left(s, t'\right)\right) dt'. \tag{4.10}$$

---

[1]The origin for t is immaterial, the intersection could be assumed to occur at $t = t_0$, say, without
affecting the resulting solution so long as we replace all occurrences of t by $t - t_0$.

Finally, to express u as a function of x and y, the equation (4.9) have to be solved to give s and t in terms of x and y:

$$t = \frac{\alpha x + \beta y}{\alpha p + \beta q}, \quad s = \frac{qx - py}{\alpha p + \beta q},$$

which are valid so long as $\alpha p + \beta q \neq 0$. This condition has a geometric interpretation. Since $\alpha p + \beta q$ is the scalar product of the vector (p, q) (which is parallel to the characteristics) and the vector (α, β) (which is orthogonal to the line ℓ), the change of variables (4.9) is one-to-one if, and only if, (p, q) is not parallel to (α, β). That is, *the initial condition should not be specified on any line parallel to the characteristics* (see Fig. 4.1). This is an important conclusion that applies more widely and, when it holds, we have a unique solution at a point P (see Fig. 4.1) that depends only on the initial value g given at the foot Q of the characteristic through P and the values of f along PQ.

Finally, note that in the simple homogeneous case the coefficients of the PDE $pu_x + qu_y = 0$ form a vector (p, q), whereas the solution (4.10) depends only on s, that is, $qx - py$ whose coefficients form a vector $(q, -p)$ that is orthogonal to (p, q). This means that boundary information is simply conveyed along characteristics. This is visually evident in the solution to Example 1.2 in Chap. 1. Since information travels a horizontal distance p/q for each unit of vertical distance, p/q is known as the *characteristic speed* of the equation. For example in the case of (**pde.1**), solved in Example 1.2, the characteristic speed is equal to the wave speed c.

Example 4.3 (Quarter-plane problem) Solve the PDEs $u_x + u_y = 0$ and $u_x - u_y = 0$ in the quarter plane $x > 0, y > 0$ given that $u(x, 0) = g_0(x)$ for $x > 0$ and $u(0, y) = g_1(y)$ for $y > 0$.

From the previous example with $p = q = 1$ we see that the PDE $u_x + u_y = 0$ has characteristics $x - y = $ constant and general solution $u(x, y) = F(x - y)$ (where F is an arbitrary function) which is constant along characteristics. The solution at any particular point thus depends on whether the characteristic through that point first intersects the x-axis or the y-axis (see Fig. 4.2, left). For example, the characteristic through a point $P_0(X_0, Y_0)$ with $X_0 > Y_0$ cuts the x-axis at $Q_0(X_0 - Y_0, 0)$ and so, using the boundary condition $u(x, 0) = g_0(x)$, we find

$$u(X_0, Y_0) = g_0(X_0 - Y_0)$$

or, more succinctly, $u(P_0) = g_0(Q_0)$. Similarly, the characteristic through $P_1(X_1, Y_1)$ with $X_1 < Y_1$ cuts the y-axis at $Q_1(0, Y_1 - X_1)$ and so, using the boundary condition $u(0, y) = g_1(y)$, we find

$$u(X_1, Y_1) = g_1(Y_1 - X_1)$$

or $u(P_1) = g_1(Q_1)$. If the boundary conditions are such that $g_0(0) = g_1(0)$ then the solution in the quarter plane will be continuous and $u(X, X) = g_0(0) = g_1(0)$ for

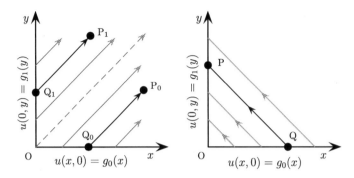

Fig. 4.2 Characteristics for Example 4.3 with $p = q = 1$ (*left*) and with $p = 1, q = -1$ (*right*)

all $X \geq 0$. Otherwise, there is a discontinuity at the origin which will propagate into the domain along the characteristic $x = y$ shown as a dashed line in Fig. 4.2 (left). To the right of the dashed line the solution is $u(x, y) = g_0(x - y)$ and to the left, $u(x, y) = g_1(y - x)$.

For $u_x - u_y = 0$ the characteristics are $x + y = $ constant and the general solution is $u(x, y) = F(x + y)$, where F is again an arbitrary function. What distinguishes this from the previous case is that any characteristic now intersects the boundary twice (see Fig. 4.2 (right)). The characteristic through the point $P(0, Y)$, for example, cuts the x-axis at $Q(Y, 0)$ and so u has the value $u = g_0(Y)$ along this characteristic. However, at P we have $u(0, Y) = g_1(Y)$ and, in general, there will be a contradiction at P between the value of u at the boundary and its value on the characteristic. This implies that the problem is not properly posed.

The moral to be taken from the two PDE problems in this example is that boundary conditions should only be imposed on boundaries along which characteristics are directed *into* the domain. The PDE $u_x - u_y = 0$ should have a boundary condition on the x-axis but not the y-axis. $\diamond$

It was shown in Example 4.2 that a boundary condition should not be applied on boundaries parallel to characteristics. Our final example illustrates the consequences of a characteristic being tangential to the curve along which a boundary condition is specified.

Example 4.4 Solve the PDE $u_x + u_y = 0$ in the domain $y > \varphi(x)$, $x \in \mathbb{R}$ given that $u = g(x)$ on the curve $y = \varphi(x)$, where $\varphi(x) = x/(1 + |x|)$.

We first observe that the curve $\varphi(x) = x/(1 + |x|)$ is continuously differentiable (see Exercise 4.1) and monotonically increases from its value $\varphi(x) = -1$ as $x \to -\infty$ to $\varphi(x) = 1$ as $x \to \infty$. The characteristics $y = x + $ constant and the initial curve are shown in Fig. 4.3 (left). Since $\varphi'(0) = 1$ the characteristic $y = x$ through the origin (shown dashed in the figure) is tangent to $\varphi(x)$ at the origin.

The general solution of the PDE is $u(x, y) = A(x - y)$, where $A(\cdot)$ is an arbitrary function. Parameterizing the initial curve by s, the boundary condition becomes

$$x = s, \quad y = \varphi(s), \quad u(s, \varphi(s)) = g(s)$$

so that the function A may be determined from the relationship

$$u(s, \varphi(s)) = A(s - \varphi(s)) = g(s).$$

This requires that $x = s - \varphi(s)$ be solved for s in terms of x so as to give $A(x) = g(s(x))$. When $\varphi(x) = x/(1 + |x|)$ this is accomplished by treating the cases $x \geq 0$ and $x < 0$ separately,

$$s(x) = \begin{cases} \frac{1}{2}\left(x + \sqrt{x^2 + 4x}\right) & x \geq 0 \\ \frac{1}{2}\left(x - \sqrt{x^2 - 4x}\right) & x < 0 \end{cases}. \tag{4.11}$$

This leads us to the explicit solution $u(x, y) = g(s(x - y))$, which uniquely defines u throughout the domain $y > \varphi(x)$.

Snapshots of the solution u in the case of the smooth boundary condition, $g(x) = e^{-x^2}$, are visualised in Fig. 4.3. An unexpected feature of the solution is that, although g is infinitely differentiable, there is a discontinuity in the slope of the solution u along the characteristic $y = x$. This is attributable to this characteristic being tangent to the initial curve at the origin.

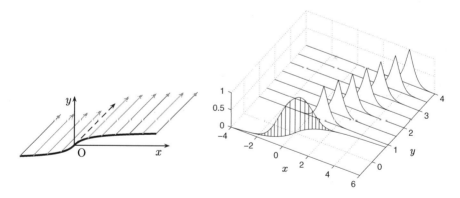

Fig. 4.3 Characteristics for Example 4.4 (*left*) and snapshots of the solution plotted at half unit intervals for $1 \leq y \leq 4$ (*right*)

4.2 Characteristics of Second-Order PDES

We now turn to the issue of boundary conditions that ensure the well-posedness of BVPs associated with the linear PDE (4.1) when the coefficients are constant. Since the generic behaviour of solutions is dependent on the terms involving the highest derivatives we will focus on the special case $p = q = r = 0$, that is, the PDE

$$\mathcal{L}u := a\partial_x^2 u + 2b\partial_x\partial_y u + c\partial_y^2 u = f, \tag{4.12}$$

where $f(x, y)$ is a given function and we draw attention to the factor of 2 multiplying the mixed derivative. For the sake of definiteness, and without loss of generality, we will assume that $a \geq 0$.

There are three alternative possibilities known as *hyperbolic*, *parabolic* and *elliptic* PDEs, named not for any connection with conic sections but because of the shape of the level curves of the associated quadratic form

$$Q(x, y) := ax^2 + 2bxy + cy^2. \tag{4.13}$$

Important geometric properties of quadratic forms are reviewed in Appendix B.4. The key point here is that the shape of the level curves can be determined by making an appropriate change of variable. Making the same change of variable to the PDE provides information about the underlying *characteristics*.[2] The geometric properties of (4.13) depend on the sign of its discriminant: $d = b^2 - ac$. The sign of this discriminant determines whether or not the second derivative operator (4.12) can be factorised into the product of real one-dimensional operators. There are three cases to consider.

Hyperbolic   ($b^2 - ac > 0$). $\mathcal{L}$ has two distinct real factors:

$$\mathcal{L}u = (\alpha\partial_x + \beta\partial_y)(\gamma\partial_x + \delta\partial_y)u \tag{4.14}$$

with real numbers α, β, γ and δ such that

$$\frac{\beta}{\alpha} \neq \frac{\delta}{\gamma}. \tag{4.15}$$

Building on the earlier discussion (see Example 4.2), This implies that a second-order hyperbolic PDE operator has two distinct characteristic speeds β/α and δ/γ.

---

[2] In practical situations BVPs are defined on domains in $\mathbb{R}^2$ and any change of variables is likely to distort the boundary of the domain and thereby complicate the imposition of boundary conditions. Such changes of variable should therefore be viewed as tools to investigate the theoretical properties of PDEs.

Parabolic ($b^2 - ac = 0$; $a > 0$). $\mathcal{L}$ is a perfect square:

$$\mathcal{L}u = (\alpha\partial_x + \beta\partial_y)^2 u$$

for real numbers α and β. This implies that a parabolic partial differential operator has a single characteristic speed β/α.

Elliptic ($b^2 - ac < 0$). This means that neither a nor c can be zero and that both have the same sign. It also implies that an elliptic operator $\mathcal{L}$ has no real factors, so the characteristics are *complex*.

Prototypical examples of the three types of PDE will be discussed below. Identifying solutions of hyperbolic PDEs turns out to be straightforward, so this case is considered first. The characterisation of solutions of parabolic and elliptic PDEs will provide more of a challenge.

4.2.1 Hyperbolic Equations

From our considerations of first-order PDEs in the previous section we recall that the operator $p\partial_x + q\partial_y$ has associated characteristics defined by (4.4) along which $qx - py$ is constant. Thus the factorisation (4.14) gives rise to two families of characteristics:

$$s = \delta x - \gamma y; \quad t = \beta x - \alpha y, \tag{4.16}$$

which we shall refer to as the s- and the t-characteristics, respectively. Making this change of variables and using the chain rule gives

$$\frac{\partial u}{\partial x} = \frac{\partial u}{\partial s}\frac{\partial s}{\partial x} + \frac{\partial u}{\partial t}\frac{\partial t}{\partial x} = \delta\frac{\partial u}{\partial s} + \beta\frac{\partial u}{\partial t}$$

$$\frac{\partial u}{\partial y} = \frac{\partial u}{\partial s}\frac{\partial s}{\partial y} + \frac{\partial u}{\partial t}\frac{\partial t}{\partial y} = -\gamma\frac{\partial u}{\partial s} - \alpha\frac{\partial u}{\partial t}.$$

It follows that

$$\alpha\partial_x + \beta\partial_y = \sigma\partial_s, \quad \gamma\partial_x + \delta\partial_y = -\sigma\partial_t,$$

where $\sigma = \alpha\delta - \beta\gamma$ and $\sigma \neq 0$ because of (4.15), thus from (4.14) we find that

$$\mathcal{L}u = -\sigma^2\partial_s\partial_t u. \tag{4.17}$$

At this point the PDE $\mathcal{L}u - f$ is readily integrated (see Example 1.1):

$$u(s, t) = F(s) + G(t) - \Phi(s, t), \quad \Phi(s, t) = \frac{1}{\sigma^2}\iint f(x(s, t), y(s, t)\,\mathrm{d}s\,\mathrm{d}t,$$

where $F(s)$ and $G(t)$ are arbitrary functions of integration. Changing back to x-y coordinates (by inverting (4.16)) then gives an explicit characterisation of the general solution to our hyperbolic PDE:

$$u(x, y) = F(\delta x - \gamma y) + G(\beta x - \alpha y) - \Phi(s(x, y), t(x, y)). \qquad (4.18)$$

Note that the solution has three components: (a) the integral Φ of the source term, (b) the function $F(\delta x - \gamma y)$ that is constant on s-characteristics and, (c) the function $G(\beta x - \alpha y)$ that is constant on t-characteristics. This structure conforms to a familiar pattern of being the general solution of the homogeneous equation (the F and G terms) combined with a particular solution (the Φ term) of the inhomogeneous equation.

Reduction of hyperbolic equations to the form $u_{st} = \phi(s, t)$ is particularly suitable for pencil-and-paper solutions. The next example shows the elegance of this approach when applied to the wave equation (**pde.5**).

Example 4.5 (The wave equation) Find a solution of the PDE $u_{tt} - c^2 u_{xx} = 0$ (where c is a constant) in the half plane $t > 0$ with initial conditions $u(x, 0) = g_0(x)$ and $u_t(x, 0) = g_1(x)$.

Note that the wave equation is expressed here in x-t coordinates since, in practice, one of the independent variables invariably denotes time. The factorisation

$$u_{tt} - c^2 u_{xx} = (\partial_t + c\partial_x)(\partial_t - c\partial_x)u$$

suggests the change of variables $y = x - ct$, $s = x + ct$ and, following the steps taken in the previous example, leads us immediately to the general solution

$$u(x, t) = F(x - ct) + G(x + ct), \qquad (4.19)$$

where the two characteristic speeds are $\pm c$. Next, knowing the value of u and its normal derivative along the x-axis, one can explicitly determine F and G (see Exercise 4.3) to show that

$$u(x, t) = \frac{1}{2}(g_0(x - ct) + g_0(x + ct)) + \frac{1}{2c} \int_{x-ct}^{x+ct} g_1(z)\,\mathrm{d}z. \qquad (4.20)$$

This is called d'Alembert's solution to the wave equation.   ◊

4.2.2 Parabolic Equations

In the case $b^2 = ac$ the operator $\mathcal{L}$ defined by (4.12) is a perfect square,

$$\mathcal{L}u = (\alpha\partial_x + \beta\partial_y)^2 u. \qquad (4.21)$$

We will draw attention to the most important properties of parabolic PDEs by looking at some distinguished examples. Parabolic equations usually arise in space-time situations and, in addition to second derivative terms that form a perfect square, these problems typically have nonzero first derivative terms. The prototypical example of a PDE of this type is the heat (or diffusion) equation (**pde.4**) derived in Sect. 3.2.1.

Example 4.6 (The heat equation) Solve the PDE $u_t = u_{xx}$ defined on the half-plane $t > 0$ with initial condition $u(x, 0) = g(x)$ and establish the well-posedness of the initial value problem.

By writing the equation as $(\partial_x + 0\,\partial_t)^2 u = u_t$, we see that its characteristics would follow paths defined by

$$\frac{dx}{dt} = \pm\infty$$

and, in this sense, the heat equation has two infinite characteristic speeds. That is to say, information is transmitted from one part of the domain to another instantaneously. Confirmation of this assertion is provided by the fundamental solution

$$u(x, t) = \int_{-\infty}^{\infty} k(x - s, t)\, g(s)ds, \quad k(x, t) = \frac{1}{\sqrt{4\pi t}} e^{-x^2/4t} \tag{4.22}$$

that was identified in Example 1.3. For given values of x and $t > 0$, the kernel function $k(x - s, t) > 0$ is strictly positive for all $s \in \mathbb{R}$, so we can immediately conclude that $u(x, t) > 0$ whenever $g(x) > 0$ for all $x \in \mathbb{R}$: positive initial data leads to a positive solution. Moreover, if we suppose that $g(x) = c$, a positive constant, in a small neighbourhood $(x_0 - \varepsilon, x_0 + \varepsilon)$ of some point x_0 and is otherwise zero, we discover that

$$u(x, t) = \frac{c}{\sqrt{4\pi t}} \int_{x_0-\varepsilon}^{x_0+\varepsilon} e^{-(x-s)^2/4t} ds > 0.$$

Thus, since u is never equal to zero, we conclude that the initial value at x_0 affects the solution at (x, t) regardless of the distance between x and x_0 and regardless of how small the value of t that is taken.

Next, since $k(x - s, t) > 0$ and using the fact that

$$\int_{-\infty}^{\infty} k(x - s, t)ds = 1 \tag{4.23}$$

(which follows from the result given in Exercise 2.3), it follows from (4.22) that

$$|u(x, t)| \le \max_{s \in \mathbb{R}} |g(s)|.$$

This establishes the well-posedness of the (initial value) problem.

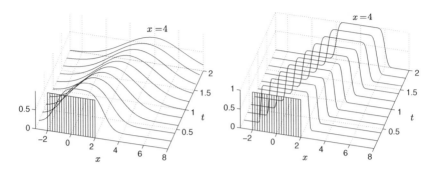

Fig. 4.4 The solution to Example 4.7 with $a = 2$, $g(x) = 1$ for $-2 \leq x \leq 2$ and with $\kappa = 1$ (*left*) and $\kappa = 0.001$ (*right*)

Example 4.7 (The advection–diffusion equation) Determine the solution of the PDE $u_t + au_x = \kappa u_{xx}$ (where a and $\kappa > 0$ are constant) on the half plane $t > 0$ with initial condition $u(x, 0) = g(x)$.

It can be shown (see Exercise 4.12) that a viable solution to this PDE is given by $u(x, t) = \phi(\xi, t)$, where $\xi = (x - at)/\sqrt{\kappa}$ and $\phi_t = \phi_{xx}$. Since $u(x, 0) = \phi(x/\sqrt{\kappa}, 0)$ it immediately follows, using the representation (4.22) for ϕ, that

$$u(x, t) = \frac{1}{\sqrt{4\pi t}} \int_{-\infty}^{\infty} \exp\left(-\frac{1}{4t}\left(\frac{x - at}{\sqrt{\kappa}} - s\right)^2\right) g(s\sqrt{\kappa})ds. \qquad (4.24)$$

For the particular case of an initial "square pulse" with $g(x) = 1$ for $|x| \leq 2$ and $g(x) = 0$ otherwise, the PDE solution u can be succinctly expressed using the *error function*

$$\mathrm{erf}(x) := \frac{2}{\sqrt{\pi}} \int_0^x e^{-s^2}ds. \qquad (4.25)$$

In particular, it may be shown that

$$u(x, t) = \frac{1}{2}\left(\mathrm{erf}\left(\frac{1}{\sqrt{4\kappa t}}(x - at + 2)\right) - \mathrm{erf}\left(\frac{1}{\sqrt{4\kappa t}}(x - at - 2)\right)\right). \qquad (4.26)$$

This solution is illustrated in Fig. 4.4. Note that the initial data is swept to the right with the advection speed a but, unlike the corresponding hyperbolic situation, discontinuities in the initial data are smeared in time. This smoothing of initial data is typical of parabolic equations. As $\kappa \to 0_+$ the solutions approach those of the one-way wave equation (**pde.1**).

4.2.3 Elliptic Equations

In the case $b^2 < ac$ the operator $\mathcal{L}$ defined by (4.12) may be written as

$$\mathcal{L} = \frac{1}{a}\left((a\partial_x + b\partial_y)^2 + (ac - b^2)\partial_y^2\right), \tag{4.27}$$

that is, as a sum of two squares. Even though $\mathcal{L}$ has no real factors it is possible to follow the procedure used for hyperbolic problems but with complex characteristics—this idea is pursued in Exercises 4.23 and 4.24.

The use of complex characteristics turns out to be of limited utility, so instead, we shall investigate an elegant change of variables that transforms $\mathcal{L}u = f$ to give Poisson's equation:

$$-(u_{ss} + u_{tt}) = \phi, \tag{4.28}$$

where $\phi(s, t)$ is the "source term" and is associated with the change of variables in the right-hand side function f. We will see that such as change of variables is given by

$$s = \alpha(bx - ay), \quad t = x$$

in which α is a scaling constant. Applying the chain rule gives

$$\partial_x = \alpha b\partial_s + \partial_t, \quad \partial_y = -\alpha a\partial_s$$

and choosing $\alpha = -1/\sqrt{(ac - b^2)}$ leads to $\mathcal{L} = a(\partial_s^2 + \partial_t^2)$, which matches the PDE operator in the Poisson equation.

For the remainder of the section we will assume that the elliptic PDE is homogeneous so that f in (4.12) is zero. Recalling Example 1.4, we know that

$$u = \frac{1}{4\pi}\log(s^2 + t^2), \quad (s, t) \neq (0, 0),$$

is a solution of Laplace's equation $u_{ss} + u_{tt} = 0$ (that is, the homogeneous version of the Poisson equation). A short calculation then shows that $s^2 + t^2 = a\widehat{Q}(x, y)$, with

$$\widehat{Q}(x, y) = \frac{cx^2 - 2bxy + ay^2}{ac - b^2},$$

and so, (provided x and y are not both zero) a solution of the elliptic equation $\mathcal{L}u = 0$ is given by the simple expression

$$u(x, y) = \frac{1}{4\pi}\log \widehat{Q}(x, y). \tag{4.29}$$

To explore the implications of (4.29), we introduce the symmetric matrix

$$A = \begin{bmatrix} a & b \\ b & c \end{bmatrix} \tag{4.30}$$

formed from the coefficients of $\mathcal{L}$ defined by (4.12). The quadratic form Q in (4.13) and its close relative $\widehat{Q}$ may be expressed in matrix-vector notation as

$$Q(x) = x^T A x, \quad \widehat{Q}(x) = x^T A^{-1} x, \quad x = [x, y]^T.$$

We now suppose that A has eigenvalues λ_1, λ_2 with corresponding eigenvectors v_1, v_2 normalised to have unit length. The eigenvalues are real by virtue of the fact that A is symmetric and, from the ellipticity condition $ac > b^2$, it follows that $\det A > 0$ and so the product $\lambda_1 \lambda_2 > 0$. Then, since[3] $\operatorname{tr} A = a + c = \lambda_1 + \lambda_2 > 0$ and, knowing that $a > 0$, we deduce that A has two positive eigenvalues.

Suppose that the two eigenvalues are ordered so that $\lambda_1 \geq \lambda_2 > 0$. We know that the principal axes of the ellipses defined by the level curves $Q = \text{constant}$ and $\widehat{Q} = \text{constant}$ are in the directions of the eigenvectors (details are given in Appendix B.4) and, moreover,

$$Q\left(\frac{1}{\sqrt{\lambda_j}} v_j\right) = 1, \quad \widehat{Q}\left(\sqrt{\lambda_j} v_j\right) = 1, \quad j = 1, 2.$$

The upshot is that general second-order elliptic PDEs model processes that act in an anisotropic manner. Diffusion is isotropic when the eigenvalues are equal, as in the special case of Laplace's equation, and the PDE is then invariant under rotation of the coordinate axes (see Exercise 4.13). Otherwise, when the two eigenvalues are not equal, there are two alternative ways of interpreting this geometrical information:

(a) The level curve $Q(x) = 1$ is an ellipse with major axis of length $1/\sqrt{\lambda_2}$ in the v_2 direction and minor axis of length $1/\sqrt{\lambda_1}$ in the v_1 direction. The diffusion coefficient is greater in the direction of v_2 than v_1.
(b) The level curve $\widehat{Q}(x) = 1$ is an ellipse with major axis of length $\sqrt{\lambda_1}$ in the v_1 direction and minor axis of length $\sqrt{\lambda_2}$ in the v_2 direction. The solution is therefore diffused (smeared) more in the direction of v_1 than v_2.

These points are illustrated in Fig. 4.5 for the elliptic operator

$$\mathcal{L}u = 13u_{xx} - 8u_{xy} + 7u_{yy}. \tag{4.31}$$

---

[3]The trace of a matrix A, denoted by $\operatorname{tr} A$, is the sum of its diagonal entries.

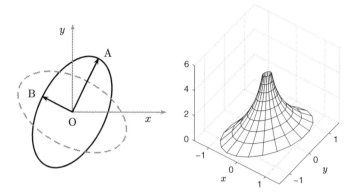

Fig. 4.5 *Left* the level curves $Q(x, y) = 1$ (*dashed*) and $\widehat{Q}(x, y) = 1$ (*solid*) related to the operator $\mathcal{L}$ defined by (4.31). OA and OB show the vectors $\sqrt{\lambda_1}\,\boldsymbol{v}_1$ and $\sqrt{\lambda_2}\,\boldsymbol{v}_2$, respectively. *Right* the solution (4.29) of $\mathcal{L}u = 0$

A discussion of the corresponding quadratic form Q is given in Example B.6, where it is shown that

$$\lambda_1 = 15, \ \lambda_2 = 5, \quad \boldsymbol{v}_1 = [1, 2]^T / \sqrt{5}, \ \boldsymbol{v}_2 = [-2, 1]^T / \sqrt{5}.$$

This example shows how the solution of Laplace's equation established in Example 1.4 can be used to generate a solution to general elliptic PDE $\mathcal{L}u = 0$.

The same solution also makes a subtle contribution to the solution of the half-plane problem for Laplace's equation in the next example—the connection is identified in Exercise 4.17. A word of caution is in order. The half-plane problem is ill posed if two initial conditions are specified on the x-axis as in Example 4.5, so a different arrangement of boundary conditions is necessary.

Example 4.8 (Laplace's equation) Explore the solution of the PDE $u_{xx} + u_{yy} = 0$ on the upper half plane $y > 0$, assuming that it satisfies the boundary condition $u(x, 0) = g(x)$ and the condition $u \to 0$ as $r = \sqrt{x^2 + y^2} \to \infty$.

A solution to this problem can be constructed that is analogous to that of the heat equation (4.22) discussed in Example 4.6. It takes the same form (see Exercise 4.16):

$$u(x, y) = \int_{-\infty}^{\infty} k(x - s, y)\, g(s)\, ds, \quad k(x, y) = \frac{1}{\pi}\left(\frac{y}{x^2 + y^2}\right), \tag{4.32}$$

except that the kernel function is defined slightly differently. The construction (4.32) allows us to deduce properties of the solution analogous to those of the heat equation. In particular,

(a) $u(x, y) > 0$ for $y > 0$ provided that $g(x) > 0$ for all $x \in \mathbb{R}$. That is, positive boundary data produces a positive solution.

Fig. 4.6 The solution to
Example 4.8 with $g(x) = 1$
for $-2 \leq x \leq 2$

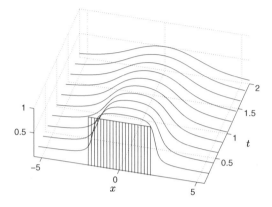

(b) If $g(x) = c$, a positive constant, in a neighbourhood of a point x_0 and is zero otherwise then $u(x, y) > 0$. That is, changing the data affects the solution *everywhere*.

(c) The boundary value problem is well posed: $|u(x, y)| \leq \max_{s \in \mathbb{R}} |g(s)|$.

Returning to the case of a "square pulse" $g(x)$ on the boundary, the integral in (4.32) can again be explicitly evaluated—this time it takes the form

$$u(x, y) = \frac{1}{\pi} \left(\tan^{-1} \left(\tfrac{x+2}{y} \right) - \tan^{-1} \left(\tfrac{x-2}{y} \right) \right). \tag{4.33}$$

Cross-sections of this function at different values of y are shown in Fig. 4.6. It can be readily seen that the discontinuous boundary data is smoothed as it diffuses into the domain. ◊

4.3 Characteristics of Higher-Order PDEs

The classification of nonlinear and higher-order PDEs can often be accomplished by factorizing the differential operators involved. Treating the general case would be overly ambitious for a textbook at this level, so we simply give some representative examples below.

Example 4.9 The PDE

$$u_{ttt} + u_{txx} + u u_{ttx} + u u_{xxx} = (\partial_t + u \partial_x)(u_{tt} + u_{xx})$$
$$= (\partial_t + u \partial_x)(\partial_t + i \partial_x)(\partial_t - i \partial_x) u = 0$$

is of mixed hyperbolic and elliptic type, since its operator factorises into one operator involving a real characteristic speed and two operators with imaginary characteristic speeds.

Example 4.10 The PDE

$$u_{ttt} - u_{txx} + uu_{ttx} - uu_{xxx} = (\partial_t + u\partial_x)(u_{tt} - u_{xx})u$$
$$= (\partial_t + u\partial_x)(\partial_t + \partial_x)(\partial_t - \partial_x)u = 0$$

is hyperbolic for $|u| \neq 1$, since its operator then factorises into operators each involving real and distinct characteristic speeds. Whenever $u = \pm 1$ two characteristic speeds coincide, making the PDE of mixed hyperbolic and parabolic type.

Example 4.11 The PDE

$$u_{tt} - u_{txx} + uu_{tx} - uu_{xxx} = (\partial_t + u\partial_x)(u_t - u_{xx})$$
$$= (\partial_t + u\partial_x)(\partial_t - \partial_{xx})u = 0$$

is of mixed hyperbolic and parabolic type, since its operator factorises into one parabolic operator and one operator involving a real characteristic speed.

4.4 Postscript

The examples in this chapter are intended to show that while the process of determining solutions to parabolic and elliptic equations is invariably a challenging task; the qualitative picture of the solutions, as revealed in Figs. 4.4 and 4.6, shows structure and simplicity. In contrast, constructing solutions to homogeneous hyperbolic problems is easier but the nature of the solutions can be relatively complex.

An important and general feature is that solutions of elliptic/parabolic problems are typically much smoother than their boundary data; whereas, for hyperbolic problems, discontinuities in boundary data are propagated into the domain with no smoothing. It is this distinction that makes the derivation of high-quality numerical methods for hyperbolic problems so much more challenging than for their elliptic/parabolic counterparts.

Exercises

4.1 If $\varphi(x) = x/(1 + |x|)$ for $x \in \mathbb{R}$, show that $\varphi'(x) = (1 + x)^{-2}$ for $x > 0$ and $\varphi'(x) = (1 - x)^{-2}$ for $x < 0$ and deduce that $\varphi(x)$ is a continuously differentiable function.

4.2 Consider the function $u(x, y) = g(s(x))$, where $g(t) = \exp(-t^2)$ and $s(x)$ is defined by (4.11) in Example 4.4. Find $u_x(x, y)$ for $x > y$ and $x < y$ and show that $u_x(x, y)$ is not continuous at $x = y$. In particular, show that $u_x(x, x_-) = 1$ while $u_x(x, x_+) = -1$ thus confirming the behaviour shown in Fig. 4.3.

4.3 Show, by applying the initial conditions given in Example 4.5, to the general solution (4.19) of the wave equation, that $F(x) + G(x) = g_0(x)$ and $G'(x) - F'(x) = g_1(x)/c$. Deduce D'Alembert's solution (4.20). [Hint: first replace x by $x \pm ct$ in the first of these relations and integrate the second over the interval $(x - ct, x + ct)$.]

4.4 Compute u_{xx} and u_{tt} from (4.20) and hence verify that u satisfies the wave equation $u_{tt} - c^2 u_{xx} = 0$ together with the initial conditions $u(x, 0) = g_0(x)$, $u_t(x, 0) = g_1(x)$, $x \in \mathbb{R}$. Sketch the characteristics.

4.5 Determine conditions under which each of the PDEs in Exercise 2.9 is elliptic, parabolic or hyperbolic.

4.6 Show that the wave equation $\partial_x^2 u - \partial_y^2 u = 0$ can be written

(a) in factored form: $(\partial_x - \partial_y)(\partial_x + \partial_y)u = 0$,
(b) as a first-order system $u_x = v_y$, $u_y = v_x$,
(c) as $u_{st} = 0$ under the change of variable $s = x + y$ and $t = x - y$.

Use the form (c) to determine the general solution.

4.7 Use the change of variables from Exercise 4.6(c) to determine the general solution of the PDE $u_{xx} - u_{yy} = x + y$ as a function of x and y. Can you find a solution that satisfies the boundary conditions $u(x, 0) = x$, $u(0, y) = -\frac{1}{2}y^3$?

4.8 Show that the PDE $u_{xx} - 2u_{xy} - 3u_{yy} = 0$ is hyperbolic and that it has the general solution $u(x, t) = F(3x + y) + G(x - y)$. Adapt the strategy outlined in Exercise 4.3 to determine the arbitrary functions so that u satisfies the initial conditions $u(x, 0) = g_0(x)$ and $u_t(x, 0) = g_1(x)$ for $x \in \mathbb{R}$.

4.9 Show that the equation $2u_{xx} + 5u_{xt} + 3u_{tt} = 0$ is hyperbolic. Use an appropriate change of variable to find the general solution of this equation. Hence, determine the solution satisfying the initial conditions $u(x, 0) = 0$, $u_t(x, 0) = xe^{-x^2}$ for $-\infty < x < \infty$.

4.10 Show that the change of dependent variable $v(r, t) = ru(r, t)$ transforms the PDE $a^2 \partial_r (r^2 u_r) = r^2 \partial_t^2 u$ to $a^2 \partial_r^2 v = \partial_t^2 v$ and that the change of independent variables $p = r - at, q = r + at$ then reduces this to the form $\partial_p \partial_q v = 0$. Deduce that the general solution can be written in the form

$$u(r, t) = \frac{1}{r}(f(r + at) + g(r - at))$$

for arbitrary functions f and g. Hence, find the solution satisfying the initial conditions $u(r, 0) = 0$, $\partial_t u(r, 0) = e^{-r^2}$, $-\infty < r < \infty$.

4.11 Show that the spherical wave equation in n dimensions

$$u_{tt} = u_{rr} + \frac{n-1}{r} u_r$$

has solutions of the form $u(r, t) = r^m f(t - r)$ provided that f satisfies a certain first-order ODE. This differential equation will be satisfied for all twice differentiable functions f provided that the coefficients of f and f' are both zero. Show that this is only possible when either $n = 1$ or $n = 3$, and determine the value of m in these cases.

4.12 Verify the derivation of the solution (4.24) of the advection–diffusion problem in Example 4.7.

4.13 Consider the change of variables

$$x = s \cos \alpha - t \sin \alpha, \quad y = s \sin \alpha + t \cos \alpha$$

corresponding to a rotation of the x-y axes through a constant angle α. Show that

$$u_{xx} + u_{yy} = u_{ss} + u_{tt},$$

so that Laplace's equation is invariant under such a transformation.

4.14 Show that the operator $\mathcal{L}$ defined by (4.12) may be written as

$$\mathcal{L} = \frac{1}{c} \left((ac - b^2) \partial_x^2 + (b\partial_x + c\partial_y)^2 \right).$$

By following the same steps as in Sect. 4.2.3 show that this leads to the same solution (4.29) as using (4.27).

4.15 Determine a solution of the PDE $2u_{xx} + 3u_{yy} = 0$ in the form (4.29) and sketch the contours $u(x, y) = $ constant.

4.16 Show, by differentiating under the integral sign, that (4.32) is a solution of Laplace's equation for $y \neq 0$.

4.17 Suppose that

$$G(x, y, t) = \frac{1}{4\pi} \left(\log(x^2 + (y + t)^2) - \log(x^2 + (y - t)^2) \right).$$

Show that the kernel $k(x, y)$ defined by (4.32) is related to G by $k(x, y) = \partial_t G(x, y, t)|_{t=0}$ and is normalised such that

$$\int_{-\infty}^{\infty} k(x - s, y) \, ds = 1.$$

4.18 Establish the properties (a)–(c) in Example 4.8.

4.19 By using the standard trigonometric formula for $\tan^{-1} a - \tan^{-1} b$, show that the curve $u(x, y) = \frac{1}{2}$, where u is given by (4.33), is a semicircle of radius 2.

Deduce that $u(x, y) \to \frac{1}{2}(g(2^+) + g(2^-))$ as $x \to 2, y \to 0$. What is the corresponding limit as $x \to -2, y \to 0$?

4.20 Consider the change of variable to polar coordinates: $x = r \cos \theta, y = r \sin \theta$. Use the chain rule to obtain u_r and u_θ in terms of u_x and u_y and hence show that

$$\partial_x = \cos \theta \partial_r - \frac{1}{r} \sin \theta \partial_\theta$$

$$\partial_y = \sin \theta \partial_r + \frac{1}{r} \cos \theta \partial_\theta.$$

Hence, by considering

$$\partial_x^2 u = (\cos \theta \partial_r - \frac{1}{r} \sin \theta \partial_\theta)(\cos \theta \partial_r - \frac{1}{r} \sin \theta \partial_\theta)u,$$

or otherwise, show that

$$u_{xx} + u_{yy} = u_{rr} + \frac{1}{r}u_r + \frac{1}{r^2}u_{\theta\theta}.$$

For what functions f and what values of n is $u = r^n f(\theta)$ a solution of Laplace's equation? Consider also $u = f(\theta) \ln r$.

4.21 Use the results of Exercise 4.20 to solve the following problems.

(a) Show that there is a function $u = F(r)$ which solves the Poisson equation $-\nabla^2 u = 1$ in the circle $r < 1$ and is zero on the boundary of the circle. (This function is used to compute the rigidity of a cylindrical column when twisted.)
(b) Show that there are two solutions of Laplace's equation in the circle $r < 1$ such that $u = \cos \theta$ on the boundary of the circle. Why should one of these be discarded?

4.22 ☆ Suppose that u and v satisfy the pair of first-order PDEs $u_x = v_y, u_y = -v_x$, known as the Cauchy–Riemann equations. Show that both u and v must also satisfy Laplace's equation.

4.23 Show that $u_{xx} + u_{yy} = u_{zz^*}$ under the complex change of variable $z = x + iy$ ($z^* = x - iy$). Hence prove that the general solution of Laplace's equation can be written as
$$u(x, y) = F(x + iy) + G(x - iy),$$

where F and G are arbitrary functions.

If F is a real function, deduce that $\Re F(x + iy)$ and $\Im F(x + iy)$ are also solutions of Laplace's equation.

4.24 Use the previous exercise to construct real families of solutions to Laplace's equation corresponding to each of the functions (a) $F(x) = x^n$ ($n = 0, 1, 2$), (b) $F(x) = \exp(-nx)$, (c) $F(x) = \sin nx$ and (d) $F(x) = \log x$ (Hint: Use polar coordinates).

Use the result of part (d) to give an alternative validation of the solution given in Example 1.4. How is this solution related to that given in Exercise 1.10?

4.25 The biharmonic equation $\nabla^4 u = f$ (see Extension 3.2.2) may be written as a coupled pair of Poisson equations $-\nabla^2 u = v$ and $-\nabla^2 v = f$. Use this, together with Exercise 4.20 to show that $u = Cr^2 \log r$ is a solution of the homogeneous biharmonic equation for $r > 0$ and any constant C.

4.26 Show that the change of variables in Exercise 4.23 reduces the homogeneous biharmonic equation to $u_{zzz^*z^*} = 0$.

Suppose that $u(x, y) = \Re\left(z^* F(z) + G(z)\right)$, where $z = x + iy$ ($z^* = x - iy$) and F, G are real functions. Show that u is a solution of the homogeneous biharmonic equation. This solution is associated with the name of Goursat.

Chapter 5
Boundary Value Problems in $\mathbb{R}^1$

Abstract This chapter focuses on one-dimensional boundary value problems. Key concepts like maximum principles, comparison principles and infinite series solutions are introduced in a one-dimensional setting. This chapter establishes the theoretical framework that is used to establish the well-posedness of PDE problems in later chapters.

A wide range of second-order ODEs may be written in the form

$$u'' = F(x, u, u'), \qquad ' \equiv \frac{d}{dx}, \tag{5.1}$$

where $u \equiv u(x)$ and F is a given function. Why might one be interested in solving such a problem? Our motivation is to introduce some key concepts and techniques relevant to PDEs (especially elliptic equations like Laplace's equation, see Sect. 4.2.3) in a rather simpler setting. To this end, we shall focus on *linear* variants of (5.1), that take the general form

$$u'' = a(x)u' + b(x)u - f(x). \tag{5.2}$$

This is called a *reaction–advection–diffusion* equation since the second derivative term represents "diffusion", the term $a(x)u''$ represents "advection" (if $a > 0$ then there is a "wind" blowing from left to right with strength a) and the term $b(x)u$ represents "reaction". The term f is often called the "source" term.

In cases where the differential equation is posed on an interval, $x_\ell < x < x_r$ say, and one piece of supplementary information is given at each end of the interval we arrive at what is known as a "second-order two–point boundary value problem". Note that any such BVP may be transformed by means of the change of independent variable $\xi = (x - x_\ell)/(x_r - x_\ell)$ to a BVP on $0 \le \xi \le 1$ since

$$\frac{d}{dx} = \frac{1}{L}\frac{d}{d\xi}, \quad \frac{d^2}{dx^2} = \frac{1}{L^2}\frac{d^2}{d\xi^2},$$

where $L = x_r - x_\ell$ is the length of the interval. We shall take as our starting point the second-order differential equation (5.2) for which it is assumed that the

D.F. Griffiths et al., *Essential Partial Differential Equations*, Springer Undergraduate Mathematics Series, DOI 10.1007/978-3-319-22569-2_5

change of variable described above has been carried out and then the variable ξ has been renamed to be x. The differential equation is therefore posed on the interval $0 < x < 1$.

Recall that the general solution of such an equation consists of

- the complementary function: that is, the general solution of the homogeneous equation[1]

$$u'' - a(x)u' - b(x)u = 0.$$

This will have two linearly independent solutions, $u_1(x)$ and $u_2(x)$ say, from which we construct the general solution

$$u(x) = Au_1(x) + Bu_2(x). \tag{5.3}$$

- A particular solution of (5.2): which we denote by $P(x)$, say.

The general solution of the inhomogeneous equation (5.2) is then a composition of the two:

$$u(x) = Au_1(x) + Bu_2(x) + P(x). \tag{5.4}$$

This leaves a solution containing two arbitrary constants that have to be found using supplementary information (that is, boundary conditions).

Boundary conditions required to complete the specification of the boundary value problem are relationships between u and its first derivative. They are typically one-dimensional analogues of the BCs (2.11) for Laplace's equation.

At $x = 0$ (and at $x = 1$) the three possibilities are:

Dirichlet: the value of u is specified, e.g., $u(0) = 3$.
Neumann: the value of u' is specified, e.g., $u'(0) = 1$.
Robin: a linear combination of u' and u is specified, e.g., $2u(0) - u'(0) = 2$.

Important questions for a mathematician include the following.

- Does a given two-point boundary value problem have a solution?
- Is the solution unique?
- What properties can be deduced about the solution without having to solve the boundary value problem?
- How can a *numerical approximation* to a solution be computed (to deal with situations where no closed form solutions exist)?
- Is the numerical solution close to the exact solution?

We will give answers to these questions in the next two chapters of the book.

---

[1] An equation in a variable u, say, is homogeneous if replacing u by cu, where c is a constant, leaves the equation unchanged—thus cu is also a solution. Choosing $c = 0$ shows that $u = 0$ must be a solution of a homogeneous equation. The equation must also be linear.

5.1 Qualitative Behaviour of Solutions

The following example illustrates some of the issues involved. The example is instructive in the sense that we can explicitly write down the solution. We will not be so lucky in general.

Example 5.1 For what values of the real constant b does the BVP

$$\left.\begin{array}{c} u'' - bu = \varepsilon, \quad 0 < x < 1 \\ u(0) = u(1) = 0, \end{array}\right\} \tag{5.5}$$

in which ε is a constant, have a unique solution?

It may be shown (see Exercise 5.2) that, for $b > 0$, the solution is

$$u(x) = -\frac{\varepsilon}{b}\left(1 - \frac{\cosh\sqrt{b}\,(x - \frac{1}{2})}{\cosh\frac{1}{2}\sqrt{b}}\right) \qquad (b > 0). \tag{5.6}$$

Though it might appear that this formula fails when $b = 0$, it is found, by taking the limit $b \to 0+$ (Exercise 5.3) or by solving (5.5) with $b = 0$, that

$$u(x) = -\tfrac{1}{2}\varepsilon x(1 - x) \qquad (b = 0). \tag{5.7}$$

When $b < 0$ we may write $b = -|b|$ in (5.6) so that $\sqrt{b} = \sqrt{-|b|} = i\sqrt{|b|}$. Then, using the result that $\cosh i\theta = \cos\theta$ (for $\theta \in \mathbb{R}$), we conclude that

$$u(x) = -\frac{\varepsilon}{b}\left(1 - \frac{\cos\sqrt{|b|}\,(x - \frac{1}{2})}{\cos\frac{1}{2}\sqrt{|b|}}\right) \qquad (b < 0). \tag{5.8}$$

This formula clearly fails, and the BVP has no solution, when $\cos\frac{1}{2}\sqrt{|b|} = 0$, that is when $b = -(2n-1)^2\pi^2$, $n = 1, 2, \ldots$. We note that

$$|u(\tfrac{1}{2})| \to \infty \text{ as } b \to -(2n-1)^2\pi^2$$

indicating that the BVP is not well posed for these values of b. Furthermore, when $b = -(2n\pi)^2$, the homogeneous BVP has the nontrivial solutions

$$u(x) = A\sin 2n\pi x,$$

where A is an arbitrary constant. In these cases, $A\sin 2n\pi x$ should be added to the right hand side of (5.8) to give the general solution. Representative solutions are plotted in Fig. 5.1 for $b = 0, \pm 100$. ◊

Fig. 5.1 Solutions to (5.5) with $\varepsilon = 1, b = 100$ (*solid*), $b = 0$ (*dotted*) and $b = -100$ (*dashed*)

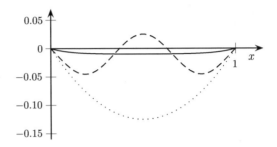

There are two conclusions that we wish to draw from this example.

(a) The BVP is well posed for $b \geq 0$ and, when $\varepsilon > 0$, the solution is negative throughout the interval $0 < x < 1$. For this reason we generally arrange for the coefficient of the second derivative in second-order BVPs to be negative so that positive data will lead to positive solutions. For example, rewriting the BVP in this example as

$$-u'' + bu = -\varepsilon, \quad 0 < x < 1 \\ u(0) = u(1) = 0,$$

would have ensured a non-negative solution when the right hand side is non-negative ($-\varepsilon > 0$). This conclusion is also valid when ε is allowed to vary with x.

(b) The situation is more complex when $b < 0$: although there is generally a unique solution (though its well posedness has not been investigated here), there are values of b, namely, $b = -(n\pi)^2, n \in \mathbb{N}$, for which the BVP has either no solution or else the solution is not unique.

The situation is analogous to those encountered when solving systems of linear algebraic equations—there is either no solution, exactly one solution or an infinity of solutions.

Conclusions were drawn in the preceding example on the basis that its solution was known. We shall be more ambitious in the next example where b is allowed to vary with x and we draw similar conclusions by applying some familiar techniques from the calculus of maxima and minima of functions of one variable.

Example 5.2 Consider the solvability of the BVP

$$-u'' + b(x)u = f(x), \quad 0 < x < 1 \\ u(0) = \alpha, \qquad u(1) = \beta. \tag{5.9}$$

If we focus on the issue of uniqueness of solution then, by virtue of Theorem 2.7, it is sufficient to consider the homogeneous problem—we therefore set $\alpha = \beta = 0$

and $f(x) \equiv 0$ for all $x \in (0, 1)$. Thus

$$u'' = b(x)u \tag{5.10}$$

and we consider any subinterval in which

(a) $b(x) > 0$ and $u(x) > 0$. This implies that $u'' > 0$ so the solution must be concave,
(b) $b(x) > 0$ and $u(x) < 0$. Then $u'' < 0$ so the solution must be convex (concave down),
(c) $b(x) < 0$ and $u(x) > 0$. Then $u'' < 0$ so the solution must be convex,
(d) $b(x) < 0$ and $u(x) < 0$. Then $u'' > 0$ so the solution must be concave.

These four possibilities are sketched in Fig. 5.2. It is evident in cases (a) and (b), where $b(x) > 0$, that, if these inequalities were to hold over the entire interval, then such a solution would not be capable of satisfying the homogeneous BCs $u(0) = u(1) = 0$ unless $u(x) \equiv 0$ for all $x \in [0, 1]$. In these cases the homogeneous problem would have a unique solution.

In cases (c) and (d), where $b(x) < 0$, there is a natural tendency for the solution to cross and recross the x-axis and therefore lead to oscillatory solutions. We would also anticipate that the more negative that $b(x)$ becomes, the stronger the curvature and so the wavelength of oscillations would become shorter. It is not possible to assert with any confidence that any particular BVP would have oscillatory solutions because this would depend on the strength of the curvature relative to the length of the interval. However, the potential for such solutions clearly exists which, in turn, opens up the possibility of nonunique solutions since there are many places where $u = 0$.   ◊

A trick allows us to relax the strict inequalities in the previous example to nonstrict inequalities. This is the topic of the next section. It opens the door to a systematic approach for showing the well-posedness of BVPs.

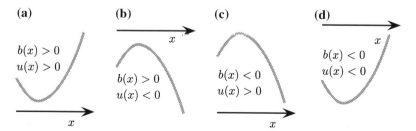

Fig. 5.2 Parts **a–d** show sketches of possible solutions for Example 5.2. The *arrows* indicate the location of the x-axis

5.2 Comparison Principles and Well-Posedness

Before plunging into details, we begin by outlining the strategy that we shall use to investigate the well-posedness of two-point BVPs. A key point is that, by expressing the essential ideas using the abstract operators introduced in Sect. 2.1, we will be able to apply the same strategy in later chapters. It will prove useful both for determining properties of PDEs and for proving convergence of numerical methods.

Thus, suppose that our BVP is written in the form (see (2.15))

$$\mathscr{L}u = \mathscr{F}, \tag{5.11}$$

where $\mathscr{L}$ contains both the differential and boundary operators and $\mathscr{F}$ the data terms comprising the right hand side of the differential equation and the boundary conditions. We will insist that $\mathscr{L}$ satisfies two properties:

① $\mathscr{L}$ is *linear*.
② $\mathscr{L}$ is *inverse monotone*: that is, $\mathscr{L}u \geq 0$ implies that $u \geq 0$.

We will also need to find (or construct) a bounded, non-negative function $\varphi(x)$, called a *comparison function*, so that

③ $\mathscr{L}\varphi(x) \geq 1$ for all $x \in [0, 1]$.

One way to interpret inverse monotonicity is that a non-negative source term is guaranteed to generate a non-negative solution. Furthermore, if u and v are two functions such that

$$\mathscr{L}u \geq \mathscr{L}v$$

then, by linearity of $\mathscr{L}$, we have $\mathscr{L}(u-v) \geq 0$ and, by virtue of inverse monotonicity, we conclude that $u \geq v$. This is known as a *comparison principle*. The two properties ① and ② ensure solution uniqueness.

Theorem 5.3 (Uniqueness) *If the linear operator $\mathscr{L}$ is inverse monotone then the equation $\mathscr{L}u = \mathscr{F}$ has a unique solution.*

Proof We suppose that there is a second solution v. Then $\mathscr{L}v = \mathscr{F}$ and so $\mathscr{L}v = \mathscr{L}u$. Thus, on the one hand $\mathscr{L}v \geq \mathscr{L}u$ from which we deduce that $v \geq u$ while, on the other hand, $\mathscr{L}v \leq \mathscr{L}u$ so that $v \leq u$. Combining these two results we conclude that $v = u$! $\qquad\square$

The additional property ③ is all that is required for a well-posed problem in the sense of Definition 2.10.

Theorem 5.4 *Suppose that the linear operator $\mathscr{L}$ is inverse monotone, it has a comparison function φ and that the norm $\| \cdot \|$ is defined such that*

$$-\|\mathscr{F}\| \leq \mathscr{F} \leq \|\mathscr{F}\|.$$

Then the problem $\mathscr{L}u = \mathscr{F}$ is well posed:

$$-\|\mathscr{F}\|\varphi \le u \le \|\mathscr{F}\|\varphi$$

at all points $x \in [0, 1]$, which means that

$$\max_x |u| \le C\|\mathscr{F}\|,$$

where $C = \max_{x \in [0,1]} \varphi$.

Proof Using ③, the upper bound follows from the sequence of inequalities

$$\mathscr{L}u = \mathscr{F} \le \|\mathscr{F}\| = \|\mathscr{F}\| \times 1 \le \|\mathscr{F}\| \times \mathscr{L}\varphi = \mathscr{L}(\|\mathscr{F}\|\varphi),$$

We thus see that $\mathscr{L}u \le \mathscr{L}(\|\mathscr{F}\|\varphi)$ which leads, via inverse monotonicity, to $u \le \|\mathscr{F}\|\varphi$. The lower bound can be proved in a similar manner and is left to Exercise 5.5. $\qquad\square$

A note of caution: when applying this result to a particular boundary value problem, care must be taken when choosing the sign of the highest derivative in both the differential operator and the boundary conditions.

Example 5.5 Show that the BVP

$$\left.\begin{array}{ll} -u'' = f(x), & 0 < x < 1 \\ u(0) = \alpha, & u(1) = \beta \end{array}\right\} \tag{5.12}$$

is well posed. Establish upper and lower bounds on the solution when $\alpha = -3$, $\beta = 2$ and $f(x) = \sin(x^2)$.

This BVP corresponds to that in Example 5.2 with $b(x) \equiv 0$. A suitable definition for $\mathscr{L}$ is, in this case,

$$\mathscr{L}u(x) = \begin{cases} u(0) & \text{for } x = 0, \\ -u''(x) & \text{for } 0 < x < 1, \\ u(1) & \text{for } x = 1. \end{cases}$$

The argument that will be used below to show that $\mathscr{L}$ is inverse monotone is more elaborate than is strictly necessary: the aim is to develop an approach that can be easily extended to more complicated problems.

Let us begin by supposing that $\mathscr{L}u > 0$ holds with strict inequality. That is,

$$u(0) > 0, \quad -u''(x) > 0 \text{ for } x \in (0, 1) \text{ and } u(1) > 0.$$

We will now show, by contradiction, that these conditions imply that $u(x) \ge 0$ for $x \in [0, 1]$. To this end, let us suppose that $u(x)$ is negative on some part of the unit interval and let $s \in (0, 1)$ be a point where u achieves a negative minimum. We know from calculus that

$$u(s) < 0, \quad u'(s) = 0 \text{ and } u''(s) > 0.$$

The last of these contradicts the requirement that $\mathscr{L}u(s) > 0$.

In order to shore up this argument to accommodate the case that $\mathscr{L}u \geq 0$, we first construct a suitable comparison function. These are usually constructed from low degree polynomials. Trying $\varphi(x) = Ax^2 + Bx + C$ we get

$$\mathscr{L}\varphi(x) = \begin{cases} C & \text{for } x = 0, \\ -2A & \text{for } 0 < x < 1, \\ A + B + C & \text{for } x = 1. \end{cases}$$

We can thus ensure that $\mathscr{L}\varphi(x) = 1$ for all $x \in [0, 1]$ by making the specific choices $C = 1$, $A = -1/2$ and $B = 1/2$, leading to the function

$$\varphi(x) = 1 + \tfrac{1}{2}x(1 - x) \tag{5.13}$$

which is non-negative for $x \in [0, 1]$, as required by the definition.

The second, key step towards establishing inverse monotonicity is to introduce a new function v according to

$$v(x) = u(x) + \varepsilon\varphi(x)$$

with $\varepsilon > 0$. Note that

$$\mathscr{L}v(x) = \mathscr{L}u(x) + \varepsilon.$$

The inequality $\mathscr{L}u \geq 0$ implies that $\mathscr{L}v \geq \varepsilon > 0$ and our earlier argument may now be applied to prove that $v(x) \geq 0$. The fact that $u(x) + \varepsilon\varphi(x) \geq 0$ for all $x \in [0, 1]$ and for all $\varepsilon > 0$ also ensures that $u(x) \geq 0$ for all $x \in [0, 1]$. (If this were not the case, then we could choose ε so small so that $u(x) + \varepsilon\varphi(x) < 0$ at some point $x \in (0, 1)$ giving, once more, a contradiction.) This establishes inverse monotonicity.

Finally, since $\mathscr{F}$ is given by

$$\mathscr{F}(x) = \begin{cases} \alpha & \text{for } x = 0, \\ f(x) & \text{for } 0 < x < 1, \\ \beta & \text{for } x = 1, \end{cases} \tag{5.14}$$

we can simply define

$$\|\mathscr{F}\| = \max\{|\alpha|, \max_{0 \leq x \leq 1} |f(x)|, |\beta|\}. \tag{5.15}$$

In the specific case $\alpha = -3$, $\beta = 2$ and $f(x) = \sin(x^2)$ we find that $\|\mathscr{F}\| = 3$ and, since $C = \max_{0 \le x \le 1} \varphi(x) = 9/8$, Theorem 5.4 immediately implies that

$$\max_{0 \le x \le 1} |u(x)| \le 27/8. \qquad \diamond$$

What makes this result so impressive is that a bound on the magnitude of the solution has been determined despite the fact that the function f was purposely chosen so that the BVP could not be solved in closed form. Theoretical bounds like this can also provide useful checks on numerical solutions.

The argument used in the above example can be extended to establish the well-posedness of the reaction–advection–diffusion problem (5.2) when $b(x) \ge 0$ (see Exercise 5.6). The next example introduces a BVP with Robin boundary conditions. In this case the sign of the derivative terms in the boundary conditions is particularly important.

Example 5.6 Show that the BVP

$$\left. \begin{array}{c} -u'' = f(x), \ 0 < x < 1 \\ u(0) - u'(0) = \alpha, \ u(1) + u'(1) = \beta \end{array} \right\} \qquad (5.16)$$

is well posed. Establish upper and lower bounds on the solution when $\alpha = 3$, $\beta = 2$ and $f(x) = \cos(x^2)$.

Setting

$$\mathscr{L}u(x) = \begin{cases} u(0) - u'(0) & \text{for } x = 0, \\ -u''(x) & \text{for } 0 < x < 1, \\ u(1) + u'(1) & \text{for } x = 1 \end{cases}$$

and with $\mathscr{F}$ given by (5.14), the argument used in the previous example carries over to this case. The only difference is the comparison function, which takes the form

$$\varphi(x) = \tfrac{1}{2} \left(3 + x - x^2 \right),$$

and the fact that we have to preclude the possibility that $v \ (= u(x) + \varepsilon \varphi(x))$ has a negative minimum at either one of the two end points of the interval.

Consider the point $x = 0$. If $v(x)$ is a negative minimum then $v(0) < 0$ and $v'(0) \ge 0$ which implies that

$$\mathscr{L}v(0) = v(0) - v'(0) < 0.$$

This cannot happen since it contradicts the assumption that $\mathscr{L}v(x) > 0$ for all $x \in [0, 1]$. A similar argument holds when $x = 1$.

When $\alpha = 3$, $\beta = 2$ and $f(x) = \cos(x^2)$, we find $\|\mathscr{F}\| = 3$ and, since $C = \max_{0 \le x \le 1} \varphi(x) = 13/8$, Theorem 5.4 implies that

$$\max_{0 \le x \le 1} |u(x)| \le 33/8,$$

However, since $\mathscr{F} \ge 0$, we deduce that $u(x) \ge 0$ for all $x \in [0, 1]$ and we have the improved bounds

$$0 \le u(x) \le \|\mathscr{F}\|\varphi(x) \le 3\varphi(x), \quad x \in [0, 1]. \qquad \diamondsuit$$

There is no first derivative term in any of the examples presented thus far. We shall rectify this omission in Sect. 5.3.2. Getting to this stage of technical sophistication will involve the notion of orthogonality in a function space. This important topic is discussed next.

5.3   Inner Products and Orthogonality

The geometrical properties of differential operators and solutions to boundary value problems may be conveniently expressed using the notion of an *inner product*. For complex-valued functions $u(x)$ and $v(x)$ defined on the interval $x \in [0, L]$, the inner product is simply the complex number given by

$$\langle u, v \rangle = \int_0^L u(x)\, v^*(x)\mathrm{d}x, \tag{5.17}$$

where $v^*(x)$ is the complex conjugate of $v(x)$. Note that $\langle u, u \rangle$ is both real and positive (unless u is identically zero). If u and v are *square integrable*, that is, if

$$\int_0^L |u(x)|^2 \mathrm{d}x < \infty \text{ and } \int_0^L |v(x)|^2 \mathrm{d}x < \infty,$$

then the integral in (5.17) is well defined because of the Cauchy–Schwarz[2] inequality:

$$|\langle u, v \rangle| \le \langle u, u \rangle^{1/2} \langle v, v \rangle^{1/2} < \infty. \tag{5.18}$$

The inner product (5.17) has a close analogy with the inner product for column vectors (see Appendix B). The geometric analogy with vectors is continued in the next definition.

---

[2]The proof of (5.18) can be found in a first course on functional analysis. See Appendix B for an equivalent statement in the case of finite-dimensional vectors.

Definition 5.7 (*Orthogonality*) Any two nontrivial functions $u(x)$ and $v(x)$ are said to be *orthogonal* if

$$\langle u, v \rangle = 0.$$

Orthogonal functions are necessarily *linearly independent* (see Exercise 5.11c).

It is usual to deal with infinite sets of mutually orthogonal functions. That is, for a set of functions $\{\phi_n(x)\}$ we have

$$\langle \phi_n, \phi_m \rangle = 0, \quad m \neq n.$$

Note that $\langle \phi_n, \phi_n \rangle > 0$.

The most celebrated set of mutually orthogonal functions is given by the complex exponentials, $\phi_n(x) = e^{2\pi i n x / L}$ ($n = 0, \pm 1, \pm 2, \ldots$), see Exercise 5.19. These functions are called *Fourier modes*. The next example shows that the functions formed by their real and imaginary parts are also orthogonal.

Example 5.8 Show that the functions

$$\{1, \sin(2\pi x/L), \cos(2\pi x/L), \ldots, \sin(2\pi n x/L), \cos(2\pi n x/L), \ldots\}$$

form a mutually orthogonal set with respect to the inner product (5.17).

To establish the orthogonality of $\varphi_n = \sin(2\pi n x/L)$ and $\varphi_m = \sin(2\pi m x/L)$, we simply use the trigonometric identity

$$\sin A \sin B = \tfrac{1}{2}(\cos(A - B) - \cos(A + B)).$$

Thus, if $m \neq n$,

$$
\begin{aligned}
\langle \varphi_n, \varphi_m \rangle &= \frac{1}{2} \int_0^L (\cos(2\pi(m - n)x/L) - \cos(2\pi(m + n)x/L)) \, dx \\
&= \frac{1}{2} \left[\frac{\sin(2\pi(m - n)x/L)}{(2\pi(m - n)/L)} - \frac{\sin(2\pi(m + n)x/L)}{(2\pi(m + n)/L)} \right]_0^L \\
&= 0,
\end{aligned}
$$

since the numerators of the terms on the right hand side vanish at both end points. Also, when $m = n$, $\langle \sin(2\pi n x/L), \sin(2\pi n x/L) \rangle = L/2$. The remaining cases are left to Exercise 5.20. ◊

An important consequence of orthogonality concerns functions $f(x)$ that can be expressed in terms of a convergent series of the form

$$f(x) = \sum_{n=1}^{\infty} a_n \phi_n(x).$$

Such formulae are easily inverted—that is, a simple expression can be obtained for the coefficients a_n—when the functions $\phi_n(x)$ are mutually orthogonal. To see this, the first step is to multiply both sides of the relation by $\phi_m^*(x)$, for any value of m. We then integrate the result to give

$$\int_0^L f(x)\,\phi_m^*(x)\mathrm{d}x = \sum_{n=1}^\infty a_n \int_0^L \phi_n(x)\,\phi_m^*(x)\mathrm{d}x$$

or, more elegantly,

$$\langle f, \phi_m \rangle = \sum_{n=1}^\infty a_n \langle \phi_n, \phi_m \rangle.$$

By virtue of orthogonality, the only nonzero term in the sum on the right-hand side is $\langle \phi_m, \phi_m \rangle$ and it follows that

$$\langle f, \phi_m \rangle = a_m \langle \phi_m, \phi_m \rangle.$$

Replacing the dummy index m by n gives a simple expression for the coefficient:

$$a_n = \frac{\langle f, \phi_n \rangle}{\langle \phi_n, \phi_n \rangle}. \tag{5.19}$$

Note that the right-hand side of (5.19):

$$\int_0^L f(x)\,\phi_n^*(x)\mathrm{d}x \ \Big/ \ \int_0^L \phi_n(x)\,\phi_n^*(x)\mathrm{d}x,$$

is called as the *projection* of the function f onto the function ϕ_n. It is particularly convenient, from a theoretical point of view, to normalise (by dividing the function ϕ_n by the square root of $\langle \phi_n, \phi_n \rangle > 0$) so that the rescaled functions satisfy $\langle \phi_n, \phi_n \rangle = 1$. If this is done then the coefficients in the expansion for f become $a_n = \langle f, \phi_n \rangle$ and the set $\{\phi_n\}$ is then known as an *orthonormal set*.

We shall explore two-point BVPs that naturally give rise to orthogonal function sets in Sect. 5.4. The material in this section is a prerequisite for Chap. 8, where it will be shown that these orthogonal functions contribute to the solution of some important PDE problems. Before this, we need to identify suitable differential operators— those with a property that is analogous to *symmetry* for a conventional matrix (see Appendix B.2).

5.3.1 Self-adjoint Operators

In this section we will focus on BVPs described, in the notation of Sect. 2.1, by differential operators $\mathcal{L}$ having the special form

$$\mathcal{L}u(x) := -(p(x)u'(x))' + q(x)u(x), \quad 0 < x < 1, \tag{5.20}$$

where $q(x)$ is be assumed to be continuous and $p(x)$ is a positive and continuously differentiable function on $(0, 1)$. These are known as Sturm–Liouville operators. We shall further suppose that the associated boundary condition operator $\mathcal{B}$ takes the form[3]

$$\mathcal{B}u(0) := a_0u(0) - b_0u'(0), \qquad \mathcal{B}u(1) := a_1u(1) + b_1u'(1), \tag{5.21}$$

where coefficients a_0, b_0, a_1, b_1 are given constants with $b_0 \geq 0$, $b_1 \geq 0$ and $|a_0| + |b_0| > 0$, $|a_1| + |b_1| > 0$ to avoid degeneracy. The pair of equations $\mathcal{L}u = f$ and $\mathcal{B}u = g$ describe a two-point boundary value problem on $[0, 1]$.

Although the coefficients in (5.20) and (5.21) will either be real constants or real-valued functions, certain situations (detailed later in the chapter) will necessitate consideration of complex dependent variables. Thus, taking complex-valued functions u and v we have, after integrating by parts twice,

$$\langle u, \mathcal{L}v \rangle = \int_0^1 u\left(-(p(x)v^{*\prime})' + q(x)v^*\right)dx \tag{5.22}$$

$$= \left[-puv^{*\prime}\right]_0^1 + \int_0^1 \left(p(x)u'v^{*\prime} + q(x)uv^*\right)dx \tag{5.23}$$

$$= \left[p(x)(u'v^* - uv^{*\prime})\right]_0^1 + \int_0^1 \left(-(p(x)u')' + q(x)u\right)v^*dx.$$

Thus,

$$\langle u, \mathcal{L}v \rangle = \left[p(x)(u'v^* - uv^{*\prime})\right]_0^1 + \langle \mathcal{L}u, v \rangle$$

which can be rearranged to read

$$\int_0^1 \left(u\mathcal{L}v^* - v^*\mathcal{L}u\right)dx = \left[p(x)(u'v^* - uv^{*\prime})\right]_0^1 \tag{5.24}$$

and is known as *Green's identity*. It can further be shown, see Exercise 5.7, that the contributions from the boundary terms vanish if u and v are both subject to

---

[3]The reason for the different signs on the derivative terms at $x = 0$ and $x = 1$ is that we take the "outward" directed derivative at each boundary point.

homogeneous boundary conditions, $\mathcal{B}u = 0$ and $\mathcal{B}v = 0$, in which case

$$\langle u, \mathcal{L}v \rangle = \langle \mathcal{L}u, v \rangle \Longleftrightarrow \int_0^1 (v^* \mathcal{L}u - u \mathcal{L}v^*) \mathrm{d}x = 0, \tag{5.25}$$

a result known as *Lagrange's identity*. It inspires the following definition.

Definition 5.9 (*Self-adjoint operator*) A differential operator $\mathcal{L}$ such that

$$\langle u, \mathcal{L}v \rangle = \langle \mathcal{L}u, v \rangle + \text{boundary terms}$$

is said to be *formally* self adjoint. The full operator, $\mathscr{L}$, defined by

$$\mathscr{L}u(x) = \begin{cases} \mathcal{L}u(x), & 0 < x < 1 \\ \mathcal{B}u(x), & x = 0, 1, \end{cases} \tag{5.26}$$

is said to be self adjoint if $\langle u, \mathcal{L}v \rangle = \langle \mathcal{L}u, v \rangle$ for all functions satisfying the homogeneous boundary conditions $\mathcal{B}u = 0$ and $\mathcal{B}v = 0$. Note that unqualified self-adjointness is a property of both the differential operator and the associated boundary conditions.

5.3.2 A Clever Change of Variables

The reaction–advection–diffusion operator in (5.2) can be expressed as a Sturm–Liouville operator by finding a suitable integrating factor. The first step is to multiply both sides of (5.2) by a function $p(x)$ to give

$$p(x)u'' - p(x)a(x)u' - p(x)b(x)u = p(x)f(x). \tag{5.27}$$

Next, using the rule for differentiating a product, we find that

$$\frac{\mathrm{d}}{\mathrm{d}x}(pu') = pu'' + p'u'$$

which leads to the identity $pu'' = (pu')' - p'u'$, so (5.27) becomes

$$(pu')' - (pa + p')u' - pbu = pf.$$

The coefficient of the first derivative term can be made to vanish by choosing $p(x)$ so that $p' = -a(x)p$. That is,

$$p(x) = \mathrm{e}^{-\int a(x)\mathrm{d}x}$$

and, making this specific choice, we find that

$$-(p(x)u')' + q(x)u = g(x),$$

(5.28)

with

$$p(x) = \exp\left(-\int a\,\mathrm{d}x\right), \quad q(x) = p(x)b(x), \quad g(x) = -p(x)f(x).$$

(5.29)

The special form (5.20) therefore includes all second-order linear ODEs whose coefficient $a(x)$ is integrable! This is a significant result. Further appreciation may be gained by making the change of independent variable $\xi = \xi(x)$ given by

$$\xi = \int_0^x \frac{1}{p(s)}\,\mathrm{d}s \iff \frac{\mathrm{d}\xi}{\mathrm{d}x} = \frac{1}{p(x)} \text{ and } \xi(0) = 0.$$

(5.30)

By the chain rule,

$$\frac{\mathrm{d}}{\mathrm{d}\xi} = \frac{\mathrm{d}x}{\mathrm{d}\xi}\frac{\mathrm{d}}{\mathrm{d}x} = p(x)\frac{\mathrm{d}}{\mathrm{d}x},$$

so we see that the Sturm–Liouville equation (5.28) is transformed to a simple (reaction-diffusion) equation:

$$-\frac{\mathrm{d}^2 u}{\mathrm{d}\xi^2} + \tilde{q}(\xi)u = \tilde{g}(\xi),$$

(5.31)

where $\tilde{q}(\xi) = p(x)q(x)$ and $\tilde{g}(\xi) = p(x)g(x)$.

Note that in order for the change of variable (5.30) to be one-to-one it is necessary that $p(x)$ be positive over the entire open interval $(0, 1)$. When $1/p(x)$ is integrable, a finite interval in x is transformed to a finite interval in ξ and the associated Sturm–Liouville operator is said to be *regular*. Otherwise, when $p(x) = x$ for example, the Sturm–Liouville operator is said to be *singular*.

Supposing that the differential equation has been transformed to one with no first derivative term then the arguments of earlier sections may be applied to deduce its qualitative properties. Thus, when homogeneous Dirichlet BCs are applied at both ends of the interval, the BVP associated with (5.31) will have a unique solution when $\tilde{q}(\xi) > 0$ throughout the interval. Since $\tilde{q}(\xi) = p^2(x)b(x)$ it follows that there will be a unique solution provided $b(x) > 0$. An alternative treatment involves making a change of dependent variable. The details can be found in Exercise 5.17.

We are now in a position to mine a rich vein of orthogonal functions, known as eigenfunctions. These functions will prove to be useful when constructing solutions to linear PDEs in later chapters.

5.4 Some Important Eigenvalue Problems

The general eigenvalue problem associated with a differential operator $\mathcal{L}$ and bound-ary condition operator $\mathcal{B}$ is to find non-trivial functions u (that is, not zero for all $x \in [0, 1]$), and corresponding constants λ, satisfying the two-point boundary value problem:

$$\left.\begin{array}{l} \mathcal{L}u(x) = \lambda w(x)u(x), \quad 0 < x < 1 \\ \mathcal{B}u(x) = 0, \quad x = 0, 1, \end{array}\right\} \tag{5.32}$$

where $w(x)$ is a given positive function known as a *weight function*.

A fundamental example of (5.32) is obtained when $\mathcal{L}$ is defined by (5.20) and $\mathcal{B}$ by (5.21). This gives

$$\left.\begin{array}{l} -(p(x)u')' + q(x)u = \lambda w(x)u, \quad 0 < x < 1 \\ a_0 u(0) - b_0 u'(0) = 0, \quad a_1 u(1) + b_1 u'(1) = 0. \end{array}\right\} \tag{5.33}$$

Problems of this type are known as *Sturm–Liouville problems*. Every value of λ for which the boundary value problem (5.33) has a nontrivial solution is called an *eigenvalue* and the corresponding solution, u, is called an *eigenfunction*.

It may be shown (Exercise 5.21) that, if u is an eigenfunction corresponding to an eigenvalue λ, then cu is also an eigenfunction for any nonzero constant c.

We present some examples below that can be solved in closed form and we also present results that apply when this is not possible. We shall assume below that

$$p(x) > 0, \quad w(x) > 0, \quad 0 < x < 1. \tag{5.34}$$

Our first example, closely related to Example 5.1, is the simplest possible yet it has properties that are typical of Sturm–Liouville problems.

Example 5.10 Determine the eigenvalues and eigenfunctions of the problem

$$-u'' = \lambda u, \quad 0 < x < 1$$

with boundary conditions $u(0) = u(1) = 0$.

The form of the general solution of this ODE will depend on whether $\lambda < 0$, $\lambda = 0$ or $\lambda > 0$. We consider each case in turn.

(a) Suppose firstly that $\lambda < 0$. We write $\lambda = -\mu^2$ ($\mu \neq 0$) then the ODE has general solution

$$u(x) = Ae^{\mu x} + Be^{-\mu x}.$$

The first BC implies that $A + B = 0$ so $u(x) = A(e^{\mu x} - e^{-\mu x})$ and the second BC gives $A(e^{\mu} - e^{-\mu}) = 0$, which may be written as $2A \sinh \mu = 0$. This implies that $A = 0$ since $\mu \neq 0$. Thus, since the only solution when $\lambda < 0$ is $u(x) \equiv 0$

(the trivial solution), there cannot be any negative eigenvalues. This conclusion could also have been reached using the arguments presented in Example 5.2.

(b) Suppose, next, that $\lambda = 0$. Then $u'' = 0$ which has general solution $u(x) = A + Bx$. The BCs imply that $A = B = 0$. This is, again, the trivial solution so $\lambda = 0$ cannot be an eigenvalue.

(c) Suppose, finally, that $\lambda > 0$. This time we write $\lambda = \omega^2$ and we have the general solution

$$u(x) = A \sin \omega x + B \cos \omega x.$$

The first BC implies that $B = 0$ while the BC $u(1) = 0$ gives $A \sin \omega = 0$. Thus, either $A = 0$ (a possibility that we ignore because it again gives the trivial solution) or else $\sin \omega = 0$, that is, $\omega = n\pi$, $n = \pm 1, \pm 2, \ldots$ leading to the infinite sequence of eigenvalues

$$\lambda_n = n^2 \pi^2, \qquad n = 1, 2, \ldots$$

with corresponding eigenfunctions $\phi_n = \sin n\pi x$. The negative values of n may also be ignored since there is only one eigenvalue and one linearly independent eigenfunction associated with each pair $+n$ and $-n$.

Note that, it can be seen that the BVP in Example 5.1 fails to have a unique solution precisely when the coefficient b is an eigenvalue of the associated homogeneous problem. ◇

It was implicitly assumed in the above example that the eigenvalues λ are real. This assumption is justified in the next theorem.

Theorem 5.11 *Suppose that $p(x) > 0$, $q(x) \geq 0$ and $w(x) > 0$ for $x \in (0, 1)$. Then the eigenvalues of the Sturm–Liouville problem*

$$\left. \begin{array}{r} -(p(x)u')' + q(x)u - \lambda w(x)u, \quad 0 < x < 1 \\ u(0) = u(1) = 0, \end{array} \right\} \tag{5.35}$$

are real and positive.

Proof To prove that all eigenvalues are real we use Lagrange's identity with $v = u$. Since $\mathcal{L}u = \lambda wu$ it follows, by taking the complex conjugate of each side, that $\mathcal{L}u^* = \lambda^* wu^*$ (recall that the coefficients in $\mathcal{L}$ are real). Then (5.25) gives

$$\int_0^1 \left(u^* \mathcal{L}u - u \mathcal{L}u^* \right) dx = -(\lambda - \lambda^*) \int_0^1 w(x) |u(x)|^2 dx = 0.$$

The integrand is strictly positive because of the assumption $w(x) > 0$ and so $\lambda^* - \lambda = 0$ from which we deduce that λ must be real.

To show that the eigenvalues are positive, we set $v = u$ in (5.23) and apply the boundary conditions $u(0) = u(1) = 0$:

$$\langle u, \mathcal{L}u \rangle = \int_0^1 \left(p(x) |u'(x)|^2 + q(x) |u(x)|^2 \right) dx.$$

The left hand side can be written as

$$\langle u, \mathcal{L}u \rangle = \langle u, \lambda w u \rangle = \lambda \int_0^1 w(x) |u(x)|^2 dx$$

and so we find that

$$\lambda = \frac{\int_0^1 \left(p(x)|u'(x)|^2 + q(x) |u(x)|^2 \right) dx}{\int_0^1 w(x) |u(x)|^2 dx} > 0$$

since both numerator and denominator are positive.   □

Corollary 5.12 *The homogeneous BVP*

$$\left. \begin{array}{r} -(p(x)u'(x))' + q(x)u(x) = 0, \quad 0 < x < 1 \\ u(0) = u(1) = 0, \end{array} \right\} \tag{5.36}$$

has only the trivial solution $u(x) \equiv 0$.

Proof This situation would correspond to a Sturm–Liouville problem with an eigenvalue $\lambda = 0$ and nontrivial eigenfunction u. This contradicts the results in Theorem 5.11.   □

Example 5.13 Find the eigenfunctions and eigenvalues of the boundary value problem

$$\left. \begin{array}{r} -u'' + 4u = \lambda u, \quad 0 < x < \pi \\ u'(0) = u(\pi) = 0. \end{array} \right\} \tag{5.37}$$

If we rewrite the ODE in the form $-u'' = (\lambda - 4)u$, then Theorem 5.11 tells us that the eigenvalues $(\lambda - 4)$ of this problem are real and positive. It can be shown that only trivial solutions exist for $\lambda \leq 4$ (Exercise 5.24). Assuming $\lambda > 4$, we write $\lambda = 4 + \omega^2$ which leads to the general solution

$$u(x) = A \sin \omega x + B \cos \omega x.$$

Next, the BC $u'(0) = 0$ gives $A = 0$ while the BC $u(\pi) = 0$ gives $B \cos \omega \pi = 0$. Since B cannot be zero (it would lead to the trivial solution), we must have

$\cos \omega \pi = 0$. This can only occur if ω is an odd multiple of $\frac{1}{2}$: $\omega = \frac{1}{2}(2n - 1)$ so the nth eigenvalue is given by

$$\lambda_n = 4 + \left(n - \tfrac{1}{2}\right)^2$$

for $n = 1, 2, \ldots$, and the corresponding eigenfunction is

$$\phi_n = \cos\left(n - \tfrac{1}{2}\right)x. \qquad \diamond$$

To complete the discussion of Sturm–Liouville problems, some other general properties of the eigenvalues and eigenfunctions are listed below:

① Real eigenvalues (and consequently eigenfunctions) also occur for the general boundary conditions, $\mathcal{B}u = 0$, where $\mathcal{B}$ is defined by (5.21). The eigenvalues will still be positive if the coefficients in (5.21) are all positive.
② The eigenfunctions corresponding to distinct eigenvalues are orthogonal with respect to the weighted inner product defined by

$$\left\langle u, v \right\rangle_w = \int_0^1 w(x)u(x)v^*(x)\mathrm{d}x. \qquad (5.38)$$

That is $\left\langle u, v \right\rangle_w = 0$. This is explored in Exercises 5.31 and 5.33.
③ All eigenvalues of regular Sturm–Liouville problems are simple—there is only one eigenfunction (up to a multiplicative constant) corresponding to each eigenvalue. It follows that the eigenvalues may be ordered so that

$$0 < \lambda_1 < \lambda_2 < \lambda_3 < \cdots . \qquad (5.39)$$

Also $\lambda_k \to \infty$ as $k \to \infty$.
For a detailed discussion and proof, see Pryce [16, Theorem 2.4].

Example 5.14 Verify that the Laplacian operator in polar coordinates is formally self adjoint with respect to a particular weighted inner product in the special case of circular symmetry.
In polar coordinates, $x = r \cos \theta$, $y = r \sin \theta$, the Laplacian operator is

$$-\nabla^2 u = -\left(\frac{\partial^2 u}{\partial r^2} + \frac{1}{r}\frac{\partial u}{\partial r} + \frac{1}{r^2}\frac{\partial^2 u}{\partial \theta^2}\right).$$

If u is a function of r only (that is, when there is circular symmetry) we get

$$\mathcal{L}u(r) = -\left(\frac{\partial^2 u}{\partial r^2} + \frac{1}{r}\frac{\partial u}{\partial r}\right) = -\frac{1}{r}(ru')',$$

where the dash here denotes differentiation with respect to r. Assuming a circular region of radius R and choosing a weighted inner product of the type (5.38) with weight $w(r) = r$ on an interval $0 \le r \le R$ we find that

$$\langle u, \mathcal{L}v \rangle_r = \int_0^R u\,(-rv^{*\prime})'\,\mathrm{d}r. \tag{5.40}$$

The result follows by integrating the right hand side by parts twice as is done in the derivation of (5.24). The right hand side of (5.40) is a special case of (5.22) with $p = r$ and $q = 0$. ◊

Example 5.15 Show that the problem $\mathcal{L}u = f$ with homogeneous boundary conditions $\mathcal{B}u = 0$, where $\mathcal{L}$ and $\mathcal{B}$ are defined by (5.20) and (5.21), respectively, is well posed.

We will need to use all the properties of Sturm–Liouville eigenvalue problems listed above. Suppose that λ_n is the eigenvalue associated with the normalised eigenvector ϕ_n of the Sturm–Liouville problem (5.33), and suppose that u can be expanded in terms of the eigenvectors as

$$u(x) = \sum_{n=1}^{\infty} \alpha_n \phi_n(x).$$

Note that u and ϕ_n both satisfy the boundary condition $\mathcal{B}u = 0$. Then

$$\mathcal{L}u = \sum_{n=1}^{\infty} \alpha_n \mathcal{L}\phi_n(x) = \sum_{n=1}^{\infty} \alpha_n \lambda_n w(x) \phi_n(x)$$

and, by virtue of Property ②, we find that

$$\langle u, u \rangle_w = \sum_{n=1}^{\infty} \alpha_n^2 \langle \phi_n, \phi_n \rangle_w = \sum_{n=1}^{\infty} \alpha_n^2,$$

and that

$$\int_0^1 u \mathcal{L}u \, \mathrm{d}x = \sum_{n=1}^{\infty} \alpha_n \lambda_n \int_0^1 w(x) u(x) \phi_n(x) \, \mathrm{d}x$$

$$= \sum_{n=1}^{\infty} \alpha_n \lambda_n \langle u, \phi_n \rangle_w = \sum_{n=1}^{\infty} \lambda_n \alpha_n^2.$$

It follows from Property ③ that

$$\sum_{n=1}^{\infty} \lambda_n \alpha_n^2 \geq \lambda_1 \sum_{n=1}^{\infty} \alpha_n^2 = \lambda_1 \|u\|_w^2$$

where $\|u\|_w = (u, u)_w^{1/2}$ is the (suitably weighted) norm of u.

Thus, given that $\mathcal{L}u = f$, and applying the Cauchy–Schwarz inequality (5.18) in the weighted inner product gives

$$\lambda_1 \|u\|_w^2 \leq \int_0^1 u \mathcal{L}u \, dx = \int_0^1 u f \, dx$$
$$= \int_0^1 w u \, (f/w) \, dx = (u, f/w)_w \leq \|u\|_w \|f/w\|_w$$

Finally, dividing by $\lambda_1 \|u\|_w > 0$ gives

$$\|u\|_w \leq \frac{1}{\lambda_1} \|f/w\|_w \tag{5.41}$$

and we deduce that the weighted norm of u is bounded provided that the same norm of f/w is bounded. We note that in situations (such as the previous example) where $w = 0$ at one or both ends of the interval we will need to insist that f vanishes at the same point. ◊

Exercises

5.1 Find the general solution of $u'' + 3u' - 4u = 1$. Hence find the solution satisfying the boundary conditions $u(0) = 1$, and $u(1) = 3$.

5.2 Verify that (5.6) and (5.8) are solutions to the BVP in Example 5.1.

5.3 Show that taking the limit $b \to 0$ in both (5.6) and (5.8) leads to (5.7).

5.4 Rewrite the following equations into the form (5.28):

(a) $-u'' + 20u' = 1$,
(b) $u'' + 3u' - 4u = x^2$,
(c) $(1 - x^2)u'' - 2xu' + ku = 0$ (Legendre's equation),
(d) $x^2 u'' + xu' + (x^2 - \nu^2)u = 0$ (Bessel's equation, see Appendix D).

5.5 Starting from the observation that $\mathcal{L}u = \mathcal{F} \geq -\|\mathcal{F}\|$, establish the lower bound $-(\|\mathcal{F}\|\varphi) \leq u$ in Theorem 5.4.

5.6 Show that the ideas in Example 5.5 may be extended to establish the well-posedness of the BVP (5.2) when $b(x) \geq 0$ for $x \in [0, 1]$.

5.7 * Given that $a_0u(0) - b_0u'(0) = 0$ and $a_0v(0) - b_0v'(0) = 0$, show that $u'(0)v(0) - u(0)v'(0)) = 0$.

5.8 Suppose that the operator $\mathcal{L}$ is defined by $\mathcal{L}u = -u'' + au' + bu$, where the coefficients a and b are real constants. Using only integration by parts, show that

$$\langle u, \mathcal{L}v \rangle = \langle \mathcal{M}u, v \rangle + \text{boundary terms}$$

where $\mathcal{M}$, the adjoint of $\mathcal{L}$, is defined by $\mathcal{M}u = -u'' - au' + bu$. This shows that the operator $\mathcal{L}$ is self-adjoint ($\mathcal{L} = \mathcal{M}$) when $a = 0$.

5.9 Express the ODE

$$-2xu'' - u' = 2f(x), \quad 0 < x < 1,$$

in the form (5.31) and show, in particular, that $\tilde{q}(\xi) \equiv 0$. Use the arguments presented in Example 5.2 to show that this equation has a unique solution when subject to Dirichlet BCs.

5.10 * Show that the functions $u(x) = 2x - 1$ and $v(x) = 4x^2 - 4x + \frac{1}{2}$ are orthogonal with respect to the inner product (5.17) over the interval $0 < x < 1$.

5.11 Show that, for integrable functions u and v, the scalar product (5.17) has the following properties:

(a) $\langle u, v \rangle = \langle v, u \rangle^*$,
(b) $\langle u, u \rangle \geq 0$,
(c) If $u(x)$ is continuous on $[0, L]$, then $\langle u, u \rangle = 0$ implies $u(x) \equiv 0$,
(d) $\langle c_1u_1 + c_2u_2, v \rangle = c_1\langle u_1, v \rangle + c_2\langle u_2, v \rangle$, where u_1 and u_2 are integrable functions on $[0, L]$ and c_1 and c_2 are constants,
(e) If u and v are continuous functions that are orthogonal with respect to the inner product $\langle u, v \rangle$, show that they are linearly independent. (Hint: assume that they are linearly dependent: that is, there are non-zero constants a, b such that $au(x) + bv(x) = 0$ and show that this leads to a contradiction.)

5.12 Solve the BVP

$$\left.\begin{array}{l} u'' + a^2u = \sin \pi x, \quad 0 < x < 1 \\ u(0) = 1, u(1) = -2, \end{array}\right\}$$

for all $a \in \mathbb{R}$. What are the solutions in the cases $a = \pm\pi$? (Hint: compare with Example 5.1.)

5.13 Show that the differential equation $-u''(x) = 4\pi^2u(x)$ $(0 < x < 1)$ with boundary conditions $u(0) = u(1) = 0$ has a nontrivial solution $\phi(x)$.

Now consider the ODE $-u'' - 4\pi^2u = f(x)$, $0 < x < 1$ with boundary conditions $u(0) = u(1) = 0$. Show, by multiplying both sides of this differential equation by ϕ and integrating over the interval $0 < x < 1$, that no solution can exist unless $\int_0^1 f(x)\phi(x)\,dx = 0$.

Illustrate this result by showing that there is no solution in the case when $f(x) = \sin 2\pi x$ and that there are an infinite number of solutions when $f(x) = 1$. (Hint: when $f(x) = \sin(2\pi x)$, show that the general solution is $u(x) = \frac{1}{4\pi}x\cos 2\pi x + A\sin 2\pi x + B\cos 2\pi x$.)

5.14 For what value(s) of a does the ODE $-u'' - 9u = x - a$, $0 < x < \pi$ have a solution with end conditions $u(0) = 0$, $u(\pi) = 0$?

5.15

(a) Determine the eigenvalues λ and corresponding eigenfunctions $\phi(x)$ of the ODE $-\phi''(x) = \lambda\phi(x)$, $0 < x < 1$, with boundary conditions $\phi(0) = \phi'(1) = 0$.
(b) Suppose that $-u''(x) = \omega^2 u(x) + f(x)$, $0 < x < 1$, with boundary conditions $u(0) = 1$, $u'(1) = -2$. If ω^2 is equal to one of the eigenvalues λ from part (a), show that a solution $u(x)$ cannot exist unless

$$\int_0^1 f(x)\phi(x)\,dx = 2\phi(1) - \phi'(0).$$

Find a constant function $f(x)$ for which this condition is satisfied when ω^2 is equal to the smallest of the eigenvalues and determine the resulting solution $u(x)$.

5.16 ☆ Use the chain rule to make the change of variables $\xi = x/L$ to convert the BVP

$$\left.\begin{array}{l} -u'' = \lambda u, \quad 0 < x < L \\ u(0) = u(L) = 0, \end{array}\right\}$$

to one involving derivatives with respect to ξ. Deduce the eigenvalues and eigenfunctions of this problem by following Example 5.10.

5.17 Show, by making the substitution $u(x) = M(x)w(x)$ into (5.2) and choosing $M(x)$ appropriately, that the ODE may be transformed to the form $-w'' + Q(x)w - G(x)$. Express the functions Q and G in terms of $a(x)$, $b(x)$ and $M(x)$.

5.18 ☆ Show that the function $u(x) = x^m$ is square integrable on the interval $(0, 1)$ for $m > -\frac{1}{2}$ but not for $m \le \frac{1}{2}$.

5.19 Show that the functions $\phi_n(x) = e^{2\pi i n x/L}$, $n = 0, \pm 1, \pm 2, \ldots$, are mutually orthogonal with respect to the inner product (5.17) on the interval $(0, L)$. Evaluate $\langle \phi_n, \phi_n \rangle$.

5.20 Complete the working out of Example 5.8.

5.21 ☆ Show that, if u is an eigenfunction corresponding to an eigenvalue λ, then cu is also an eigenfunction for any constant c.

5.22 * Show that the Sturm–Liouville problem (5.33) has an eigenvalue $\lambda = 0$ when $a_0 = a_1 = 0$, $b_0 \neq 0$, $b_1 \neq 0$ and $q(x) \equiv 0$ by identifying a suitable eigenfunction.

5.23 Check that the eigenfunctions in Example 5.13 are orthogonal with respect to the inner product (5.17).

5.24 * Show that the choices $\lambda = 4$ and $\lambda = 4 - \mu^2 < 4$ in Example 5.13 both lead to trivial solutions.

5.25 Repeat Example 5.13 with the alternative BCs: $u(0) = u(\pi) = 0$.

5.26 Determine the eigenvalues and corresponding eigenfunctions of the differential equation

$$-u''(x) = \lambda u(x), \quad 0 < x < \pi$$

with boundary conditions $u(0) = u'(0)$ and $u(\pi) = u'(\pi)$.
Show, in particular, that there is one negative eigenvalue and express the corresponding eigenfunction in its simplest form.

5.27 Consider the eigenvalue problem

$$\left. \begin{array}{c} -x^2 u'' + 2xu' - 2u = \lambda x^2 u, \quad 0 < x < 1, \\ u'(0) = 0, \quad u(1) = u'(1). \end{array} \right\}$$

Determine a function $M(x)$ so that, under the change of variable $u(x) = M(x)w(x)$, it can be transformed to the standard form

$$-w'' = \lambda w, \quad 0 < x < 1$$

and establish the appropriate boundary conditions for w.
Hence determine all eigenvalues and eigenfunctions of the original problem.

5.28 Show that $\lambda = \mu^2$ is an eigenvalue of the BVP

$$\left. \begin{array}{c} -u''(x) = \lambda u(x), \quad 0 < x < 1 \\ u(0) = 0, u(1) = u'(1), \end{array} \right\}$$

provided that μ is a root of the equation $\tan \mu = \mu$. By considering the graphs of $\tan \mu$ and μ, show that there are an infinite number of eigenvalues λ_n and that $\lambda_n \to (n + \frac{1}{2})^2 \pi^2$ as $n \to \infty$.

5.29 Consider the eigenvalue problem

$$-u''(x) = \lambda w(x)u(x), \quad 0 < x < 1,$$
$$u'(0) = 0, \quad u(1) + u'(1) = 0,$$

where $w(x)$ is a given positive function. Prove that all eigenvalues λ are real and positive.

5.30 Extend Theorem 5.11 to prove that all eigenvalues of the Sturm–Liouville problem (5.33) are real and positive provided that a_0, b_0, a_1 and b_1 are positive real numbers.

5.31 Show that the properties in Exercise 5.11 also hold for the weighted inner product defined by (5.38).

5.32 Suppose that all members of the set of functions $\{\phi_n(x)\}_{n\in\mathbb{N}}$ are mutually orthogonal with respect to the weighted inner product $\langle u, v\rangle_w$. If $f(x)$ is given by the convergent sum $f(x) = \sum_{n=1}^{\infty} a_n\phi_n(x)$, show that the coefficients satisfy

$$a_n = \frac{\langle f, \phi_n\rangle_w}{\langle \phi_n, \phi_n\rangle_w}.$$

5.33 Suppose that $\phi_m(x)$ and $\phi_n(x)$ are eigenfunctions of the Sturm–Liouville problem (5.33) corresponding to distinct eigenvalues λ_m and λ_n. Use Lagrange's identity to prove that $\phi_m(x)$ and $\phi_n(x)$ are orthogonal with respect to the weighted inner product (5.38), that is, $\langle \phi_m, \phi_n\rangle_w = 0$ for $m \neq n$.

Chapter 6
Finite Difference Methods in $\mathbb{R}^1$

Abstract This chapter is an introduction to finite difference approximation methods. Key concepts like local truncation error, numerical stability and convergence of approximate solutions are developed in a one-dimensional setting. This chapter establishes the theoretical framework that is used to analyse the convergence of finite difference approximations in later chapters.

Our aim in this chapter is to introduce a simple technique for constructing numerical solutions to two-point boundary value problems based on the use of *finite differences*. Whereas the exact solution is a function $u(x)$ defined on an interval $x \in [0, L]$, say, our numerical solution is sought only at a finite set of *grid points*

$$x_0 = 0 < x_1 < \cdots < x_M = L.$$

For the sake of simplicity, we will assume that these grid points are equally spaced, so that $x_m = mh$ ($m = 0, 1, \ldots, M$), where $h = L/M$ is the grid size. The values of the exact solution of the BVP on the grid are

$$u(x_0), u(x_1), \ldots, u(x_M).$$

In the approximation process the given BVP will be replaced by a set of $M + 1$ algebraic equations (known as finite difference equations) in $M + 1$ unknowns

$$U_0, U_1, \ldots, U_M.$$

We shall often refer to U as a grid function, one whose domain is the set of grid points $\{x_0, x_1, \ldots, x_M\}$ and whose value at the mth grid point x_m is denoted by U_m. A typical situation is depicted in Fig. 6.1.

It is essential that the approximation be convergent in the sense that each item in the list $U_0, U_1, \ldots, U_M$ should converge to the corresponding item in the list $u(x_0), u(x_1), \ldots, u(x_M)$ as $M \to \infty$. This will mean that any desired level of accuracy can be achieved by choosing M appropriately. Having to take impractically large values of M in order to meet some desired accuracy tolerance is what drives

© Springer International Publishing Switzerland 2015   85
D.F. Griffiths et al., *Essential Partial Differential Equations*, Springer
Undergraduate Mathematics Series, DOI 10.1007/978-3-319-22569-2_6

Fig. 6.1 A grid of points x_m ($m = 0, 1, \ldots, M$), the exact solution $u(x)$ (*solid curve*), its restriction to the grid u_m ($m = 0, 1, \ldots, M$, *crosses*) and a notional numerical solution U_m ($m = 0, 1, \ldots, M$, dots)

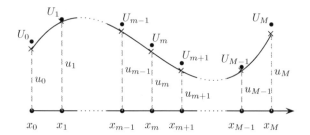

the search for more efficient numerical methods. There are two main ways in which convergence can be problematic:

(a) there may *not be a limit* as $M \to \infty$—this we attribute to a lack of stability of the approximation method,

(b) there may be convergence to the *wrong limit* as $M \to \infty$—this we attribute to an inconsistency with the original boundary value problem.

These key ideas of consistency, stability and convergence are central to the development and analysis of numerical methods and much of this chapter is devoted to exploring the relationships between these concepts.

The replacement of differential equations and their associated boundary conditions by algebraic equations will be accomplished through Taylor series expansions. Other means are also possible and some of these are explored in exercises at the end of this chapter. The building blocks of the approximation process are developed in the following section and their deployment and effectiveness is studied in subsequent sections.

6.1 The Approximation of Derivatives

We will suppose throughout this chapter that v is a smooth function, by which we mean that it is a continuous function that possesses sufficiently many continuous derivatives in order that various Taylor expansions are valid.

Fig. 6.2 The gradients of the chords AB (*backward*), BC (*forward*) and AC (*central*) are three possibilities for approximating the gradient of the tangent to the function $v(x)$ (*solid curve*) at B

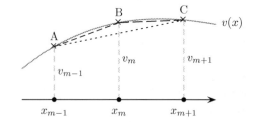

Referring to Fig. 6.2, the gradient of the function $v(x)$ at B ($x = x_m$) may be approximated by the gradients of any of the chords AB, BC or AC. We shall formalise the connections in this section and provide a means for assessing the degree of approximation involved in each case.

We shall begin with the slope of the forward chord at B(x_m, v_m). In the Taylor expansion with remainder

$$v(x + h) = v(x) + hv'(x) + \tfrac{1}{2}h^2 v''(\xi), \tag{6.1}$$

for some number $\xi \in (x, x + h)$, we choose $x = x_m$, and rearrange to give

$$v'(x_m) = h^{-1}\big(v(x_{m+1}) - v(x_m)\big) - \tfrac{1}{2}hv''(\xi_m), \quad \xi_m \in (x_m, x_{m+1}), \tag{6.2}$$

where we have written ξ_m with a subscript to reflect its dependence on x_m. We shall adopt a notation whereby $v_m, v'_m, v''_m, \ldots$ are used to denote $v(x_m), v'(x_m), v''(x_m), \ldots$, respectively. Thus,

$$v'_m = h^{-1}\big(v_{m+1} - v_m\big) + R_m.$$

The remainder term R_m is known as the *local truncation error*. It corresponds to truncating the Taylor series underlying (6.2) and is given by $R_m = -\tfrac{1}{2}hv''(\xi_m)$. Although it cannot be evaluated (except in very special cases) we observe that it can be made arbitrarily small by choosing h to be sufficiently small—provided that v'' is bounded in the neighbourhood of x_m. If we use the "big O" notation whereby $R_m = \mathcal{O}(h)$ signifies that the remainder is roughly proportional[1] to h as $h \to 0$, then we can write

$$v'_m = h^{-1}\big(v_{m+1} - v_m\big) + \mathcal{O}(h). \tag{6.3}$$

When the remainder term is omitted this provides an estimate, known as the forward difference approximation, of v' at $x = x_m$. The forward difference operator $\triangle^+$ is defined by

$$\triangle^+ v_m := v_{m+1} - v_m \tag{6.4}$$

and so $v'_m \approx h^{-1}\triangle^+ v_m$ with an error of $\mathcal{O}(h)$.

In a similar fashion, starting with the Taylor series

$$v(x - h) = v(x) - hv'(x) + \tfrac{1}{2}h^2 v''(\xi), \quad \xi \in (x - h, x) \tag{6.5}$$

and the backward difference operator $\triangle^-$ defined by

$$\triangle^- v_m :- v_m - v_{m-1}, \tag{6.6}$$

---

[1] More precisely, suppose that $z(h)$ is a quantity that depends on h. We say that $z(h) = \mathcal{O}(h^p)$ if there is a constant C, independent of h, such that $|z(h)| \leq Ch^p$ as $h \to 0$.

it may be shown that

$$v'_m = h^{-1}\triangle^- v_m + \mathcal{O}(h).\tag{6.7}$$

Thus $v'_m \approx h^{-1}\triangle^- v_m$, again with an error proportional to h.

The two approximations introduced above are known as first-order approximations since the error is proportional to (the first power of) h. A more accurate approximation may be obtained by including one more term in each of the expansions (6.1) and (6.5), so that

$$v(x \pm h) = v(x) \pm hv'(x) + \tfrac{1}{2}h^2 v''(x) \pm \tfrac{1}{6}h^3 v'''(\xi_\pm).$$

Subtracting one from the other gives

$$v'(x) = \tfrac{1}{2}h^{-1}\big(v(x+h) - v(x-h)\big) + \tfrac{1}{12}h^2\big(v'''(\xi_+) + v'''(\xi_-)\big)$$

and, assuming that $v'''(x)$ is continuous, the intermediate value theorem implies that there is a point $\xi_m \in (\xi_-, \xi_+)$ such that

$$v'_m = \tfrac{1}{2}h^{-1}\big(v_{m+1} - v_{m-1}\big) + \tfrac{1}{6}h^2 v'''(\xi_m), \qquad \xi_m \in (x_{m-1}, x_{m+1}).\tag{6.8}$$

Defining the second-order central difference operator $\triangle$ by

$$\triangle v_m := \tfrac{1}{2}\big(v_{m+1} - v_{m-1}\big),\tag{6.9}$$

we find that $v'_m = h^{-1}\triangle v_m + \mathcal{O}(h^2)$, which represents a second-order, central difference approximation of v'_m.

In summary, there are three basic ways of approximating v'_m using the three grid values v_{m-1}, v_m and v_{m+1}. The second-order approximation $(h^{-1}\triangle v_m)$ will always be more accurate than the first-order approximations $(h^{-1}\triangle^+ v_m$ and $h^{-1}\triangle^- v_m)$ provided that h is sufficiently small and the underlying function v is sufficiently smooth. Moreover, any weighted average of the form

$$h^{-1}\big(\theta\triangle^+ + (1-\theta)\triangle^-\big)v_m$$

will also give a first-order approximation to v'_m except for the special case $\theta = 1/2$, which coincides with the central difference approximation $h^{-1}\triangle v_m$.

Before leaving this topic, let us introduce another central difference,

$$\delta v(x) := v(x + \tfrac{1}{2}h) - v(x - \tfrac{1}{2}h),$$

which corresponds to the use of $\triangle$ with h replaced by $h/2$. When $x = x_m$, a point on the grid, the right-hand side involves values of v midway between grid points, while, if $x = x_{m+1/2}$ (by which we mean $x = x_m + 1/2h$) we have that

$$\delta v_{m+1/2} := v_{m+1} - v_m.\tag{6.10}$$

Exploiting the connection with $\triangle$ we can deduce from (6.8) that

$$v'(x) = h^{-1}\delta v(x) + \tfrac{1}{24}h^2 v'''(\xi), \qquad \xi \in (x - \tfrac{1}{2}h, x + \tfrac{1}{2}h). \tag{6.11}$$

Although δ cannot be used on its own (because it requires values at points that are not on the grid), it is nevertheless indispensable in the approximation of even derivatives. Indeed, the customary approximation for second derivatives is based on $\delta^2 v_m = \delta(\delta v_m)$, that is,

$$
\begin{aligned}
\delta^2 v_m &= \delta(\delta v_m) \\
&= \delta(v_{m+1/2} - v_{m-1/2}) = \delta v_{m+1/2} - \delta v_{m-1/2} \\
&= (v_{m+1} - v_m) - (v_m - v_{m-1}) = v_{m+1} - 2v_m + v_{m-1}.
\end{aligned} \tag{6.12}
$$

To see how this works (see Exercise 6.1 for a formal derivation), consider the two Taylor series expansions

$$v(x \pm h) = v(x) \pm hv'(x) + \tfrac{1}{2}h^2 v''(x) \pm \tfrac{1}{6}h^3 v'''(x) + \tfrac{1}{24}h^4 v''''(x) + \cdots. \tag{6.13}$$

Adding these series together we obtain

$$v(x + h) + v(x - h) = 2v(x) + h^2 v''(x) + \tfrac{1}{12}h^4 v''''(x) + \cdots$$

and setting $x = x_m$ leads to

$$v_m'' = h^{-2}\left(v_{m+1} - 2v_m + v_{m-1}\right) - \tfrac{1}{12}h^2 v_m'''' + \cdots \tag{6.14}$$

Thus, from (6.12) we get a centered finite difference approximation to the second derivative,

$$v_m'' \approx h^{-2}\delta^2 v_m,$$

which is second order since the remainder term in (6.14) is proportional to h^2.

The main results of this section are summarised in Table 6.1. Here the remainder terms are expressed in the shorthand form $\mathcal{O}(h^p)$, which is less cumbersome than using precise expressions like (6.8). Note that, in general, there are a variety of different ways of combining approximations of first derivatives to approximate higher derivatives.

Table 6.1 Finite difference operators: definitions and Taylor expansions

Forward difference operator	$\triangle^+ v_m := v_{m+1} - v_m = hv_m' + \tfrac{1}{2}h^2 v_m'' + \mathcal{O}(h^3)$
Backward difference operator	$\triangle^- v_m := v_m - v_{m-1} = hv_m' - \tfrac{1}{2}h^2 v_m'' + \mathcal{O}(h^3)$
Central difference operator	$\triangle v_m := \tfrac{1}{2}\left(v_{m+1} - v_{m-1}\right) = hv_m' + \tfrac{1}{6}h^3 v_m''' + \mathcal{O}(h^5)$
Central difference operator	$\delta v_m := v_{m+1/2} - v_{m-1/2} = hv_m' + \tfrac{1}{24}h^3 v_m''' + \mathcal{O}(h^5)$
Second-order centered difference operator	$\delta^2 v_m := v_{m+1} - 2v_m + v_{m-1} = h^2 v_m'' + \tfrac{1}{12}h^4 v_m'''' + \mathcal{O}(h^6)$

6.2   Approximation of Boundary Value Problems

We begin by considering the two-point BVP, see (5.9),

$$\left.\begin{aligned}-u''(x) + r(x)u(x) = f(x), \quad 0 < x < 1 \\ u(0) = \alpha, \quad u(1) = \beta,\end{aligned}\right\} \tag{6.15}$$

where $r(x)$, $f(x)$ are given continuous functions on $[0, 1]$. The boundary conditions in this example do not require approximation (more complicated boundary conditions will be covered in Sect. 6.4.1) so we can fix our attention on the differential equation. When this is evaluated at the mth grid point we get

$$- u''(x_m) + r(x_m)u(x_m) = f(x_m), \tag{6.16}$$

and, when the finite difference (6.14) (with v replaced by u) is used to represent the second derivative term, we find that

$$- h^{-2}\delta^2 u_m + \mathcal{O}(h^2) + r_m u_m = f_m, \tag{6.17}$$

where $r_m = r(x_m)$ and $f_m = f(x_m)$. The order term, representing the remainder terms in the Taylor expansions, is retained in (6.17), so this equation continues to be satisfied identically by the exact solution of the BVP. When the order term is neglected, equation (6.17) will no longer be satisfied by u but by some grid function U, say, which we hope will be close to u. This process leads to a finite difference representation of (6.15), namely

$$\left.\begin{aligned}U_0 = \alpha, \\ -h^{-2}\delta^2 U_m + r_m U_m = f_m, \quad m = 1, 2, \ldots, M - 1, \\ U_M = \beta.\end{aligned}\right\} \tag{6.18}$$

We will henceforth refer to (6.18) as a *discrete* BVP. Writing it explicitly gives

$$- \frac{1}{h^2}[U_{m-1} - 2U_m + U_{m+1}] + r_m U_m = f_m, \tag{6.19}$$

for $m = 1, 2, \ldots, M-1$, which is a linear relationship between three consecutive grid values of U. This is the fewest number of grid points that can be used to approximate the second derivative of a function. Taken together with the boundary conditions $U_0 = \alpha$, $U_M = \beta$, there are $M + 1$ linear algebraic equations with which to determine the $M + 1$ grid values of U.

It is convenient for developing the theory of finite difference equations to introduce notation that resembles that used for differential equations (see Sects. 2.1 and 5.2). Thus, a finite difference operator $\mathcal{L}_h$ (the subscript h acts as a reminder that it involves a grid of size h) that represents the discretized differential equation is defined via[2]

$$\mathcal{L}_h U_m := -h^{-2}\delta^2 U_m + r_m U_m, \qquad m = 1, 2, \ldots, M - 1. \qquad (6.20)$$

Similarly, we let $\mathcal{F}_h$ represent the corresponding source term—its mth component being, in this case, $\mathcal{F}_{h,m} = f_m$. Equation (6.19) may then be written succinctly as $\mathcal{L}_h U = \mathcal{F}_h$.

In keeping with the convention introduced in Sect. 2.1, a script font is used to represent differential/difference operators together with their boundary conditions. Accordingly, we define

$$\mathscr{L}_h U_m = \begin{cases} U_m & \text{for } m = 0, M, \\ \mathcal{L}_h U_m & \text{for } m = 1, 2, \ldots, M - 1, \end{cases} \qquad (6.21)$$

and

$$\mathscr{F}_{h,m} = \begin{cases} \alpha & \text{for } m = 0, \\ f_m & \text{for } m = 1, 2, \ldots, M - 1, \\ \beta & \text{for } m = M. \end{cases} \qquad (6.22)$$

The discrete BVP (6.18) can then be succinctly written as

$$\mathscr{L}_h U = \mathscr{F}_h. \qquad (6.23)$$

This may be clearly identified as being the discrete analogue of the BVP in (5.11), namely, $\mathscr{L} u = \mathscr{F}$.

It is also useful to express the discrete BVP in matrix–vector notation. To this end, setting $m = 1$ in (6.19) and using the left-end condition $U_0 = \alpha$ gives

$$\left(\frac{2}{h^2} + r_1\right) U_1 - \frac{1}{h^2} U_2 = f_1 + \frac{\alpha}{h^2}.$$

Similarly, for $m = 1, 2, \ldots, M - 1$ we get

$$-\frac{1}{h^2} U_{m-1} + \left(\frac{2}{h^2} + r_m\right) U_m - \frac{1}{h^2} U_{m+1} = f_m$$

and finally, setting $m = M - 1$, gives

$$-\frac{1}{h^2} U_{M-2} + \left(\frac{2}{h^2} + r_{M-1}\right) U_{M-1} = f_{M-1} + \frac{\beta}{h^2},$$

---

[2]The shorthand version $\mathcal{L}_h U_m$ to denote the value of $\mathcal{L}_h U$ at the mth grid point.

using the right-end condition $U_M = \beta$.

Next, if we let $\boldsymbol{u} \in \mathbb{R}^{M-1}$ denote the vector

$$\boldsymbol{u} = [U_1, U_2, \ldots, U_{M-1}]^\mathsf{T},$$

which contains the unknown grid values of U in their natural order, then (6.19) may be expressed in the matrix form

$$A\boldsymbol{u} = \boldsymbol{f}, \tag{6.24}$$

where

$$A = \frac{1}{h^2} \begin{bmatrix} a_{1,1} & -1 & & & \\ -1 & a_{2,2} & -1 & & \\ & \ddots & \ddots & \ddots & \\ & & & & -1 \\ & & & -1 & a_{M-1,M-1} \end{bmatrix} \tag{6.25}$$

(only the nonzero entries are shown), $a_{m,m} = 2 + r_m h^2$ and

$$\boldsymbol{f} = \begin{bmatrix} f_1 + \alpha/h^2 \\ f_2 \\ \vdots \\ f_{M-2} \\ f_{M-1} + \beta/h^2 \end{bmatrix}$$

It should be observed that not only does the dimension of the matrix A grow without bound as $h \to 0$ but its nonzero elements also tend to $\pm\infty$. The matrix A can be expressed more neatly by defining T to be the $(M - 1) \times (M - 1)$ symmetric tridiagonal matrix

$$T = \frac{1}{h^2} \begin{bmatrix} 2 & -1 & & & \\ -1 & 2 & -1 & & \\ & \ddots & \ddots & \ddots & \\ & & -1 & 2 & -1 \\ & & & -1 & 2 \end{bmatrix} \tag{6.26}$$

which represents the approximation of the second derivative with Dirichlet boundary conditions. Then $A = T + D$, where

$$D = \mathrm{diag}(r_1, r_2, \ldots, r_{M-1})$$

is a diagonal matrix. The algebraic system (6.24) will have a unique solution provided A is nonsingular; the following lemma establishes a somewhat stronger result.

Lemma 6.1 *The matrix A of equation (6.24) is symmetric and positive definite if* $r(x) \geq 0$ *or equivalently if* $r_m \geq 0$ *for all m.*

Proof The symmetry is obvious so we need to prove that $v^{\mathsf{T}} A v > 0$ for all nonzero vectors $v \in \mathbb{R}^{M-1}$. The condition $r_m \geq 0$ ensures that $v^{\mathsf{T}} D v \geq 0$ and hence

$$v^{\mathsf{T}} A v = v^{\mathsf{T}} (T + D) v = v^{\mathsf{T}} T v + v^{\mathsf{T}} D v \geq v^{\mathsf{T}} T v.$$

The positive definiteness of A will then follow from the positive definiteness of T. To establish this, let L denote the $(M-1) \times M$ matrix

$$L = \frac{1}{h} \begin{bmatrix} 1 & -1 & & & \\ & \ddots & \ddots & & \\ & & 1 & -1 & \\ & & & 1 & -1 \end{bmatrix}$$

A direct computation reveals that $LL^{\mathsf{T}} = T$ and so, defining $w = L^{\mathsf{T}} v$,

$$v^{\mathsf{T}} T v = v^{\mathsf{T}} L L^{\mathsf{T}} v = w^{\mathsf{T}} w \geq 0.$$

Clearly, $w^{\mathsf{T}} w = 0$ if and only if $w = 0$ but in this case, solving $L^{\mathsf{T}} v = w = 0$ leads to $v = 0$. $\qquad\square$

Note that the matrix L is a representation of the difference operator $-h^{-1} \Delta^+$ in that $Lv_m = -(U_{m+1} - U_m)/h$. Likewise, L^{T} (apart from its first and last rows which are influenced by BCs) represents $h^{-1} \Delta^-$ (see Exercise 6.2).

Example 6.2 Use the method (6.18) to solve the differential equation

$$-u''(x) + 2\sigma^2 \sec^2(\sigma x) u(x) = 4\sigma^2 \cos^2(\sigma x) \qquad (6.27)$$

on the interval $0 < x < 1$ with $\sigma = 3/2$ and the boundary conditions $u(0) = 1$, $u(1) = -\sin^2 \sigma$.

The matrix A and right hand side f of the linear algebra system (6.24) are readily calculated for $r(x) = 2\sigma^2 \sec^2(\sigma x)$, $f(x) = 4\sigma^2 \cos^2(\sigma x)$, $\alpha = 1$ and $\beta = -\sin^2 \sigma$. Current software is capable of solving such systems reliably—because A is positive definite—and usually in less time than it takes to calculate the elements of A and f. The numerical solutions with $M = 8$, 16 and 32 are shown by the dots in Fig. 6.3 along with the exact solution $u(x) = \cos^2(\sigma x) - \tan(\sigma x)/\tan \sigma$. $\qquad\Diamond$

It is observed that the greater the number of grid points the closer the numerical solution is to the exact solution—an indication, but no more, that the numerical solutions converge to the exact solution as $h \to 0$. The difference $E = u - U$, known as the *global error*, is regarded as a grid function since it is defined only at

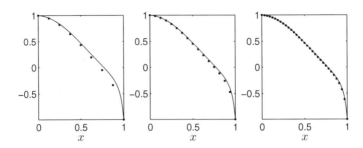

Fig. 6.3 The exact solution $u(x)$ (*solid line*) and the numerical solution U (*dots*) for Example 6.2 when $M = 8$, 16 and 32

Table 6.2 The maximum global error as a function of h for Example 6.2

M	h	$\|u - U\|_{h,\infty}$	Ratio
8	0.125	0.13	–
16	0.0625	0.071	1.86
32	0.03125	0.026	2.71
64	0.015625	0.0078	3.38
128	0.0078125	0.0021	3.78
256	0.00390625	0.00052	3.93
512	0.001953125	0.00013	3.98

The final column shows the ratio of the global error with grid size $2h$ to the global error obtained using a grid size h

grid points. A natural measure of its magnitude (and one that we shall use throughout this chapter) is provided by the ℓ_∞ or *maximum norm*, defined by

$$\|u - U\|_{h,\infty} := \max_m |u_m - U_m|, \qquad (6.28)$$

where we have included a subscript h as a reminder that the maximum is taken over all points of the grid, rather than the entire interval.

Definition 6.3 (*Convergence*) A numerical method is said to converge if $\|u - U\|_{h,\infty} \to 0$ as $h \to 0$. It is said to be convergent of order p if $\|u - U\|_{h,\infty} = \mathcal{O}(h^p)$, for some $p > 0$.[3]

Sample results for Example 6.2 are shown in Table 6.2. The final column shows the ratio by which the error is reduced when h is halved and suggests that the global error is reduced by a factor of about four whenever the grid size h is halved—provided that h is sufficiently small. The results suggest that $\|u - U\|_{h,\infty} \propto h^2$ so that it converges at a second-order rate. Thus reducing h by a factor of 3 will cause a reduction in the global error of $1/9$ and improve the accuracy of the numerical solution

---

[3]The order of convergence p is usually an integer but exceptions to this rule are not uncommon, so beware.

Fig. 6.4 Log-log plot of the maximum global error versus h for Example 6.2

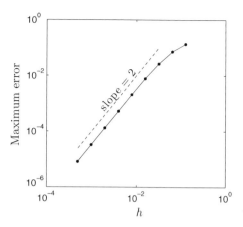

by almost one decimal digit. The convergence can be visualised by supposing that $\|u - U\|_{h,\infty} = \mathcal{O}(h^p)$ so that

$$\|u - U\|_{h,\infty} \approx Ch^p,$$

if h is small enough that the higher-order terms are negligible. Taking logarithms,

$$\log \|u - U\|_{h,\infty} \approx p \log h + \log C$$

so a plot of $\log \|u - U\|_{h,\infty}$ versus $\log h$ should reveal a straight line of gradient p. Figure 6.4 shows such a log–log plot for Example 6.2. The maximum global errors for $M = 8, 16, \ldots, 2048$ are shown by dots and a dashed line of slope two has been included for comparison. This provides compelling evidence that the global error is proportional to h^2 when h is sufficiently small.   ◊

Numerical experimentation can often expose methods that are not convergent but a mathematical analysis is necessary in order to establish convergence. This aspect will be addressed in the next section.

6.3 Convergence Theory

The theory in this section can be applied quite broadly: it revolves around the notation introduced earlier whereby the BVP $\mathscr{L}u = \mathscr{F}$ is approximated by a discrete BVP $\mathscr{L}_h U = \mathscr{F}_h$. The aim is to follow the principles used in the previous chapter (see Sect. 5.2) to establish the well posedness of the underlying BVP. The first step in the analysis is the determination of the local truncation error (LTE). This provides a measure of how well the discrete BVP matches the original BVP.

Definition 6.4 (*Local truncation error*) The local truncation error, denoted by $\mathscr{R}_h$, is defined to be the residual when the exact solution u of the BVP is substituted into the discrete equations. Thus,

$$\mathscr{R}_h := \mathscr{L}_h u - \mathscr{F}_h.$$

When attention is restricted to the approximation of the differential equation, the local truncation error is defined to be[4] $\mathcal{R}_h := \mathcal{L}_h u - \mathcal{F}_h$, so that $\mathscr{R}_h = \mathcal{R}_h$ except at boundary points.

Definition 6.5 (*Consistency*) The approximation $\mathscr{L}_h U = \mathscr{F}_h$ is said to be consistent with $\mathscr{L} u = \mathscr{F}$ if $\mathscr{R}_h := \mathscr{L}_h u - \mathscr{F}_h \to 0$ as $h \to 0$. It is consistent of order p if $\mathscr{R}_h = \mathcal{O}(h^p)$ with $p > 0$.

Returning to the discrete BVP (6.18), the Dirichlet boundary conditions are replicated exactly so the LTE at $x = x_m$ is given by

$$\mathcal{R}_m = -h^{-2}\delta^2 u_m + r_m u_m - f_m, \tag{6.29}$$

where u is the solution of (6.15). Since $f = -u'' + r\,u$, the result (6.14) leads to

$$\begin{aligned}
\mathcal{R}_m &= -\left(u_m'' + \tfrac{1}{12}h^2 u_m'''' + \mathcal{O}(h^4)\right) + r_m u_m - \left(-u_m'' + r_m u_m\right) \\
&= -\tfrac{1}{12}h^2 u_m'''' + \mathcal{O}(h^4)
\end{aligned} \tag{6.30}$$

so $\mathscr{R}_h = \mathcal{O}(h^2)$ and (6.18) is consistent with (6.15) of order two provided, of course, that $u''''(x)$ is bounded on the interval $(0, 1)$.

For a consistent approximation we have the situation where the numerical solution satisfies $\mathscr{L}_h U = \mathscr{F}_h$ whereas the exact solution u satisfies the nearby relation $\mathscr{L}_h u = \mathscr{F}_h + \mathscr{R}_h$. Thus the issue is whether or not this implies that U is close to u. Defining the global error (as in the previous section) by $E = u - U$, we see, by linearity of $\mathscr{L}_h$, that $\mathscr{L}_h E = \mathscr{L}_h u - \mathscr{L}_h U$ and so

$$\mathscr{L}_h E = \mathscr{R}_h, \tag{6.31}$$

which is actually the same as the equation for U with the right hand side $\mathscr{F}_h$ replaced by $\mathscr{R}_h$. The requirement that E should tend to zero whenever $\mathscr{R}_h$ tends to zero leads to the following definition.

Definition 6.6 (ℓ_∞ *stability*) The discrete operator is said to be stable (with respect to the maximum norm $\| \cdot \|_{h,\infty}$) if there is a constant $C > 0$, *independent of h*, such that the solution of the equation $\mathscr{L}_h U = \mathscr{F}_h$ satisfies

$$\|U\|_{h,\infty} \le C\|\mathscr{F}_h\|_{h,\infty},$$

where C is known as the stability constant.

---

[4]The subscript h will often be omitted and we write $\mathcal{R}$ and $\mathscr{R}$ since they have no continuous counterparts and to avoid the notation becoming too onerous.

This mirrors the definition of a well-posed BVP given in Sect. 2.2.3 with the added requirement that the upper bound C be independent of h (so as to preclude the possibility that $C \to \infty$ as $h \to 0$). Two further points are worthy of mention. The first is that stability of $\mathcal{L}_h$ makes no reference to the original differential operator and, secondly, stability is established with respect to a particular norm, in this case the maximum norm. The next theorem is often paraphrased as "consistency and stability implies convergence".

Theorem 6.7 (Convergence) *Suppose that discrete BVP $\mathcal{L}_h U = \mathcal{F}_h$ is a consistent approximation of $\mathcal{L}u = \mathcal{F}$ and that $\mathcal{L}_h$ is stable in the sense of Definition 6.6. Then $\|u - U\|_{h,\infty} \to 0$ as $h \to 0$. Moreover, if the order of consistency is $p > 0$ then the order of convergence is also p.*

Proof Stability of $\mathcal{L}_h$, together with (6.31) give

$$\|E\|_{h,\infty} \leq C\|\mathcal{R}_h\|_{h,\infty}$$

so that $\|\mathcal{R}_h\|_{h,\infty} = \mathcal{O}(h^p)$ implies that $\|E\|_{h,\infty} = \mathcal{O}(h^p)$ provided that C is bounded independently of h.   □

We have already seen how consistency of a finite difference scheme can be established. As for stability, we exploit its relationship with well-posedness (see Sect. 5.2) which was built around the ideas of inverse monotonicity ($\mathcal{L}u \geq 0$ implies $u \geq 0$) and the existence of a comparison function $\varphi > 0$ such that $\mathcal{L}\varphi \geq 1$. The analogue for discrete operators is given by the following lemma.

Lemma 6.8 (Stability) *Suppose that the operator $\mathcal{L}_h$ is inverse monotone and that there is a comparison function $\Phi > 0$ such that[5] $\mathcal{L}_h \Phi \geq 1$. Then $\mathcal{L}_h$ is stable, with stability constant $C = \max_m \Phi_m$, provided that Φ is bounded independently of h.*

Proof See Theorem 5.4.   □

The discrete operators encountered so far have been such that $\mathcal{L}_h U_m$ has involved a linear combination of three consecutive values of U, namely U_{m-1}, U_m and U_{m+1}. The next definition identifies some simple inequalities that are sufficient (though not necessary) for operators of this type to be inverse monotone when accompanied by Dirichlet boundary conditions. The definition will later be extended to include approximations of other boundary conditions as well as approximations of partial differential equations.

Definition 6.9 (*Positive type operator*) A finite difference operator of the form

$$\mathcal{L}_h U_m := -a_m U_{m-1} + b_m U_m - c_m U_{m+1} \quad (m = 1, 2, \dots, M-1) \qquad (6.32)$$

is said to be of *positive type* if the coefficients satisfy the inequalities

---

[5]Inequalities of the form $V \geq 0$, where V is a grid function, mean that $V_m \geq 0$ for all $m = 0, 1, \dots, M$ (or $m = 1, 2, \dots, M-1$, depending on context).

$$a_m \geq 0, \quad c_m \geq 0, \quad b_m \geq a_m + c_m \qquad (6.33)$$

and $b_m > 0$.

Theorem 6.10 (Inverse monotonicity) *Suppose that the difference operator $\mathscr{L}_h$ is defined by*

$$\mathscr{L}_h U_m = \begin{cases} U_m & \text{for } m = 0, M, \\ L_h U_m & \text{for } m = 1, 2, \ldots, M - 1, \end{cases} \qquad (6.34)$$

where L_h is of positive type, then $\mathscr{L}_h U \geq 0$ implies that $U \geq 0$.

Proof Suppose, contrary to the statement of the theorem, that there is at least one grid point where U is negative. This means that U attains its negative minimum at an internal grid point x_j, say ($1 \leq j \leq M - 1$, since we know that U_0 and U_M are both nonnegative from $\mathscr{L}_h U \geq 0$). Thus,

$$U_{j-1} \geq U_j \text{ and } U_{j+1} \geq U_j$$

and it follows from (6.32) and (6.33) that

$$L_h U_j \leq (a_j - b_j + c_j)U_j \leq 0. \qquad (6.35)$$

If the inequality happens to be strict ($L_h U_j < 0$) it would contradict the assumption $\mathscr{L}_h U \geq 0$ and the theorem would be proved.

We suppose therefore that equality holds. This can occur if and only if $a_j - b_j + c_j = 0$ and $U_{j-1} = U_j = U_{j+1}$, that is, the same negative minimum value is taken at three consecutive grid points. In this case, we apply the same reasoning with either $m = j - 1$ or $m = j + 1$. Continuing in this way, we either obtain a contradiction because of strict inequality or we reach a stage when the same negative minimum holds at $m = 0$ or $m = M$ where we know, by hypothesis, that $U_m \geq 0$. This again leads to a contradiction.   □

By applying the theorem to $-U$ we may also prove that $\mathscr{L}_h U \leq 0$ implies that $U \leq 0$. A comparison principle follows as in the continuous case: $\mathscr{L}_h U \geq \mathscr{L}_h V$ implies that $U \geq V$.

Corollary 6.11 (Uniqueness) *Suppose that $\mathscr{L}_h$ satisfies the conditions of Theorem 6.10. Then the discrete equation $\mathscr{L}_h U = \mathscr{F}_h$ has a unique solution.*

Proof See Theorem 5.3.

                                                                              □

Example 6.12 Show that the finite difference operator defined by (6.21) and (6.20) is inverse monotone when $r(x) \geq 0$.

Using (6.20) and (6.12) we find

$$\mathcal{L}_h U_m = h^{-2}\left(-U_{m-1} + (2 + r_m h^2)U_m - U_{m-1}\right)$$

and so, comparing with (6.32), $a_m = c_m = h^{-2}$ and $b_m = 2h^{-2} + r_m$. The condition $b_m \geq a_m + c_m$ required by Definition 6.9 is therefore satisfied provided that $r_m \geq 0$, for all m. Thus, not only will the solution of the discrete BVP $\mathcal{L}_h U = \mathcal{F}_h$ be unique, it will also be nonnegative provided that the data $\mathcal{F}_h$ is nonnegative.   ◊

The final step required in order use Lemma 6.8 to ascertain stability of a discrete BVP is to identify a suitable comparison function. The good news is that a comparison function φ that has previously been found for the continuous problem $\mathcal{L}u = \mathcal{F}$ can also be used for $\mathcal{L}_h U = \mathcal{F}_h$, provided that $\mathcal{L}_h$ is consistent with $\mathcal{L}$ in the sense that $\mathcal{L}_h \varphi \to \mathcal{L}\varphi$, as $h \to 0$. In many cases φ is a polynomial of low degree and we have $\mathcal{L}_h \varphi \equiv \mathcal{L}\varphi \geq 1$ and we may therefore choose $\varPhi = \varphi$. More generally, for any constant $c > 1$, we have

$$\mathcal{L}_h(c\varphi) = c\mathcal{L}_h\varphi \to c\mathcal{L}\varphi \geq c > 1,$$

as $h \to 0$. Thus $\varPhi = c\varphi$ may be used as a comparison function for $\mathcal{L}_h$ (provided, of course, that c is independent of h). In both cases the discrete operator $\mathcal{L}_h$ is seen to inherit a comparison function from $\mathcal{L}$. This contrasts with inverse monotonicity, which has to be established independently for $\mathcal{L}$ and $\mathcal{L}_h$.

Example 6.13 Show that the discrete BVP (6.18), repeated here for convenience,

$$\left.\begin{array}{r}U_0 = \alpha, \\ -h^{-2}\delta^2 U_m + r_m U_m = f_m, \quad m = 1, 2, \ldots, M - 1, \\ U_M = \beta,\end{array}\right\} \qquad (6.36)$$

is stable when $r(x) \geq 0$.

Inverse monotonicity of the corresponding operator $\mathcal{L}_h$ (defined by (6.21) and (6.20)) was established in Example 6.12 and it remains to find a suitable comparison function. With $\mathcal{L}_h$ defined by (6.20),

$$\mathcal{L}_h \varphi_m = -h^{-2}\delta^2 \varphi_m + r_m \varphi_m \geq -h^{-2}\delta^2 \varphi_m = -\varphi_m'' - \tfrac{1}{12}h^2 \varphi''''(\xi_m) \qquad (6.37)$$

where we have used $r_m \geq 0$, $\varphi_m \geq 0$ together with the result of Exercise 6.1 (a more precise version of (6.14)). The comparison function

$$\varphi(x) = 1 + \tfrac{1}{2}x(1 - x)$$

was shown in Example 5.5 to satisfy $-\varphi''(x) \geq 1$ for $x \in (0, 1)$ along with $\varphi(x) \geq 1$ at $x = 0, 1$. Since φ is a polynomial of degree less than four the remainder term in (6.37) vanishes so

$$\mathcal{L}_h \varphi_m = -\varphi_m'' \geq 1$$

and, because φ has the correct behaviour at the boundary points, $\mathscr{L}_h\varphi \geq 1$. We conclude that $\mathscr{L}_h$ is stable with stability constant $C = \max \varphi = 9/8$.   ◊

Thus, by virtue of Theorem 6.7, the solution of (6.36) will converge to the solution u of (6.15) provided that $u''''(x)$ is bounded on the interval $(0, 1)$, a property that can usually be deduced from the differential equation. For example, by differentiating the ODE $u'' = ru - f$ twice, we find that

$$u''''(x) = r''(x)u(x) + 2r'(x)u'(x) + r(x)u''(x) - f''(x),$$

in which the right hand side is bounded provided that r and f and their first and second derivatives are continuous on $(0, 1)$. Convergence of the finite difference method used in Example 6.2 can be verified by a particular instance of this argument (see Exercise 6.4). It is worth noting that finite difference methods of the type considered in this section usually converge (albeit at a slower rate) even if $u''''(x)$ is not bounded, but the analysis is considerably more delicate than that given in Theorem 6.7.

The next example illustrates the new issues that occur when the differential equation contains a first derivative term.

Example 6.14 (Advection–Diffusion) Write down a finite difference approximation that is consistent of order two with the BVP

$$\left. \begin{aligned} -\varepsilon u''(x) + 2u'(x) = f(x), \quad 0 < x < 1 \\ u(0) = \alpha, \quad u(1) = \beta, \end{aligned} \right\} \tag{6.38}$$

where $\varepsilon > 0$. Prove that the numerical solution is convergent of order two.

The approximations (6.8) and (6.14) suggest the numerical method

$$\left. \begin{aligned} -\varepsilon h^{-2}\delta^2 U_m + 2h^{-1}\Delta U_m = f_m, \quad m = 1, 2, \ldots, M - 1, \\ U_0 = \alpha, \quad U_M = \beta, \end{aligned} \right\} \tag{6.39}$$

which leads to the definition of the operator $\mathcal{L}_h$,

$$\mathcal{L}_h U_m = -\varepsilon h^{-2}\delta^2 U_m + 2h^{-1}\Delta U_m \tag{6.40a}$$

$$= -\left(\varepsilon h^{-2} + h^{-1}\right) U_{m-1} + 2\varepsilon h^{-2} U_m - \left(\varepsilon h^{-2} - h^{-1}\right) U_{m+1} \tag{6.40b}$$

for $m = 1, 2, \ldots, M - 1$. The form (6.40a) is most convenient for determining consistency. Thus, using $f(x) = -\varepsilon u''(x) + 2u'(x)$ and appropriate results from Table 6.1, we find

$$\begin{aligned} \mathcal{R}_m &= \mathcal{L}_h u_m - f_m \\ &= -\varepsilon h^{-2}\delta^2 u_m + 2h^{-1}\Delta u_m - \left(-\varepsilon u''_m + 2u'_m\right) \\ &= -\tfrac{1}{12}\varepsilon h^2 u''''_m + \tfrac{1}{3}h^2 u'''_m + \mathcal{O}(h^4). \end{aligned}$$

Accordingly, $\mathcal{R} = \mathcal{O}(h^2)$ when u and its derivatives up to order four are continuous and bounded.

When the finite difference equations are organised as a linear algebraic system $Au = f$ (cf. (6.24)) the coefficient matrix A is no longer symmetric—see Exercise 6.10—and so its nonsingularity cannot be deduced from Lemma 6.1. However, utilising Definition 6.9 and Theorem 6.10 (this is where the second form (6.40b) of the definition of $\mathcal{L}_h$ is useful) with

$$a_m = \varepsilon h^{-2} + h^{-1}, \quad b_m = 2\varepsilon h^{-2}, \quad c_m = \varepsilon h^{-2} - h^{-1},$$

we find that the operator is of positive type whenever $h \leq \varepsilon$, in which case the corresponding operator $\mathcal{L}_h$ (formed by $\mathcal{L}_h$ with Dirichlet boundary conditions) is inverse monotone. The existence of a unique numerical solution then follows from Corollary 6.11.

In order to use Lemma 6.8 to establish stability of $\mathcal{L}_h$, a suitable comparison function must be found. As in the previous example, this can be achieved by finding a comparison function for the BVP. Thus, trying a linear function $\varphi(x) = Ax + B$ where A and B are constants, gives

$$\mathcal{L}\varphi(x) = \begin{cases} \varphi(0) = B, & x = 0, \\ -\varepsilon\varphi''(x) + 2\varphi'(x) = 2A, & 0 < x < 1, \\ \varphi(1) = A + B, & x = 1, \end{cases}$$

and so $\mathcal{L}\varphi \geq 1$ when $B \geq 1, 2A \geq 1$ leading to $\varphi(x) = \frac{1}{2}x + 1$. Since the consistency error was shown earlier to involve third and fourth derivatives, $\mathcal{L}_h\varphi = \mathcal{L}\varphi \geq 1$ and we may choose $\Phi(x) = \varphi(x)$ as a discrete comparison function. Hence $\mathcal{L}_h$ is stable with stability constant $C = 3/2$. Convergence of order two is thus guaranteed whenever u and its first four derivatives are continuous and bounded on $(0, 1)$.

The condition $h \leq \varepsilon$ required for inverse monotonicity does not affect the issue of convergence, since this concerns the situation when $h \to 0$. The condition does have important practical significance however. Indeed, if the condition is not satisfied,

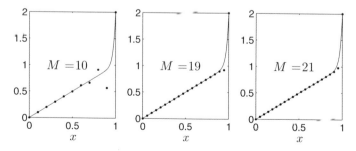

Fig. 6.5 The exact solution (*solid line*) and the numerical solution (*dots*) for Example 6.14 when $\varepsilon = 1/20$, $f(x) = 2$ and $h = 1/10, 1/19$ and $1/21$

that is if $h > \varepsilon$, then the numerical solutions may be prone to so-called wiggles (that is, they exhibit grid dependent oscillations that are not present in the exact solution). This is illustrated in Fig. 6.5 where we show the results of numerical experiments when $\varepsilon = 1/20$, $f(x) = 2$ and the boundary values are $\alpha = 0$, $\beta = 2$. The exact solution is given by

$$u(x) = x - \frac{e^{-2/\varepsilon} - e^{-2(1-x)/\varepsilon}}{1 - e^{-2/\varepsilon}},$$

which is a monotonic increasing function. When $M = 10$ (so that $h/\varepsilon = 2$) there are significant oscillations that become barely perceptible when $M = 19$ ($h/\varepsilon = 20/19 > 1$) and are not visible when $M = 21$ ($h/\varepsilon = 20/21 < 1$). When ε becomes significantly smaller, the requirement $h \leq \varepsilon$ becomes impractical and more sophisticated numerical methods are needed (see Morton [15] or Roos, Stynes and Tobiska [19]). We shall return to this in a later chapter of the book.   ◊

When presented with a finite difference scheme for the first time, some care is usually required in the definition of the operators $\mathcal{L}_h$ and $\mathscr{L}_h$ in order to ensure that they are correctly scaled with respect to h. This point is illustrated by the following example.

Example 6.15 Show that the finite difference method

$$- (1 + h^2)U_{m-1} + 2U_m - (1 + h^2)U_{m+1} = 0, \tag{6.41}$$

for $m = 1, 2, \ldots, M - 1$ with end conditions $U_0 = 2$ and $U_M = -1$ is a convergent approximation of the BVP,

$$\left. \begin{array}{l} -u''(x) + 2u(x) = 0, \quad 0 < x < 1 \\ u(0) = 2, \quad u(1) = -1. \end{array} \right\} \tag{6.42}$$

To check on consistency, we substitute $U = u$ into the left side of (6.41)

$$- (1 - h^2)u_{m-1} + 2u_m - (1 - h^2)u_{m+1} =$$
$$2u(x_m) - (1 - h^2)\big(u(x_m - h) + u(x_m + h)\big)$$

and, from the Taylor expansions (6.13) with $x = x_m$, we have

$$u(x_m - h) + u(x_m + h) = 2h^2 u(x_m) + h^2 u''(x_m) + \tfrac{1}{12}h^4 u''''(x_m) + \cdots$$

so that

$$-(1 - h^2)u_{m-1} + 2u_m - (1 - h^2)u_{m+1} = h^2 \mathcal{L}u(x_m) + \mathcal{O}(h^4),$$

where $\mathcal{L}u(x) = -u''(x) + 2u(x)$. The presence of the factor h^2 on the right hand side shows that the given finite difference equations needs to be divided by h^2 in order to define $\mathcal{L}_h$:

$$\mathcal{L}_h U_m := h^{-2} \left(-(1 - h^2)U_{m-1} + 2U_m - (1 - h^2)U_{m+1} \right).$$

The remaining issues for this example are left to Exercise 6.14. ◇

6.4 Advanced Topics and Extensions

The material of this section is not essential in a first course on the numerical solution of boundary value problems but it does lead to a variety of challenging exercises for the diligent student. The first extension we consider are boundary value problems associated with the general boundary conditions described at the start of Chap. 5.

6.4.1 Boundary Conditions with Derivatives

The family of boundary value problems of interest to us initially is

$$\left. \begin{array}{l} -u''(x) + r(x)u(x) = f(x), \quad 0 < x < 1 \\ u(0) = \alpha, \quad au(1) + bu'(1) = \beta, \end{array} \right\} \tag{6.43}$$

having a Dirichlet BC at $x = 0$ and a Robin BC at $x = 1$. (The abbreviation BC for boundary condition will be used throughout this section.) As in Sect. 6.2, we shall assume that $r(x)$ and $f(x)$ are continuous functions on the interval $[0, 1]$. We shall also assume that $b \neq 0$ since the situation when $b = 0$ has already been dealt with.

The differential equation will be approximated using the finite difference method used for the BVP (6.15), that is, $\mathcal{L}_h U = f$, where (see (6.20))

$$\mathcal{L}_h U_m = -h^{-2}\delta^2 U_m + r_m U_m, \quad m = 1, 2, \ldots, M - 1,$$

and $\mathcal{L}u(x) = -u''(x) + r(x)u(x)$.

In order to approximate the derivative $u'(1) = u'(x_M)$ by values of u lying on the grid $\{x_0, x_1, \ldots, x_M\}$ a backward difference approximation must be used. That is (see Table 6.1),

$$u'_M = h^{-1}\triangle^- u_M + \mathcal{O}(h)$$
$$= h^{-1}(u_M - u_{M-1}) + \tfrac{1}{2}hu''_M + \mathcal{O}(h^2). \qquad (6.44)$$

On neglecting the last two terms on the right hand side, the Robin BC $au(1) + bu'(1) = \beta$ is seen to lead to the numerical boundary condition

$$aU_M + bh^{-1}\triangle^- U_M = \beta, \qquad (6.45)$$

that is, $aU_M + b(U_M - U_{M-1})/h = \beta$ (see Fig. 6.6).

We therefore have the discrete BVP $\mathscr{L}_h U = \mathscr{F}_h$, where

$$\mathscr{L}_h U_m = \begin{cases} U_0, \\ L_h U_m, \\ aU_M + bh^{-1}\triangle^- U_M, \end{cases} \qquad \mathscr{F}_{h,m} = \begin{cases} \alpha & \text{for } m = 0, \\ f_m & \text{for } m = 1, 2, \ldots, M - 1, \\ \beta & \text{for } m = M. \end{cases}$$
$$(6.46)$$

This gives $M + 1$ linear algebraic equations for the values U_m $(m = 0, 1, \ldots, M)$. The system may be expressed in matrix-vector form $A\boldsymbol{u} = \boldsymbol{f}$ by defining $\boldsymbol{u} = [U_1, U_2, \ldots, U_M]^\mathsf{T}$, with an $M \times M$ tridiagonal matrix A

$$A = \frac{1}{h^2} \begin{bmatrix} a_{1,1} & -1 & & & \\ -1 & a_{2,2} & -1 & & \\ & \ddots & \ddots & \ddots & \\ & & & & -1 \\ & & & -1 & a_{M,M} \end{bmatrix}, \qquad (6.47)$$

where $a_{m,m} = 2 + h^2 r_m$ $(m < M)$, $a_{M,M} = 1 + ha/b$, and

$$\boldsymbol{f} = \left[f_1, f_2, \ldots, f_{M-1}, f_M^\beta\right]^\mathsf{T}, \qquad (6.48)$$

with $f_M^\beta = \beta/(hb)$. Note that equation (6.45) representing the boundary condition has been divided by bh so as to preserve the symmetry of A.

The components of the LTE $\mathscr{R}_h = \mathscr{L}_h u - \mathscr{F}_h$ at $x = x_m$ $(m = 0, 1, \ldots, M - 1)$ are $\mathcal{O}(h^2)$, as in the previous section. At $x = x_M$ we have, from (6.46),

Fig. 6.6 The finite difference approximation replaces the Robin boundary condition at $x = x_M = 1$ by a relationship between U_M and U_{M-1}

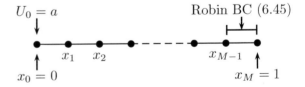

$$\begin{aligned}
\mathscr{R}_{h,M} &= \mathscr{L}_h u_M - \mathscr{F}_{h,M} \\
&= a u_M + b h^{-1} \triangle^- u_M - \beta \\
&= a u_M + b\left(u'_M - \tfrac{1}{2} h u''_M\right) - \beta + \mathcal{O}(h^2)
\end{aligned} \tag{6.49}$$

and, since u satisfies $a u_M + b u'_M = \beta$, we find that $\mathscr{R}_{h,M} = \mathcal{O}(h)$ giving an order of consistency of $p = 1$. We will demonstrate presently that the rate of convergence is also first order but, before doing so, we shall describe how second-order consistency can be recovered.

The reduced order consistency is clearly caused by the term $-\tfrac{1}{2} b h u''_M$ in the LTE (6.49), which originates from the leading error term in (6.44). Applying the ODE $u'' = r u - f$, the approximation (6.44) can be written as

$$u'_M = h^{-1} \triangle^- u_M + \tfrac{1}{2} h\left(r_M u_M - f_M\right) + \mathcal{O}(h^2).$$

so that the Robin BC $a u_M + b u'_M = \beta$ now leads to

$$a U_M + b\left(h^{-1} \triangle^- U_M + \tfrac{1}{2} h r_M U_M\right) = \beta + \tfrac{1}{2} b h f_M. \tag{6.50}$$

This suggests a modified discrete BVP $\widehat{\mathscr{L}}_h U = \widehat{\mathscr{F}}_h$ in which the quantities $\widehat{\mathscr{L}}_h, \widehat{\mathscr{F}}_h$ differ from $\mathscr{L}_h, \mathscr{F}_h$ only at $x = x_M$, where (compare with (6.46))

$$\widehat{\mathscr{L}}_h U_M = (a + \tfrac{1}{2} b h r_M) U_M + b h^{-1} \triangle^- U_M, \quad \widehat{\mathscr{F}}_{h,M} = \beta + \tfrac{1}{2} b h f_M. \tag{6.51}$$

This modified method is consistent of order two with the BVP (6.43) (see Exercise 6.15). When written in matrix–vector notation $A\boldsymbol{u} = \boldsymbol{f}$, the matrix A and vector $\boldsymbol{f}$ are still defined as in (6.47) and (6.48) except that

$$a_{M,M} = 1 + \frac{ha}{b} + \frac{1}{2} h^2 r_M, \quad f_M^\beta = \frac{\beta}{hb} + \frac{1}{2} f_M.$$

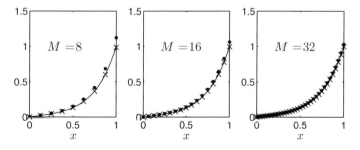

Fig. 6.7 The exact solution (*solid line*) and the numerical solutions for Example 6.16 using $\sigma = 4$ with the first-order approximation (*dots*) and second-order approximation (*crosses*) of the Robin BC

Fig. 6.8 Log–log plot of
maximum global error
versus h for the first and
second-order accurate BCs
(*dots* and +, respectively) of
Example 6.16. *Two dashed
lines* with slopes of one and
two are included for
reference

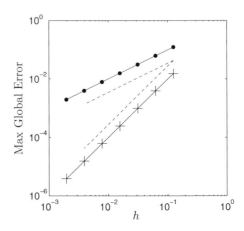

These represent minor changes to the original equations—the gain in order of consistency is obtained at essentially no extra cost.

Example 6.16 Use the methods described in this section to solve the differential equation $-u''(x)+\sigma^2 u(x) = 0$ on $0 < x \leq 1$ with the boundary conditions $u(0) = 1$ and $\sigma u(1)+u'(1) = \sigma e^\sigma / \sinh \sigma$, where σ is a positive constant. The exact solution, for comparison purposes, is $u(x) = \sinh(\sigma x)/\sinh(\sigma)$.

The solutions with $\sigma = 4$, $M = 8$, 16 and 32 are shown in Fig. 6.7 for both methods (6.45) and (6.50) for approximating the Robin BC at $x = 1$. There is a discernible difference between the first- and second-order methods at the two smaller values of M and they both merge with the exact solution when $M = 32$. The log–log plot shown in Fig. 6.8 provides a quantitative measure of the accuracy of the two methods for $M = 2^k$ ($k = 3, 4, \ldots, 9$). On the finest grid ($M = 512$) the first-order method has a maximum error of 2×10^{-3} compared with 4×10^{-6} for the second-order method—a significant increase in accuracy in return for the additional care taken in the design of the numerical boundary condition.   ◇

The analysis of problems that involve Robin BCs require generalisations of Definition 6.9 and Theorem 6.10.

Definition 6.17 (*Positive type operator–2*) A difference operator $\mathcal{L}_h$ defined by

$$\mathcal{L}_h U_m = -a_m U_{m-1} + b_m U_m - c_m U_{m+1}, \quad m = 0, 1, \ldots, M, \qquad (6.52)$$

with $a_0 = 0$ and $c_M = 0$ is said to be of *positive type* if

$$a_m \geq 0, \quad b_m > 0, \quad c_m \geq 0, \quad b_m \geq a_m + c_m, \quad m = 0, 1, \ldots, M \qquad (6.53)$$

and also $b_m > a_m + c_m$ for **at least one** value of m.[6]

Theorem 6.18 (Inverse monotonicity–2) *If $\mathcal{L}_h$ is of positive type then, $\mathcal{L}_h U \geq 0$ implies that $U \geq 0$.*

Proof See Exercise 6.19.   □

Example 6.19 Prove that the methods used in Example 6.16 are convergent and establish the rate of convergence in each of the two schemes.

The problem addressed in Example 6.16 is a special case of the BVP (6.43) and so consistency of the two methods follows from the earlier discussion. The discrete operator for the first method is

$$
\mathcal{L}_h U_m =
\begin{cases}
U_0, & m = 0, \\
-h^{-2} U_{m-1} + (2h^{-2} + \sigma^2) U_m - h^{-2} U_{m+1}, & m = 1, 2, \ldots, M - 1, \\
-h^{-1} U_{M-1} + (\sigma + h^{-1}) U_M, & m = M,
\end{cases}
$$

and so $a_0 = c_0 = 0$, $b_0 = 1$,

$$
a_m = c_m = h^{-2}, \quad b_m = 2h^{-2} + \sigma^2, \quad m = 1, 2, \ldots, M - 1,
$$

$a_M = h^{-1}$, $b_M = \sigma + h^{-1}$ and $c_M = 0$. Since σ is a positive constant, these fulfil the inequalities required by Definition 6.17 so $\mathcal{L}_h$ is inverse monotone by Theorem 6.18. The operator $\widehat{\mathcal{L}}_h$ representing the second-order method is similarly defined except that

$$
\widehat{\mathcal{L}}_h U_M = -h^{-1} U_{M-1} + (\sigma + h^{-1} + \tfrac{1}{2}\sigma^2 h) U_M.
$$

Since $b_M = \sigma + h^{-1} + \tfrac{1}{2}\sigma^2 h > h^{-1} = a_M + c_M$, this operator is also of positive type. Next, to show that these discrete operators are stable we have to find a suitable comparison function. This can be achieved by looking at the underlying continuous problem. Thus, if $\mathcal{L}$ is the operator representing the BVP (6.43) and $\varphi(x) = C$ is a constant function, then

$$
\mathcal{L}\varphi(x) =
\begin{cases}
\varphi(0) = C, & x = 0, \\
-\varphi''(x) + \sigma^2 \varphi(x) = \sigma^2 C, & 0 < x < 1, \\
\varphi'(1) + \sigma \varphi(1) = \sigma C, & x = 1,
\end{cases}
$$

and so $\mathcal{L}\varphi \geq 1$ by choosing $C = \max\{1, 1/\sigma^2\}$. Since $\mathcal{L}_h \varphi = \widehat{\mathcal{L}}_h \varphi = \mathcal{L}\varphi$ when $\varphi(x) = C$, it follows that $\Phi = C$ will also be a comparison function for both $\mathcal{L}_h$ and $\widehat{\mathcal{L}}_h$.

The methods $\mathcal{L}_h U = \mathcal{F}_h$ and $\widehat{\mathcal{L}}_h U = \widehat{\mathcal{F}}_h$ in Example 6.16 are stable and consistent of order one and two, respectively, and they therefore converge at first

---

[6]This condition is satisfied if, for example, a Dirichlet BC is applied at one or both ends of the interval: $\mathcal{L}_h U_0 := U_0$ would lead to $b_0 = 1 > a_0 + c_0 = 0$.

and second-order rates provided that u and its first four derivatives are continuous. This is easily checked in this example since we know that the exact solution is $u(x) = \sinh(\sigma x)/\sinh \sigma$.   ◊

In the final example additional practical and theoretical issues are introduced by replacing the Dirichlet BC at $x = 0$ by a Neumann BC.

Example 6.20 Find a second-order consistent finite difference approximation of the BVP

$$-u''(x) + \sigma^2 u(x) = 0, \quad 0 < x < 1 \atop u'(0) = -1, \quad \sigma u(1) + u'(1) = 2e^\sigma,} \tag{6.54}$$

and examine the convergence of the scheme when $\sigma > \frac{1}{2}$.

The starting point for approximating the BC at the left endpoint is the forward difference operator (see Table 6.1),

$$u'(0) = h^{-1}\Delta^+ u_0 - \tfrac{1}{2}hu_0'' + \mathcal{O}(h^2).$$

To upgrade this to a second-order approximation, the ODE at $x = 0$ gives $u_0'' = \sigma^2 u_0$ and so

$$u'(0) = h^{-1}\Delta^+ u_0 - \tfrac{1}{2}h\sigma^2 u_0 + \mathcal{O}(h^2).$$

Thus, the BC $u'(0) = -1$ leads to the numerical boundary condition

$$h^{-1}\Delta^+ U_0 - \tfrac{1}{2}h\sigma^2 U_0 = -1,$$

that is

$$- \left(h^{-1} + \tfrac{1}{2}h\sigma^2\right)U_0 + h^{-1}U_1 = -1. \tag{6.55}$$

This, together with (6.19) and (6.50) defines our discrete BVP. To analyse the method we need to express it in the operator form, $\mathscr{L}_h U = \mathscr{F}_h$, where $\mathscr{L}_h$ is inverse monotone. We observe that the signs of the coefficients of U_0 and U_1 in (6.55) are the opposite of those required by Theorem 6.18 (at $m = 0$), so both sides of the numerical BC (6.55) should be multiplied by -1 in order to define a suitable operator $\mathscr{L}_h$:

$$\mathscr{L}_h U_m = \begin{cases} -h^{-1}\Delta^+ U_0 + \tfrac{1}{2}h\sigma^2 U_0, & m = 0, \\ \mathcal{L}_h U_m, & m = 1, 2, \ldots, M-1, \\ (\sigma + \tfrac{1}{2}\sigma^2 h)U_M + h^{-1}\Delta^- U_M, & m = M, \end{cases} \tag{6.56}$$

with $\mathcal{L}_h$ defined by (6.20) and $\mathscr{F}_h$ defined as in (6.46) except that $\mathscr{F}_{h,0} = 1$. This accords with our observation in Sect. 5.2 that derivatives in BCs should be based on outward-going derivatives. Confirmation that this approximation is both second-order consistent with the BVP (6.54) and inverse monotone is left as an exercise (Exercise 6.25).

A comparison function needs to be found in order to complete the proof of stability and, for this, we turn again to the continuous problem for which the operator $\mathscr{L}$ is given by

$$\mathscr{L}u(x) = \begin{cases} -u'(0), & x = 0, \\ -u''(x) + \sigma^2 u(x), & 0 < x < 1, \\ u'(1) + \sigma u(1), & x = 1. \end{cases} \tag{6.57}$$

A constant comparison function is not satisfactory since $\mathscr{L}\varphi(0) = 0$ so we look for a function of the form $\varphi(x) = A + Bx$. This will satisfy $\mathscr{L}\varphi(x) \geq 1$ if $B = -1$ and both conditions $\sigma^2(A - 1) \geq 1$ and $\sigma(A - 1) \geq 2$ hold. Since $\sigma > \frac{1}{2}$ these are satisfied for $A = 5$ so $\varphi(x) = 5 - x$. Clearly $\varphi(x) \geq 0$ for $x \in [0, 1]$. Because $\varphi(x)$ is a polynomial of degree ≤ 2, $\mathscr{L}_h\varphi = \mathscr{L}\varphi \geq 1$ and we have a function that fulfils all the criteria for a comparison function for both $\mathscr{L}$ and $\mathscr{L}_h$. Since the finite difference approximation that has been constructed is stable whenever $\sigma > \frac{1}{2}$ and is consistent of second order, it converges at a second-order rate for any value of σ in this range. $\diamond$

The approach we have used to increase the order of convergence of derivative BCs can be employed more widely to generate high-order finite difference methods. This is explored in the concluding section.

6.4.2 A Fourth-Order Finite Difference Method

To illustrate the derivation of finite difference methods with a convergence rate that is potentially higher than second-order, we return to the Dirichlet problem discussed earlier, see (6.15),

$$\left. \begin{array}{c} -u''(x) + r(x)u(x) = f(x), \quad 0 < x < 1 \\ u(0) = \alpha, \quad u(1) = \beta. \end{array} \right\} \tag{6.58}$$

The basis of the approximation in Sect. 6.2 was the following Taylor expansion (see Table 6.1)

$$u''_m = h^{-2}\delta^2 u_m - \frac{1}{12}h^2 u''''_m + \mathcal{O}(h^4). \tag{6.59}$$

An approximation of the leading term $\frac{1}{12}h^2 u''''_m$ in the truncation error must be incorporated into the finite difference method so as to increase its order of consistency. A direct approximation of this term requires the five consecutive grid values $u_{m\pm2}, u_{m\pm1}$ and u_m (see Exercise 6.32) and is rather cumbersome to implement, especially in the context of generalisations to PDEs. We shall describe an alternative approach that fits conveniently into the framework that we have established. The leading term in the truncation error in (6.59) is proportional to $u''''_m(x)$ which, via the differential equation, can be expressed as

$$u_m''''(x) = \left(r(x) u(x) - f(x) \right)''$$

then, using $h^{-2} \delta^2 v_m = v_m'' + \mathcal{O}(h^2)$ from Table 6.1 with $v = ru - f$,

$$u_m'''' = h^{-2} \delta^2 \left(r_m u_m - f_m \right) + \mathcal{O}(h^2).$$

When this is combined with (6.59) to approximate the ODE at $x = x_m$, we obtain the method

$$-h^{-2} \delta^2 U_m + \left(r_m U_m + \tfrac{1}{12} \delta^2 (r_m U_m) \right) = f_m + \tfrac{1}{12} \delta^2 f_m \tag{6.60a}$$

for $m = 1, 2, \ldots, M - 1$, with boundary conditions $U_0 = \alpha$, $U_M = \beta$. Written explicitly in terms of grid values, this becomes,

$$h^{-2} \left(-U_{m-1} + 2U_m - U_{m+1} \right) +$$
$$\tfrac{1}{12} \left(r_{m-1} U_{m-1} + 10 r_m U_m + r_{m+1} U_{m+1} \right)$$
$$= \tfrac{1}{12} \left(f_{m-1} + 10 f_m + f_{m+1} \right). \tag{6.60b}$$

The difference scheme is commonly known as Numerov's method (because it was conceived in a 1924 paper by *Boris Numerov*). It is an example of a *compact* difference scheme in that it achieves a high-order of consistency (fourth-order, in fact) while using a minimal number of consecutive grid values.

The discrete BVP may be written in the operator form $\mathscr{L}_h U = \mathscr{F}_h$, where

$$\mathscr{L}_h U_m = \begin{cases} U_m & \text{for } m = 0, M, \\ \mathcal{L}_h U_m & \text{for } m = 1, 2, \ldots, M - 1, \end{cases} \tag{6.61}$$

with $\mathcal{L}_h$ being defined by (6.32) with coefficients

$$a_m = \frac{1}{h^2} - \frac{1}{12} r_{m-1}, \quad b_m = \frac{2}{h^2} + \frac{5}{6} r_m, \quad c_m = \frac{1}{h^2} - \frac{1}{12} r_{m+1}, \tag{6.62}$$

and

$$\mathscr{F}_{h,m} = \begin{cases} \alpha & \text{for } m = 0, \\ \tfrac{1}{12} \left(f_{m-1} + 10 f_m + f_{m+1} \right) & \text{for } m = 1, 2, \ldots, M - 1, \\ \beta & \text{for } m = M. \end{cases} \tag{6.63}$$

The operator $\mathscr{L}_h$ is of positive type (and therefore inverse monotone by virtue of Theorem 6.10) so long as $r(x) \geq 0$ and $12 h^2 r(x) \leq 1$ for all $x \in [0, 1]$ (which is always achievable if h is sufficiently small and $r(x)$ is bounded). The comparison function used in Example 6.13 also applies here and so the method is stable. Consequently, Numerov's method is convergent of fourth-order whenever the solution is

Table 6.3 The maximum global error for the second-order method in Example 6.2 versus the error for Numerov's method as a function of h

M	h	Conventional method		Numerov method	
		$\|u - U\|_{h,\infty}$	Ratio	$\|u - U\|_{h,\infty}$	Ratio
8	0.125	0.13	–	0.35	–
16	0.0625	0.071	1.86	0.062	5.51
32	0.03125	0.026	2.71	0.0075	8.21
64	0.015625	0.0078	3.38	0.00060	12.61
128	0.0078125	0.0021	3.78	0.000042	14.36
256	0.00390625	0.00052	3.93	0.0000027	15.60
512	0.001953125	0.00013	3.98	0.00000017	15.89

smooth enough (certainly if u and the first six derivatives are continuous functions over $(0, 1)$).

Example 6.21 Compare the performance of Numerov's method with the second-order finite difference method used to solve the BVP in Example 6.2.

Results obtained using Numerov's method are shown in Table 6.3 and are appended to the original results in Table 6.2. Note that the second-order method has a smaller global error on the coarsest grid; but that the global error for Numerov's method can be reduced by up to a factor of 16 (commensurate with a fourth-order method) when the grid spacing is halved, compared to a maximum factor of 4 for the second-order method. The superiority of Numerov's method (at least when computing a smooth solution) is pretty self-evident.   ◊

Exercises

6.1 Derive the result

$$v_m'' = h^{-2}\left(u_{m+1} - 2u_m + u_{m-1}\right) - \tfrac{1}{12}h^2 v^{(4)}(\xi_m), \quad \xi_m \in (x_{m-1}, x_{m+1})$$

by mimicking the derivation of (6.8).

6.2 Show that $\Delta^+ \Delta^- = \delta^2$, in the sense that $\Delta^+ \Delta^- U_m = \delta^2 U_m$ for any grid function U. Show also that (a) $\Delta^- \Delta^+ = \delta^2$, (b) $\Delta = \tfrac{1}{2}(\Delta^+ + \Delta^-)$ and (c) $\delta^2 = \Delta^+ - \Delta^-$.

6.3 Verify that $\Delta^+ \Delta^+ v_m = v_{m+2} - 2v_{m+1} + v_m$, where $v_m := v(x_m)$. Show, by use of suitable Taylor expansions, that $h^{-2}\Delta^+ \Delta^+ v_m = v_m'' + \mathcal{O}(h)$, providing a first-order approximation of v_m''.
  What are the corresponding results for $\Delta^- \Delta^- v_m$?

6.4 By verifying that $u''''(x)$ is continuous and bounded in $(0, 1)$ establish that the finite difference method in Example 6.2 is convergent and that the rate is second-order.

6.5 $\star$ With $\mathscr{L}_h$ as defined in Theorem 6.10 and

$$\mathscr{F}_{h,m} = \begin{cases} \alpha, & m = 0, \\ f_m, & m = 1, 2, \ldots, M - 1, \\ \beta, & m = M, \end{cases}$$

write the equations $\mathscr{L}_h U = \mathscr{F}_h$ in the form $A\boldsymbol{u} = \boldsymbol{f}$, where A is an $(M-1) \times (M-1)$ matrix and $\boldsymbol{u}, \boldsymbol{f} \in \mathbb{R}^{M-1}$.

6.6 Show that the finite difference method

$$-h^{-2}\delta^2 U_m + x_m^2 U_m = x_m, \quad m = 1, 2, \ldots, M - 1, \\ U_0 = 1, \quad U_M = -2. \Bigg\}$$

is an approximation of the BVP

$$-u''(x) + x^2 u(x) = x, \quad 0 < x < 1 \\ u(0) = 1, \quad u(1) = -2 \Bigg\}$$

that is consistent of second order. When $M = 4$, write the finite difference equations in matrix–vector form $A\boldsymbol{u} = \boldsymbol{f}$.

6.7 $\star$ Write the finite difference approximation in Exercise 6.6 at x_m in the form

$$-a_m U_{m-1} + b_m U_m - c_m U_{m+1} = d_m$$

for $m = 1, 2, \ldots, M - 1$. Give expressions for a_m, b_m, c_m and d_m. Hence show that the corresponding finite difference operator is inverse monotone when combined with Dirichlet BCs.

6.8 Consider the discrete BVP

$$-h^{-2}\delta^2 U_m = 1, \quad m = 1, 2, \ldots, M - 1,$$

with BCs $U_0 = 0$ and $2h^{-1}(U_M - U_{M-1}) = 1$.

(a) With which BVP is it consistent? What is the order of consistency?
(b) Verify that the discrete BVP satisfies the conditions of Theorem 6.18. Deduce that the numerical solution is non-negative.

6.9 Suppose that $\mathscr{L}_h U = \mathscr{F}_h$ where $\mathscr{L}_h$ and $\mathscr{F}_h$ are defined by the previous exercise. Find values of a and c so that $\Phi(x) = cx(x - a)$ is a comparison function for $\mathscr{L}_h$, i.e.,

$$\begin{cases} \Phi_0 \geq 0, \\ -h^{-2}\delta^2 \Phi_m \geq 1, \quad m = 1, 2, \ldots, M - 1, \\ h^{-1}(\Phi_M - \Phi_{M-1}) \geq 1. \end{cases}$$

Verify that the solution of the discrete BVP in the previous exercise is given by $U_m = \frac{1}{2}x_m(3 - x_{m-1})$. Deduce an expression for the global error and thus confirm the theoretical rate of convergence is actually achieved in this case.

6.10 * Show that the finite difference equations (6.39) lead to an algebraic system $Au = f$ in which the matrix A is not symmetric.

6.11 Use the central difference operators $\triangle$ and δ^2 to construct a finite difference approximation to the BVP

$$-u''(x) + 20u'(x) = x^2, \quad 0 < x < 1 \atop u(0) = 0, \quad u(1) = 0, \Big\}$$

on a grid $\{x_j = jh, \ j = 0, 1, 2, \ldots, M\}$ and write it in the form

$$-aU_{m-1} + bU_m - cU_{m+1} = f_m.$$

For what values of M is the resulting difference operator of positive type?

Show that the discrete BVP has a linear comparison function Φ and hence prove that the finite difference approximation is stable if h is sufficiently small.

6.12 Show that the finite difference equations

$$-\varepsilon h^{-2}\delta^2 U_m + 2h^{-1}\triangle U_m = 0, \quad m = 1, 2, \ldots, M - 1,$$

(see (6.39)) have solutions $U_m = A$ and $U_m = B\rho^m$, where A, B are constants and $\rho = (\varepsilon - h)/(\varepsilon + h)$. The equations therefore have a general solution $U_m = A + B\rho^m$. Explain why this implies that solutions are generally oscillatory when $h > \varepsilon$.

6.13 Define appropriate quantities $\mathscr{L}_h$ and $\mathscr{F}_h$ corresponding to the upwind finite difference method

$$-\varepsilon h^{-2}\delta^2 U_m + 2h^{-1}\triangle^- U_m = f_m, \quad m = 1, 2, \ldots, M - 1, \atop U_0 = \alpha, \quad U_M = \beta, \Big\}$$

where $\varepsilon > 0$, and determine its order of consistency for solving the BVP (6.38). Prove, by first showing that $\mathscr{L}_h$ is of positive type, that the numerical solution converges. What is the convergence rate?

6.14 Complete the investigation of consistency, stability and convergence for Example 6.15. [Hint: show that the constant $\Phi_m = 1$ is a possible comparison function.]

6.15 Suppose that $\widehat{\mathscr{L}}U_M$ and $\widehat{\mathscr{F}}_{h,M}$ are defined by equation (6.51). Verify that the discrete BVP $\widehat{\mathscr{L}}_h U = \widehat{\mathscr{F}}_h$ is consistent of order two with the BVP (6.43) at $x = x_M = 1$.

6.16 Show that the standard central difference approximation (6.19) of the differential equation together with the numerical boundary condition $aU_0 - bh^{-1}\Delta^+ U_0 = \alpha$ leads to a discrete BVP that is consistent of order one with the BVP defined by $-u''(x) + r(x)u(x) = f(x)$ for $0 < x < 1$ with a Robin BC $au(0) - bu'(0) = \alpha$ at $x = 0$ and Dirichlet BC $u(1) = \beta$ at $x = 1$.

6.17 Explain how the approximation of the Robin BC in the previous exercise can be modified to give rise to a discrete BVP that is consistent of order two.

6.18 * With $\mathscr{L}_h$ given in Definition 6.17 and

$$
\mathscr{F}_h = \begin{cases} \alpha, & m = 0 \\ f_m, & m = 1, 2, \ldots, M-1 \\ \beta, & m = M, \end{cases}
$$

write the equations $\mathscr{L}_h U = \mathscr{F}_h$ in the form $Au = f$, where A is an $(M+1) \times (M+1)$ matrix and $u, f \in \mathbb{R}^{M+1}$.

6.19 Prove Theorem 6.18 by generalising the proof of Theorem 6.10.

6.20 * Verify that the operators $\mathscr{L}_h$ defined by (6.46) and its modification $\widehat{\mathscr{L}}_h$ defined with the BC (6.51) are both of positive type provided that the coefficients a and b in the Robin BC have the same sign.

6.21 * Show that $\mathscr{L}_h$ defined by

$$
\mathscr{L}_h U_m = \begin{cases} -h^{-1}(U_1 - U_0), & m = 0, \\ -h^{-2}\delta^2 U_m, & m = 1, 2, \ldots, M-1, \\ h^{-1}(U_M - U_{M-1}), & m = M, \end{cases}
$$

does not satisfy the conditions of Theorem 6.18.
Verify that $\mathscr{L}_h V = 0$ for any constant grid function V. Deduce that if $\mathscr{L}_h U = \mathscr{F}_h$ has a solution U, then it also has a family of solutions $U + V$ for every constant function V. (Thus, if a solution exists then there are infinitely many solutions.)

Suppose that $\mathscr{F}_h$ is given by (6.22). Show that the BVP with which the numerical method is consistent also has infinitely many solutions.

Under what conditions on $\mathscr{F}$ and $\mathscr{F}_h$ do solutions to the continuous and discrete BVPs $\mathscr{L}u = \mathscr{F}$ and $\mathscr{L}_h U = \mathscr{F}_h$, respectively, exist? [Hint: integrate the ODE over the interval $0 < x < 1$ and, using $\delta^2 = \Delta^+ \Delta^-$, sum the finite difference equations for $m = 1, 2, \ldots, M-1$.]

6.22 * Suppose that the operator $\mathscr{L}_h$ in the previous exercise is modified so that its value at $m = 0$ reads $\mathscr{L}_h U_0 = hU_0 - h^{-1}(U_1 - U_0)$. Show that the resulting operator is inverse monotone but *unstable*. [Hint: repeat the summation process used in the previous exercise.]

6.23 Show that $\mathcal{L}_h U_m := -8U_{m-1} + 65U_m - 8U_{m+1}$ defines an inverse monotone operator $\mathcal{L}_h$.

Verify that the homogeneous equations $\mathcal{L}_h U_m = 0$ have solutions $U_m = A 8^m$ and $U_m = B 8^{-m}$ for arbitrary constants A, B, and that its general solution is the sum of these two sequences. Determine the coefficients in this linear combination from the boundary conditions $U_0 \doteq \alpha$ and $U_M = \beta$.

Sketch the solutions for $m = 0, 1, \ldots, M$ when $M = 10$ and $\alpha = \pm 1$, $\beta = 2$. [Hint: Show that $U_m \approx \alpha 8^{-m} + \beta 8^{M-m}$.] Verify that, in both cases, $\min(0, \alpha, \beta) \leq U_m \leq \max(0, \alpha, \beta)$.

6.24 Suppose that $\mathcal{L}_h$ is as defined by (6.32) and that U satisfies the homogeneous equations $\mathcal{L}_h U_m = 0$, for $m = 1, 2, \ldots, M - 1$, with BCs $U_0 = \alpha$, $U_M = \beta$. Prove that $\min(0, \alpha, \beta) \leq U_m \leq \max(0, \alpha, \beta)$ when $\mathcal{L}_h$ inverse monotone.

6.25 Verify that the operator $\mathscr{L}_h$ defined by (6.56) is inverse monotone when $\sigma > 0$.

6.26 Given a finite difference operator $\mathscr{L}_h$ and an associated grid function $\mathscr{F}_h$ defined by

$$\mathscr{L}_h U_m = \begin{cases} U_0, \\ -h^{-2}\delta_x^2 U_m + U_m, & m = 1, 2, \ldots, M - 1, \\ h^{-1}\left((1 + \tfrac{1}{2}h^2)U_M - U_{M-1}\right), \end{cases}$$

$$\mathscr{F}_{h,m} = \begin{cases} 0, \\ f(mh), & m = 1, 2, \ldots, M - 1, \\ \tfrac{1}{2}hf(Mh), \end{cases}$$

where $Mh = 1$, show that $\mathscr{L}_h U = \mathscr{F}_h$ is second-order consistent with the BVP

$$\left. \begin{array}{c} -u''(x) + u(x) = f(x), \quad 0 < x < 1 \\ u(0) = 0, \quad u'(1) = 0. \end{array} \right\}$$

Prove, from first principles, that $\mathscr{F}_h \geq 0$ implies that $U \geq 0$.

6.27 Determine the order of consistency of the finite difference methods

$$\begin{cases} -h^{-1}\Delta^+ U_0 = 1, \\ -h^{-2}\delta^2 U_m + x_m U_m = 1, & m = 1, 2, \ldots, M - 1, \\ U_M = 2 \end{cases}$$

and

$$\begin{cases} -h^{-1}\Delta^+ U_0 = 1 - \tfrac{1}{2}h, \\ -h^{-2}\delta^2 U_m + x_m U_m = 1, & m = 1, 2, \ldots, M - 1, \\ U_M = 2 \end{cases}$$

for solving the BVP

$$\left.\begin{array}{rl} -u''(x) + xu(x) = 1, & 0 < x < 1 \\ -u'(0) = 0, & u(1) = 2. \end{array}\right\}$$

6.28 Develop a finite difference approximation of the BVP

$$\left.\begin{array}{rl} -u''(x) = x^2, & 0 < x < 1 \\ -u'(0) = 1, & u'(1) + u(1) = 2 \end{array}\right\}$$

that is consistent of order two.

6.29 Suppose that $u(x)$ is a solution of the differential equation

$$-u'' + 4u' + x^2 u = \sin \pi x, \quad 0 < x < 1$$

with the BCs $u(0) = 0$ and $u'(1) = 2$. Show that

$$-h^{-2}\delta^2 U_m + 4h^{-1}\Delta U_m + x_m^2 U_m = \sin \pi x_m,$$

where $x_m = mh$ ($m = 0, 1, 2, \ldots, M, h = 1/M$), is second-order consistent with the differential equation. Derive a second-order approximation of the derivative boundary condition at $x = 1$.

6.30 Suppose that $u(x)$ satisfies $-u'' + xu' = \cos \pi x$ for $0 < x \leq 1$. Determine constants A and B so that $h^{-1}(U_1 - U_0) = AU_0 + B$ is a finite difference approximation of the boundary condition $u'(0) = 2u(0)$ that is second-order consistent.

6.31 Let $L(\rho)$ denote the $M \times (M + 1)$ matrix

$$L(\rho) = \frac{1}{h} \begin{bmatrix} -1 & 1 & & & \\ & \ddots & \ddots & & \\ & & -1 & 1 \\ & & & -1 & \rho \end{bmatrix}$$

and let A_0 denote the $M \times M$ matrix obtained from (6.47) by setting $r_1 = r_2 = \cdots = r_{M-1} = 0$. If $a_{M,M} \geq 1$, show that a real value of ρ may be found such that $L(\rho)L(\rho)^\mathsf{T} = A_0$. Hence, extend the proof of Lemma 6.1 to prove that A is positive definite provided that the coefficients a and b in the Robin boundary condition in (6.43) have the same sign.

6.32 Show that $h^{-4}\delta^4 u_m = u_m'''' + \mathcal{O}(h^2)$ and combine it with (6.59) to give the approximation

$$u_m'' = h^{-2}\left(\delta^2 - \tfrac{1}{12}\delta^4\right)u_m + \mathcal{O}(h^4).$$

Hence, construct a fourth-order approximation of the ODE (6.58) that can be applied at grid points $x = x_m$, $(m = 2, 3, \ldots, M - 2)$. Verify that

$$\left(\delta^2 - \tfrac{1}{12}\delta^4\right) u_m = \tfrac{1}{12}(u_{m-2} - 16u_{m-1} + 30u_m - 16u_{m+1} + u_{m+2}),$$

to show that the finite difference equation necessarily involves values of u at five consecutive grid points.

6.33 Suppose that Numerov's method from Sect. 6.4.2 is applied to the BVP

$$\left. \begin{array}{l} -u''(x) + 12u(x) = 12x, \quad 0 < x < 1 \\ u(0) = 3, \quad u(1) = -5. \end{array} \right\}$$

Write the corresponding finite difference equations in matrix–vector form $A\boldsymbol{u} = \boldsymbol{f}$ when $M = 4$.

6.34 Suppose that $\mathscr{L}_h$ and $\mathscr{F}_h$ are defined by (6.61) and (6.63), respectively. Use appropriate Taylor expansions to verify that the local truncation error $\mathscr{R}_h = \mathscr{L}_h u - \mathscr{F}_h$ for Numerov's method is $\mathcal{O}(h^4)$ if $u \in C^6[0, 1]$ and $r, f \in C^4[0, 1]$. [Hint: defining $\mathcal{L}_h$ by $\mathcal{L}_h u_m = -h^{-2}\delta^2 u_m + \left(r_m u_m + \tfrac{1}{12}\delta^2(r_m u_m)\right)$ simplifies the algebra.]

6.35 * Suppose that $\mathscr{L}_h U = \mathscr{F}_h$, where $\mathscr{L}_h$ is inverse monotone. Suppose that, in addition, Φ is a discrete comparison function such that $\Phi \geq 0$ and that

$$\mathscr{L}_h \Phi_m \geq 1 \text{ when } \mathscr{F}_{h,m} \neq 0,$$
$$\mathscr{L}_h \Phi_m \geq 0 \text{ when } \mathscr{F}_{h,m} = 0.$$

Prove that $\|U_m\|_{h,\infty} \leq C\|\mathscr{F}_h\|_{h,\infty}$ where $C = \|\Phi\|_{h,\infty}$.

6.36 In the light of the previous exercise, show that $\Phi(x) = \tfrac{1}{2}x(1 - x)$ could be used as a comparison function for the Dirichlet problem in Example 6.13.

6.37 An alternative approach to constructing finite difference methods is the *method of undetermined coefficients*. The idea is to assume a particular form for $\mathcal{L}_h$. As an example, let us suppose that $\mathcal{L}_h$ is to involve three consecutive grid values as in (6.32), where the unknown coefficients (a_m, b_m, c_m) are determined by requiring, for each m, that $\mathcal{L}_h v_m = \mathcal{L} v_m$ for $v(x) = 1$, $(x - x_m)$ and $(x - x_m)^2$. Show that this implies that $\mathcal{L}_h v_m = \mathcal{L} v_m$ for all polynomials of degree at most two.

In the case that $\mathcal{L}v(x) = -v''(x) + r(x)v(x)$, show that this process leads to the standard finite difference approximation (6.20).

6.38 * Use the method of undetermined coefficients outlined in the previous exercise to develop a finite difference approximation of the operator $\mathcal{L}u = -\varepsilon u''(x) + 2u'(x)$. Compare the resulting approximation method with the scheme in (6.40b).

6.39 Consider the BVP

$$\left.\begin{array}{c} -u''(x) = f(x), \quad 0 < x < 1 \\ u(0) = 0, \quad u(1) = 0, \end{array}\right\}$$

with the following right-hand side functions:

(a) $f(x) = 192(x - \frac{1}{2})^2$,

(b) $f(x) = 0$ for $0 < x \le \frac{1}{2}$ and $f(x) = 384(x - \frac{1}{2})^2$ for $\frac{1}{2} < x < 1$,

(c) $f(x) = 0$ for $0 < x \le \frac{1}{2}$ and $f(x) = 96(x - \frac{1}{2})$ for $\frac{1}{2} < x < 1$.

Determine the exact solution in each case and show that they satisfy $u(1/2) = 1$.
[Hint: In cases (b) and (c) it will be necessary to determine the general solution of
the ODEs separately in the intervals $(0, \frac{1}{2})$ and $(\frac{1}{2}, 1)$ which will give rise to four
arbitrary constants. These may be determined by applying the BCs and by insisting
that both $u(x)$ and $u'(x)$ be continuous at $x = \frac{1}{2}$.]

Conduct numerical experiments using the finite difference approximation given
in equation (6.18) to ascertain whether or not the numerical solutions converge in
each of these three cases. If so, estimate the rate of convergence.

Chapter 7
Maximum Principles and Energy Methods

Abstract This chapter extends the ideas in earlier chapters and identifies two concepts that are useful for checking the well-posedness of boundary value problems. These concepts play a fundamental role in establishing the stability of finite difference solutions in later chapters.

The well-posedness of boundary value problems associated with general linear second-order PDEs was assessed in Chap. 4 by considering *fundamental solutions*, or more generally, integral formulae expressing the solution in terms of the boundary data. Such a procedure has limited utility however. Fundamental solutions only exist for a small number of problems, so more general techniques are needed. In this chapter, we develop two powerful tools—maximum principles (based on inverse monotonicity) and energy methods (based on the use of inner products)—that are applicable more widely when studying properties of PDEs.

7.1 Maximum Principles

We first show how the concept of inverse monotonicity that was introduced in Sect. 5.2 in the context of two-point boundary value problems may be extended to parabolic and elliptic BVPs. This framework requires that the BVPs be written in the form

$$\mathscr{L}u = \mathscr{F}, \tag{7.1}$$

where the operator $\mathscr{L}$ is inverse monotone and possesses a suitable comparison function. Once this is done the main theorems in Sect. 5.2 are immediately applicable in a PDE context. We present a number of examples in the sequel. Before doing so however, we will discuss some iconic results from the theory of parabolic and elliptic PDEs.

As discussed in Chap. 4, the heat equation (**pde.4**) is the quintessential parabolic PDE. We begin with a celebrated result that has an intuitive physical interpretation: if no heat is applied to the interior of an insulated rod of unit length, its temperature cannot exceed the larger of its initial temperature and that at its ends.

© Springer International Publishing Switzerland 2015
D.F. Griffiths et al., *Essential Partial Differential Equations*, Springer
Undergraduate Mathematics Series, DOI 10.1007/978-3-319-22569-2_7

Fig. 7.1 The domain Ω_τ for
the heat equation (**pde.4**)
(*shaded*) and the boundary
Γ_τ (*thick line*)

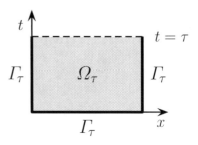

Theorem 7.1 (Maximum principle for the heat equation) *Let $\kappa > 0$, suppose that the function $u(x, t)$ satisfies the inequality*

$$-\kappa u_{xx} + u_t \le 0$$

for $(x, t) \in \Omega_\tau$, where

$$\Omega_\tau = \{(x, t) : 0 < x < 1, 0 < t \le \tau\},$$

then $u(x, t)$ is either constant or else attains its maximum value on Γ_τ, which is the boundary of Ω_τ excluding the side $t = \tau$ (see Fig. 7.1).

Proof We begin the proof as in Example 5.5 by assuming that strict inequality holds: $-\kappa u_{xx} + u_t < 0$ throughout Ω_τ. If u were to achieve a maximum at some point $(x^*, t^*) \in \Omega_\tau$ this would require

$$u_{xx}(x^*, t^*) \le 0 \text{ and } u_t(x^*, t^*) \ge 0$$

(we would have $u_t(x^*, t^*) = 0$ if the maximum occurred for $t^* < \tau$, but $u_t(x^*, t^*) \ge 0$ if it occurred at $t^* = \tau$) meaning that $-\kappa u_{xx} + u_t \ge 0$ at (x^*, t^*), contradicting our hypothesis.

Returning to the original premise that $-\kappa u_{xx} + u_t \le 0$, we go back to the comparison function $\varphi(x) = 1 + \frac{1}{2}x(1 - x)$ used in Example 5.5 and we define the function v by

$$v(x, t) = u(x, t) + \varepsilon\varphi(x). \tag{7.2}$$

It is easily shown that $-\kappa v_{xx} + v_t = -\kappa u_{xx} + u_t - \kappa\varepsilon$ so that $-\kappa v_{xx} + v_t < 0$ for all positive values of ε. The earlier argument then implies that v achieves its maximum on Γ_τ since it cannot occur in Ω_τ. The proof is completed by allowing $\varepsilon \to 0^+$. □

Corollary 7.2 *If $u_t = \kappa u_{xx}$ in Ω_τ then u attains its maximum and minimum values on Γ_τ.*

Proof See Exercise 7.2.   □

The next result shows that the operator corresponding to the heat equation with an initial condition and Dirichlet boundary conditions is inverse monotone, that is, $\mathscr{L}u \geq 0$ implies that $u \geq 0$.

Corollary 7.3 (Inverse monotonicity for the heat equation) *The operator $\mathscr{L}$ defined by*

$$\mathscr{L}u(x,t) = \begin{cases} -\kappa u_{xx} + u_t & \text{for } (x,t) \in \Omega_\tau, \\ u(x,t) & \text{for } (x,t) \in \Gamma_\tau \end{cases}$$

is inverse monotone.

Proof See Exercise 7.3. □

The well-posedness of the inhomogeneous heat equation with Dirichlet boundary conditions is the focus of the following example.

Example 7.4 Show that the BVP for the heat equation $u_t = \kappa u_{xx} + f(x,t)$, for $(x,t) \in \Omega_\tau$ is well posed for the initial/boundary condition: $u(x,t) = g(x,t)$ for $(x,t) \in \Gamma_\tau$.

With $\mathscr{L}$ in Corollary 7.3 and the source term defined by

$$\mathscr{F}(x,t) = \begin{cases} f(x,t), & (x,t) \in \Omega_\tau \\ g(x,t), & (x,t) \in \Gamma_\tau, \end{cases} \tag{7.3}$$

the BVP can be written in the requisite form (7.1). The inverse monotonicity of $\mathscr{L}$ is thus sufficient to ensure uniqueness of the solution (see Theorem 5.3). For well-posedness, we adapt the argument used in Example 5.5 to the present situation. To this end the comparison function $\varphi(x) = 1 + \frac{1}{2}x(1-x)$ satisfies $\mathscr{L}u \geq 1$ everywhere in Ω_τ as well as on the boundary Γ_τ. A suitable norm is[1]

$$\|\mathscr{F}\| = \max\left\{ \max_{(x,t)\in\Omega_\tau} |f(x,t)|, \max_{(x,t)\in\Gamma_\tau} |g(x,t)| \right\}, \tag{7.4}$$

so well-posedness follows immediately (from Theorem 5.4), provided that the functions f and g are bounded on their respective domains. ◊

The Poisson equation is the quintessential *elliptic* PDE, see (4.28). In two dimensions it can be written in the form $\mathcal{L}u = f$, with

$$\mathcal{L}u := -u_{xx} - u_{yy} \tag{7.5}$$

and is assumed to hold on a domain Ω such as that shown in Fig. 7.2. Like the heat equation, it has a well-defined maximum principle.

---

[1] The alternative norm $\|\mathscr{F}\| = \max_{(x,t)\in\Omega_\tau} |f(x,t)| + \max_{(x,t)\in\Gamma_\tau} |g(x,t)|$ leads to a slightly larger upper bound.

Fig. 7.2 The domain Ω for
the Poisson equation which
is contained between the
lines $x = 0$ and $x = a$

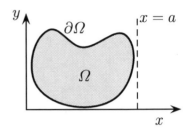

Theorem 7.5 (Maximum principle for the Poisson equation) *Suppose that the function $u(x, y)$ satisfies the inequality*

$$-u_{xx} - u_{yy} \le 0$$

for $(x, y) \in \Omega$ then $u(x, y)$ is either constant or else attains its maximum value on $\partial\Omega$, the boundary of Ω.

Proof We shall suppose that the domain Ω is contained in the strip $0 \le x \le a$ (see Fig. 7.1) then apart from cosmetic changes (replacing $u(x, t)$ by $u(x, y)$, Ω_τ by Ω and Γ_τ by $\partial\Omega$) the only alteration needed to the proof of Theorem 7.1 is to define

$$v(x, y) = u(x, y) + \varepsilon\varphi(x), \quad \varphi(x) = 1 + \tfrac{1}{2}x(x - a) \tag{7.6}$$

instead of (7.2). Note that $v_{xx} \le 0$ and $v_{yy} \le 0$ at a maximum of v in Ω.   □

Corollary 7.6 (Laplace's equation) *If $u_{xx} + u_{yy} = 0$ in Ω then u attains its maximum and minimum values on the boundary $\partial\Omega$.*

Corollary 7.7 (Inverse monotonicity for the Poisson equation) *The operator $\mathscr{L}$ defined by*

$$\mathscr{L}u(x, y) = \begin{cases} -u_{xx} - u_{yy} & \text{for } (x, t) \in \Omega, \\ u(x, y) & \text{for } (x, t) \in \partial\Omega \end{cases}$$

is inverse monotone.

The well-posedness of the Poisson equation with Dirichlet boundary conditions is established in the following example by combining the result of Theorem 7.5 and its corollaries.

Example 7.8 Show that the BVP for the Poisson equation $-u_{xx} - u_{yy} = f(x, y)$, for $(x, y) \in \Omega$ is well posed for the boundary condition: $u(x, y) = g(x, y)$ for $(x, y) \in \partial\Omega$.

To get to the requisite form (7.1), we let $\mathscr{L}$ be as in Corollary 7.7 and set

$$\mathscr{F}(x, y) = \begin{cases} f(x, y), & (x, y) \in \Omega \\ g(x, y), & (x, y) \in \partial\Omega. \end{cases} \tag{7.7}$$

The inverse monotonicity of $\mathcal{L}$ ensures uniqueness of the solution. Next, we extend the argument used in Example 7.4 to the present situation. Thus, assuming that the domain Ω fits into the strip $0 \le x \le a$, we simply take $\varphi(x) = 1 + \frac{1}{2}x(a - x)$ as the comparison function. Well-posedness immediately follows from Theorem 5.4 whenever the functions f and g are bounded on their respective domains. $\Diamond$

One needs to proceed with a lot more care if Neumann boundary conditions are substituted for the Dirichlet conditions in the previous example since a solution may not even exist in this case! This is will be addressed in the next example. First, we recall from Chap. 2 that $\vec{n} := \vec{n}(x, y)$ denotes the outward unit normal vector to a point (x, y) on $\partial\Omega$ so that $u_n := \vec{n} \cdot \text{grad } u$ represents the outward normal derivative to $\partial\Omega$.

Example 7.9 (Neumann boundary conditions) Consider the BVP for the Poisson equation $-u_{xx} - u_{yy} = f(x, y)$, for $(x, y) \in \Omega$ when subject to the BC $u_n(x, y) = g(x, y)$ for $(x, y) \in \partial\Omega$.

To make progress, we use the relation $u_{xx} + u_{yy} = \text{div grad } u$ and integrate the PDE over Ω. Applying the divergence theorem then gives

$$\int_\Omega \text{div grad } u \, d\Omega = \int_{\partial\Omega} \vec{n} \cdot \text{grad } u \, dS = \int_{\partial\Omega} u_n \, dS.$$

Since div grad $u = -f$ in Ω and $u_n = g$ on $\partial\Omega$, we find that the given functions f and g are required to satisfy a compatibility condition:

$$\int_\Omega f \, d\Omega + \int_{\partial\Omega} g \, dS = 0, \tag{7.8}$$

otherwise a solution cannot exist. Assuming that (7.8) holds, we observe that if $u(x, y)$ is a solution to the given problem, then $u(x, y) + c$ is also a solution, for any constant c.[2] This means that there are infinitely many solutions! An additional condition is required to restore uniqueness in this case. For example, one could choose to find the specific solution satisfying $u(x^*, y^*) = 0$ for some point $(x^*, y^*) \in \Omega \cup \partial\Omega$, or else one could find the solution satisfying

$$\int_\Omega u \, d\Omega = 0. \tag{7.9}$$

(In the context of u representing a pressure, this latter condition means that the net force on the domain is zero). $\Diamond$

The concept of inverse monotonicity is not so helpful when considering hyperbolic PDEs. The next example illustrates why an alternative construction is needed in such cases.

---

[2]This might occur, for instance, when u is the pressure in a fluid. In such systems it is usually only the difference in pressure between two points and not the absolute pressure that can be measured.

Example 7.10 (The wave equation) Suppose that u satisfies the wave equation $u_{tt} = a^2 u_{xx}$ in the quarter-plane $x > 0$, $t > 0$ and is subject to the boundary condition $u(0, t) = 0$ and initial conditions $u(x, 0) = g(x)$ and $u_t(x, 0) = 0$, where $g(x)$ is an odd function of x. Show that positive initial data ($g_0(x) \geq 0$, for $x \in [0, \infty)$) does not necessarily lead to a positive solution for $t > 0$.

The general solution of the wave equation is given by d'Alembert's formula (equation (4.20) with $g_1 = 0$):

$$u(x, t) = \frac{1}{2} (g(x - at) + g(x + at)).$$ (7.10)

When $x = 0$ this reduces to $u(0, t) = \frac{1}{2}(g(-at) + g(at)) = 0$ (since g is an odd function) so (7.10) is a valid solution to the quarter-plane problem. There are many examples of odd functions $g(x)$ that are positive on $(0, \infty)$ but lead to solutions u that are negative in certain parts of the positive quadrant. For example, taking $g(x) = x/(1 + x^2)$ it can be shown that $u(1, 2/a) = -1/10 < 0$. Thus, in this case, positive data does not lead to a positive solution.   ◇

7.2 Energy Methods

Energy methods provide an alternative way of establishing well-posedness. They can, moreover, be applied to hyperbolic PDEs.

To demonstrate the basic idea of energy methods we take the heat equation (**pde.4**) ($u_t = \kappa u_{xx}$) with zero Neumann end conditions as a prototypical BVP. Integrating the heat equation over the spatial interval gives

$$\int_0^1 \frac{\partial u}{\partial t} \, dx = \int_0^1 \kappa \frac{\partial^2 u}{\partial x^2} \, dx = \kappa \frac{\partial u}{\partial x} \Big|_{x=0}^{1} = 0,$$ (7.11)

where the right hand side is zero because of the homogeneous Neumann BCs. The left hand side of (7.11) can be identified with $\frac{d}{dt} \int_0^1 u(x, t) \, dx$ and the quantity $\int_0^1 u(x, t) \, dx$, which is a function of t only, may be interpreted as the total heat content of the bar at time t. Since its derivative is zero the heat content remains constant in time so is always equal to its initial value:

$$\int_0^1 u(x, t) \, dx = \int_0^1 u(x, 0) \, dx = \int_0^1 g(x) \, dx.$$ (7.12)

This is consistent with the bar being kept insulated.

In a similar vein, if both sides of the PDE are multiplied by u before integrating, we have

$$\int_0^1 u \frac{\partial u}{\partial t} \, dx = \kappa \int_0^1 u \frac{\partial^2 u}{\partial x^2} \, dx.$$

The integrand on the left hand side may be written as $\frac{1}{2} \partial_t(u^2)$ and, by integrating the right hand side by parts, we find that

$$\frac{1}{2} \frac{d}{dt} \int_0^1 u^2 \, dx = \kappa u \frac{\partial u}{\partial x} \bigg|_{x=0}^{1} - \kappa \int_0^1 \left(\frac{\partial u}{\partial x} \right)^2 dx$$

The first term on the right hand side vanishes when the ends are subject to either homogeneous Dirichlet or Neumann BCs and so, in these cases,

$$\frac{d}{dt} \int_0^1 u^2 \, dx = -2\kappa \int_0^1 \left(\frac{\partial u}{\partial x} \right)^2 dx \leq 0.$$

Therefore $\int_0^1 u^2(x, t) \, dx$, which is a measure of the magnitude of the solution at time t, is a nonincreasing quantity, so it cannot exceed its initial value:

$$\int_0^1 u^2(x, t) \, dx \leq \int_0^1 g^2(x) \, dx. \tag{7.13}$$

This implies *stability* —the solution to the heat equation cannot blow up in time! This is not all—it also implies that the problem has a unique solution (by Theorem 2.7, since $u(x, t) \equiv 0$ whenever $g(x) \equiv 0$) and is therefore well posed. Inequalities like that in (7.13) have acquired the sobriquet of "energy inequalities" despite the fact that the integral on the left hand side does not actually represent a physical energy. The key point is that it is the integral of a nonnegative quantity. ◊

Example 7.11 (The wave equation) The energy E associated with the wave equation $u_{tt} = c^2 u_{xx}$ for $0 < x < 1, t > 0$ is given by

$$E(t) = \int_0^1 \left((u_t)^2 + c^2(u_x)^2 \right) dx.$$

Show that the energy E is constant in time when the equation is supplemented with homogeneous Neumann boundary conditions.

By differentiating with respect to t, we find,

$$\frac{dE}{dt} = 2 \int_0^1 \left(u_t u_{tt} + c^2 u_x u_{xt} \right) dx.$$

Next, integrating the second term on the right hand side by parts gives

$$\frac{\mathrm{d}E}{\mathrm{d}t} = 2 \int_0^1 u_t \left(u_{tt} - c^2 u_{xx} \right) \mathrm{d}x + c^2 u_x \, u_t \Big|_0^1.$$

The imposed boundary conditions make the boundary term zero, so the right hand side is identically zero. Thus $E(t)$ does not vary with time.

The same conclusion may be drawn if the Neumann boundary conditions are replaced by Dirichlet conditions of the form $u = $ constant at either, or both, $x = 0, 1$. For example, if $u(0, t) = $ constant then $u_t(0, t) = 0$ and the corresponding boundary term above vanishes.   ◊

Exercises

7.1 Show that Theorem 7.1 can also be established by making the choice $v(x, t) = u(x, t) + \varepsilon(\tau - t)$ for $0 \le t \le \tau$.

7.2 Prove Corollary 7.2.

7.3 Prove Corollary 7.3.

7.4 Suppose that u satisfies the advection–diffusion equation $u_t + 2u_x = u_{xx}$ for $0 < x < 1$ and $t > 0$ together with homogeneous boundary conditions[3] $u_x = 2u$ at $x = 0$ and $x = 1$, and the initial condition $u(x, 0) = 6x$, for $0 < x < 1$. Show that the total mass $M(t) := \int_0^1 u(x, t) \, \mathrm{d}x$ satisfies $M'(t) = 0$ and deduce that $M(t) = 3$ for all $t \ge 0$.

Show also, by employing the energy method, that $E(t) := \int_0^1 u^2(x, t) \, \mathrm{d}x$ satisfies $E'(t) \le 0$ and deduce that $E(t) \le 12$.

7.5 Suppose that u satisfies the PDE $u_t = x u_{xx} + u_x$ together with homogeneous Dirichlet boundary conditions and the initial condition $u(x, 0) = \sin \pi x$ for $0 < x < 1$.

Use the energy method to prove that $\int_0^1 u^2(x, t) \, \mathrm{d}x \le 1/2$.

[Hint: show that $x u_{xx} + u_x = \partial_x(a u_x)$ for a suitable function $a(x)$.]

7.6 Suppose that $u(r, t)$ satisfies the heat equation in polar coordinates with circular symmetry, that is, $r u_t = (r u_r)_r$ (see Example 5.14), in the region $1 < r < 2$, $t > 0$ with BCs $u(1, t) = 0$, $u(2, t) = 0$ for $t > 0$ and initial condition $u(r, 0) = (r - 1)(2 - r)$ for $1 \le r \le 2$.

---

[3]The PDE may be written in the form of a conservation law $u_t + f(u)_x = 0$ with a flux function $f(u) = 2u - u_x$. The boundary conditions are then seen to be zero-flux conditions.

Show, using an energy argument, that

$$\frac{d}{dt} \int_1^2 ru^2 \, dr = -2 \int_1^2 r \, (u_r)^2 \, dr.$$

Deduce that $\int_1^2 ru(r, t)^2 \, dr \leq 1/20$ for all $t > 0$.

7.7 Suppose that the solution $u(x, t)$ of the heat equation is subject to homogeneous (Dirichlet or Neumann) boundary conditions and has initial condition $u(x, 0) = g(x)$. Establish the energy inequality

$$\int_0^1 u_x^2(x, t) \, dx \leq \int_0^1 (g'(x))^2 \, dx$$

by differentiating the integral on the left with respect to t.

7.8 Suppose that $u(x, y)$ satisfies the Poisson equation $-\nabla^2 u = f(x, y)$ for $(x, y) \in \Omega$ with homogeneous Dirichlet boundary conditions. By first multiplying both sides of the PDE by u and then proceeding as in Example 7.9, show that

$$\int_\Omega \left((u_x)^2 + (u_y)^2 \right) d\Omega = \int_\Omega u f \, d\Omega.$$

Deduce that the boundary value problem consisting of Poisson's equation with Dirichlet boundary conditions has a unique solution.
[Hint: div $\alpha \vec{v} = \alpha$ div $\vec{v} + \vec{v} \cdot$ grad α.]

7.9 Explore the consequences of replacing the homogeneous Dirichlet boundary condition in the previous exercise by the homogeneous Neumann boundary condition $u_n = 0$. Explain why it cannot be concluded that the problem has a unique solution.

7.10 Suppose that the wave equation of Example 7.11 is subject to the Robin boundary conditions $a_0 u(0, t) - b_0 u_x(0, t) = 0$ and $a_1 u(1, t) + b_1 u_x(1, t) = 0$ for constants a_0, b_0, a_1, b_1 with $b_0 \neq 0$ and $b_1 \neq 0$. Calculate the derivative $E'(t)$ of the energy given in the example.

How should $E(t)$ be modified in order to remain constant in time with these new boundary conditions?

What constraints should be imposed on a_0, b_0, a_1, b_1 to ensure the modified energy is a nonnegative function of u?

7.11 Consider the Korteweg-de Vries equation $u_t + 6uu_x + u_{xxx} = 0$ on the interval $-\infty < x < \infty$ for $t > 0$ (see (**pde.9**)). If u and its derivatives decay to zero sufficiently rapidly as $x \to \pm\infty$, show that the total mass $m(t)$ and the momentum $M(t)$, defined by

$$m(t) = \int_{-\infty}^{\infty} u(x,t)\,dx, \qquad M(t) = \int_{-\infty}^{\infty} u^2(x,t)\,dx,$$

are both constant in time.

7.12 Suppose that u is a solution of the Korteweg-de Vries equation as in the previous exercise. Show that the energy

$$E(t) = \int_{-\infty}^{\infty} \left(\tfrac{1}{2}(u_x)^2 - u^3 \right) dx$$

is also constant in time.[4]

---

[4] An indication of the special nature of the KdV equation is that it has an infinite number of such conserved quantities.

Chapter 8
Separation of Variables

Abstract This chapter describes a classical technique for constructing series solutions of linear PDE problems. Classical examples like the heat equation, the wave equation and Laplace's equations are studied in detail.

Separation of variables is an elegant technique for finding solutions of *linear* PDEs. It is founded on the notions of inner products, Sturm–Liouville theory and eigenvalue problems discussed in Chap. 5. Although the range of problems that can be solved is limited, the technique is generally useful; first, for revealing the nature of solutions and second, for providing exact solutions with which to test the correctness and accuracy of numerical methods.

In this chapter we will apply the technique to the model elliptic, hyperbolic and parabolic problems that were identified in Chap. 4. We study the easiest problem first.

8.1 The Heat Equation Revisited

Example 8.1 Solve the (homogeneous) heat equation (**pde.4**)

$$\frac{\partial u}{\partial t} = \kappa \frac{\partial^2 u}{\partial x^2}, \qquad (\kappa > 0), \tag{8.1}$$

in the semi-infinite strip $S = \{(x, t) : 0 < x < 1, t > 0\}$, with a nonhomogeneous initial condition, $u(x, 0) = g(x)$, $0 < x < 1$, and (Dirichlet) boundary conditions: $u(0, t) = u(1, t) = 0$, $t > 0$. Note that we will not insist that $g(0) = g(1) = 0$, so that there could be a discontinuity in the solution when we take the limit $t \to 0$.

The essence of separation of variables it to look for a solution of the form

$$u(x, t) = X(x)T(t), \tag{8.2}$$

© Springer International Publishing Switzerland 2015
D.F. Griffiths et al., *Essential Partial Differential Equations*, Springer
Undergraduate Mathematics Series, DOI 10.1007/978-3-319-22569-2_8

that is, as a product of two functions, each depending on only one of the independent variables. First, substituting the ansatz (8.2) into the boundary conditions gives

$$u(0, t) = X(0)T(t), \quad u(1, t) = X(1)T(t).$$

Since we are seeking nontrivial solutions, we require $T(t) \not\equiv 0$, thus

$$X(0) = X(1) = 0.$$

Next, we substitute (8.2) into the PDE. Rearranging the result gives

$$\frac{1}{\kappa T(t)} \frac{dT(t)}{dt} = \frac{1}{X(x)} \frac{d^2 X(x)}{dx^2}.$$

Note that the left hand side is a function of t only, while the right side is a function of x only. Thus both sides of the equation must be equal to a constant (known as the *separation constant*),

$$\frac{1}{\kappa} \frac{1}{T(t)} \frac{dT(t)}{dt} = \frac{1}{X(x)} \frac{d^2 X(x)}{dx^2} = \text{constant}. \tag{8.3}$$

Writing the constant as $-\lambda$ (a negative number, following the reasoning given in Example 5.10) the BVP for X becomes

$$\left. \begin{array}{l} -X''(x) = \lambda X(x), \quad 0 < x < 1 \\ X(0) = X(1) = 0. \end{array} \right\} \tag{8.4}$$

Working through the details of Example 5.10 leads to a countable set of eigenfunctions $X(x) := X_n(x)$, where

$$X_n(x) = \sin n\pi x,$$

associated with eigenvalues $\lambda := \lambda_n = n^2 \pi^2$ ($n = 1, 2, \ldots$). Moreover, for these distinct values, (8.3) gives

$$T'(t) = -\kappa \lambda_n T(t)$$

whose solution is $T_n(t) = A \exp(-\kappa \lambda_n t)$, where A is an arbitrary constant. Combining X_n with T_n gives an infinite family of fundamental solutions

$$u_n(x, t) = e^{-\kappa n^2 \pi^2 t} \sin n\pi x, \quad n = 1, 2, \ldots \tag{8.5}$$

where, for simplicity, we normalize each eigenfunction by setting the arbitrary constant so that $A = 1$. The first three solutions are illustrated in Fig. 8.1. As n increases the wavelength decreases, as does the penetration of the solution into the domain in the t-direction.

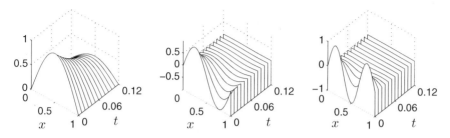

Fig. 8.1 The first three fundamental solutions u_n, $n = 1, 2, 3$ (see (8.5)) of the heat equation with $\kappa = 1$

As discussed in Sect. 2.2.1, the fact that the PDE and associated boundary conditions are linear and homogeneous means that any linear combination of fundamental solutions (8.5), that is

$$u(x, t) = \sum_{n=1}^{\infty} A_n e^{-\kappa n^2 \pi^2 t} \sin n\pi x, \qquad (8.6)$$

must also be a solution (to the combination of the PDE plus BCs), with arbitrary coefficients $\{A_n\}$.

We should take a step back at this point: Corollary 7.3 together with Theorem 5.3 ensure that our BVP has a *unique* solution. This implies that the coefficients $\{A_n\}$ in (8.6) have to be uniquely determined by problem data that we have not yet used; namely the (nonhomogeneous) initial condition $u(x, 0) = g(x)$. Thus, to find these coefficients we simply need to set $t = 0$ in (8.6). Doing so gives

$$g(x) = \sum_{n=1}^{\infty} A_n \sin n\pi x, \qquad (8.7)$$

and hence, using the construction in (5.19), we find that[1]

$$A_n = \frac{\langle g, X_n \rangle}{\langle X_n, X_n \rangle}, \qquad (8.8)$$

where $\langle u, v \rangle = \int_0^1 uv \, dx$. To summarise, (8.6) and (8.8) characterise the solution of our boundary value problem for any initial data g that possesses a convergent expansion of the form (8.7). ◇

To give a concrete illustration, let us suppose that $g(x) = x$ and $\kappa = 1$. The boundary value problem is then a model of the temperature in a rod that is initially in equilibrium

---

[1] The eigenfunctions $\{X_n(x)\}_{n=1}^{\infty}$ are orthogonal with respect to the inner product (5.17) with $L = 1$, see Example 5.8.

with a temperature $u = 0$ at $x = 0$ and $u = 1$ at $x = 1$, but with the right end plunged into an ice bath at time $t = 0$ (the boundary condition is $u = 0$ for $t > 0$ at $x = 1$). With this specific initial condition the numerator in (8.8) can be evaluated by integration by parts,

$$\langle g, X_n \rangle = \int_0^1 x \sin n\pi x \, dx = -\left[\frac{x}{n\pi} \cos n\pi x \right]_0^1 + \frac{1}{n\pi} \int_0^1 \cos n\pi x \, dx$$

$$= \frac{1}{n\pi}(-1)^{n+1},$$

while the denominator $\langle X_n, X_n \rangle = 1/2$. Thus, from (8.7), we find that

$$u(x, t) = \sum_{n=1}^{\infty} (-1)^{n+1} \frac{1}{n\pi} e^{-n^2\pi^2 t} \sin n\pi x. \tag{8.9}$$

The series solution (8.9) is visualised for $0 \le t \le 0.1$ in increments of 0.005 on the left of Fig. 8.2. The discontinuity at $x = 1$, $t = 0$, between the initial condition and the boundary condition instantaneously vanishes and the function u becomes progressive smoother as t increases. When t is relatively large, successive terms in the series decay rapidly to zero so the long-term solution looks like the leading term, that is,

$$u(x, t) \approx \frac{1}{\pi} e^{-\pi^2 t} \sin \pi x, \quad \text{when } t \to \infty.$$

This asymptotic behaviour is evident in Fig. 8.2.

Example 8.2 Characterise the solution to the heat equation (**pde.4**) when modelling the case that both ends of the rod are insulated for $t > 0$.

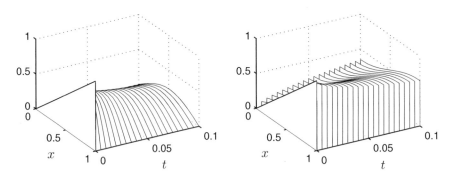

Fig. 8.2 The solution of Example 8.1 (*left*) and Example 8.2 (*right*) with $\kappa = 1$ and initial condition $g(x) = x$, shown at time increments of 0.005

Appropriate conditions for insulated ends are $u_x(0, t) = u_x(1, t) = 0$. Setting $u(x, t) = X(x)T(t)$, leads to $X'(0) = X'(1) = 0$. The general solution satisfying the combination of PDE plus BCs can be shown (see Exercise 8.1) to be

$$u(x, t) = \frac{1}{2} A_0 + \sum_{n=1}^{\infty} A_n e^{-\kappa n^2 \pi^2 t} \cos n\pi x.$$

The coefficients $\{A_n\}_{n=0}^{\infty}$ are still determined by the initial data at $t = 0$:

$$g(x) = \frac{1}{2} A_0 + \sum_{n=1}^{\infty} A_n \cos n\pi x.$$

Once more, by virtue of the orthogonality of the functions $\{\cos n\pi x\}_{n=0}^{\infty}$ we use (8.8) with $X_n(x) = \cos n\pi x$, to give

$$A_n = 2 \int_0^1 g(x) \cos n\pi x \, dx, \quad n = 0, 1, 2, \ldots.$$

For the specific example with $g(x) = x$ and $\kappa = 1$, it may be shown that

$$u(x, t) = \frac{1}{2} - \sum_{\substack{n=1 \\ n \text{ odd}}}^{\infty} \frac{4}{\pi^2 n^2} e^{-n^2 \pi^2 t} \cos n\pi x,$$

which is visualised in Fig. 8.2 (right). In this case, for relatively large times t, the first two terms dominate and

$$u(x, t) = \frac{1}{2} - \frac{4}{\pi^2} e^{-\pi^2 t} \cos \pi x + \cdots$$

so $u \to 1/2$ as $t \to \infty$ in contrast to the case of homogeneous Dirichlet boundary conditions where $u \to 0$ as $t \to \infty$. ◇

It might appear on intuitive grounds that the initial data $g(x)$ used in the above examples should be continuous and have continuously differentiable first and second space derivatives in order to lead to bona fide solutions of the heat equation. This is far from being the case—the function g need only be smooth enough that the inner product $\langle g, X_n \rangle$ in the formula (8.8) is well defined and that the resulting coefficients lead to a convergent series (8.7). The following definition makes this more precise.

Definition 8.3 (*Piecewise continuity*) A function $f(x)$ defined on an interval $[a, b]$ is said to be piecewise continuous if there are a finite number of points $a = x_0 < x_1 < \cdots < x_N = b$ such that

① f is continuous on each open interval $x_{n-1} < x < x_n$,
② f has a finite limit at each end of each subinterval,

An important point is that, while the value of f at the interior points x_n must be finite, this value need not be the same as either of the limits of f as $x \to x_n$.

If $g(x)$ and $g'(x)$ are piecewise continuous functions on $[0, 1]$ then the infinite series defined by

$$g(x) = \sum_{n=1}^{\infty} A_n \sin n\pi x, \quad A_n = \frac{\langle g, \sin n\pi x \rangle}{\langle \sin n\pi x, \sin n\pi x \rangle} = 2 \int_0^1 g \sin n\pi x \, dx \quad (8.10)$$

converges to the function $g(x)$ at all points where it is continuous, except possibly at end points where there are Dirichlet boundary conditions that $g(x)$ does not satisfy (see Example 8.1). At interior points where $g(x)$ is discontinuous, the series converges to $\frac{1}{2}(g(x^+) + g(x^-))$, the average of its left and right limits at x. If the eigenfunctions satisfy a homogeneous Dirichlet boundary condition at an end point then the series expansion in (8.10) will also be zero at that end point, regardless of the value of g there.

Tutorial exercises typically involve piecewise polynomials. That is, the domain of interest is divided into subintervals on each of which g has a polynomial expression. Such situations naturally lead to piecewise continuous functions. This is illustrated in the next example.

Example 8.4 Solve the diffusion equation $u_t = \kappa u_{xx}$ on the domain $\{(x, t) : 0 \le x \le 1, t > 0\}$ with homogeneous Dirichlet boundary conditions $u(0, t) = u(1, t) = 0$ and initial conditions

(a) $g(x) = 1$ for $3/8 < x < 5/8$ and equal to zero otherwise,
(b) $g(x) = \max\{0, 1 - |8x - 4|\}$.

From Example 8.1 we have eigenfunctions $X_n(x) = \sin n\pi x$ with corresponding eigenvalues $\lambda_n = \kappa(n\pi)^2$. In case (a), where $g(x)$ is piecewise continuous but not continuous, we find,

$$\langle g, X_n \rangle = \int_{3/8}^{5/8} \sin n\pi x \, dx$$

$$= -\frac{1}{n\pi}(\cos \tfrac{5}{8}n\pi - \cos \tfrac{3}{8}n\pi) = \frac{2}{n\pi} \sin \tfrac{1}{2}\pi n \sin \tfrac{1}{8}n\pi.$$

Since $\langle X_n, X_n \rangle = 1/2$, it is seen that the coefficients A_n given by (8.10) tend to zero as a rate inversely proportional to n, as $n \to \infty$. This slow decay is shared by all piecewise continuous functions that are not continuous. This also leads to slow convergence of the corresponding series for $g(x)$, which implies that many terms may be needed in order to obtain an accurate representation of g. Also, according to our earlier remarks, the series converges at the points of discontinuity ($x = 3/8, 5/8$) to $1/2$, which is the average of left and right hand limits. The solution is given by

$$u(x, t) = \sum_{\substack{n=1 \\ n \text{ odd}}}^{\infty} \frac{4}{n\pi} e^{-\kappa n^2 \pi^2 t} \sin \tfrac{1}{2} n\pi \sin \tfrac{1}{8} n\pi \sin n\pi x, \qquad (8.11)$$

and we observe that, for $t > 0$, the terms decay to zero very rapidly due to the presence of the exponential factor. The coefficients of the even numbered terms are all zero (the reason for which can be found in Exercise 8.3) and so the summation is altered accordingly.

Next, for case (b) where $g(x) = \max\{0, 1 - |8x - 4|\}$, it can be shown (see Exercise 8.2) that the solution is given by (8.6) with coefficients

$$A_n = \left(\frac{8}{\pi n}\right)^2 (\sin \tfrac{1}{16} n\pi)^2 \sin \tfrac{1}{2} n\pi \qquad (8.12)$$

that decay to zero like $1/n^2$ as $n \to \infty$ (compared to $1/n$ in case (a) reflecting the additional smoothness of the initial data—see Theorem E.2). The even numbered coefficients are again zero.

Finally, we note that in case (a) $A_1 = (4/\pi) \sin(\pi/8)$, while in case (b) we have $A_1 = (8/\pi)^2 \sin^2(\pi/16)$. The ratio of these two quantities is given by $(\pi/8) \cot(\pi/16) \approx 1.97$, from which it follows that the long-term amplitude in case (a) should be approximately twice that in case (b). This is borne out by a closer inspection of the computed solutions shown in Fig. 8.3. ◊

A number of features of the examples so far are also relevant to more general parabolic equations. In contrast to ordinary differential equations, where the general solution of an nth-order equation is a linear combination of n fundamental solutions, the general solutions of PDEs contain an infinite number of terms. We also note that, within the set of solutions (8.5), the terms with shorter wavelengths decay faster than those with longer wavelengths. These shorter wavelengths are responsible for the fine detail in the initial data: in general, this means that solutions to parabolic PDE problems inevitably become progressively smoother over time.

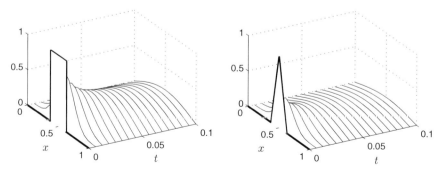

Fig. 8.3 The solutions of the initial boundary value problem for the heat equation in Example 8.4 for $0 \le t \le 0.1$

8.1.1 Extension to Other Parabolic Problems ...

A key ingredient of separation of variables is the identification of an eigenvalue
problem in one of the independent variables with (two) homogeneous boundary
conditions. The Sturm–Liouville (S–L) eigenvalue problems introduced in Sect. 5.4
can be included in this category. As will be shown shortly, S–L eigenvalue problems
naturally arise when solving *linear* parabolic BVPs of the following form:

$$\left. \begin{array}{ll} u_t + \frac{1}{w(x)}\mathcal{L}u = 0, & 0 < x < a \\ \mathcal{B}u(0,t) = \mathcal{B}u(a,t) = 0, & u(x,0) = g(x). \end{array} \right\} \tag{8.13}$$

Generality comes from different choices of *weight function* $w(x) > 0$. The main
constraint is that neither spatial derivative operators $\mathcal{L}$ nor $\mathcal{B}$ is allowed to depend on t.

Applying separation of variables to (8.13) we look for a solution of the form
$u(x,t) = X(x)T(t)$, with general boundary conditions,

$$\mathcal{B}X(0) = \mathcal{B}X(a) = 0.$$

The linearity of $\mathcal{L}$ implies that $\mathcal{L}\big(X(x)T(t)\big) = T(t)\big(\mathcal{L}X(x)\big)$ and so the PDE in
(8.13) becomes, on dividing by $X(x)T(t)$,

$$\frac{T'(t)}{T(t)} = -\frac{\mathcal{L}X(x)}{w(x)X(x)}.$$

Thus, if we le λ_n be the nth eigenvalue and X_n the corresponding eigenfunction
solving the problem

$$\left. \begin{array}{ll} \mathcal{L}X(x) = \lambda w(x)X(x), & 0 < x < a, \\ \mathcal{B}X(0) = \mathcal{B}X(a) = 0, \end{array} \right\} \tag{8.14}$$

then $T'(t) = \lambda T(t)$. This immediately leads us to a set of fundamental solutions

$$u_n(x,t) = X_n(x)e^{-\lambda_n t}, \quad n = 1, 2, \ldots \tag{8.15}$$

and a general solution

$$u(x,t) = \sum_{n=1}^{\infty} A_n X_n(x)e^{-\lambda_n t}. \tag{8.16}$$

Given the initial condition $u(x,0) = g(x)$, the coefficients $\{A_n\}$ can be directly
determined from

$$g(x) = \sum_{n=1}^{\infty} A_n X_n(x)e^{-\lambda_n t}$$

by multiplying both sides of this equation by $w(x)X_m(x)$ and using the orthogonality of the eigenfunctions with respect to the weighted inner product (5.38). The result is

$$A_m = \frac{\langle g, X_m \rangle_w}{\langle X_m, X_m \rangle_w}. \tag{8.17}$$

The combination (8.17) and (8.16) is the solution to the general boundary value problem (8.13). Two specific examples are worked out below.

Example 8.5 (The heat equation with circular symmetry) Determine the solution of the heat equation $u_t = \nabla^2 u$ in a circular disc of radius a, when the temperature at the boundary satisfies $u = 0$ and the initial temperature distribution, $u = g$ at $t = 0$, has circular symmetry.

In polar coordinates $x = r \cos \theta$, $y = r \sin \theta$, the circular symmetry of the initial data means that g is a function of radial distance r from the centre of the disc. This then implies that the solution must also be radially symmetric, that is $u := u(r, t)$. In this case, as shown in Example 5.14, the heat equation can be expressed in the form (8.13) with weight function $w = r$ (the independent variable is now r, rather then x) and with

$$\mathcal{L}u = -\frac{\partial}{\partial r}\left(r\frac{\partial u}{\partial r}\right). \tag{8.18}$$

We now follow the general procedure outlined above for separation of variables, which is to say that we look for solutions in the form $u(r, t) = R(r)T(t)$. The boundary condition $u = 0$ at $r = a$ requires that $R(a) = 0$ while circular symmetry implies that $R'(0) = 0$ (see Exercise 8.7). Thus we are led to the eigenvalue problem (8.14), which can be written as

$$\left.\begin{array}{c} R''(r) + \frac{1}{r}R'(r) + \lambda R(r), \quad 0 < r < a, \\ R'(0) = R(a) = 0. \end{array}\right\} \tag{8.19}$$

The change of variable[2] $x = \sqrt{\lambda}r$ transforms this ODE to

$$\frac{d^2 R}{dx^2} + \frac{1}{x}\frac{dR}{dx} + R = 0, \quad 0 < x < a\sqrt{\lambda},$$

which is of the form (D.1) with $\nu = 0$. Using results from Appendix D we deduce that the general solution is $R(r) = AJ_0(x) + BY_0(x)$, where J_0 and Y_0 are zero-order Bessel functions of the first and second kinds, respectively, with A, B being arbitrary constants. Since $Y_0(x) \to -\infty$ as $r \to 0$, it follows that $B = 0$ in order to give a bounded solution. Furthermore, since all eigenfunctions are unique only up to a multiplicative constant, we may choose $A = 1$, so $R(r) = J_0(x)$. The boundary condition at $r = 0$ is automatically satisfied since $J_0'(0) = 0$. The remaining

---

[2]The eigenvalues of S–L problems are necessarily positive, see Theorem 5.11, so $\sqrt{\lambda}$ is a positive real number.

boundary condition requires that $u = 0$ at $x = \sqrt{\lambda}a$. The zeros of $J_0(x)$ are labelled $\xi_0 = 0 < \xi_1 < \xi_2 < \cdots$ (see Table D.1) which means that the eigenvalues must be

$$\lambda_k = \left(\frac{\xi_k}{a}\right)^2, \quad k = 1, 2, \ldots \tag{8.20}$$

Making use of (8.16), the general solution is

$$u(r, t) = \sum_{n=1}^{\infty} A_n J_0(x) e^{-\lambda_n t}, \quad A_n = \frac{\langle g, J_0 \rangle_r}{\langle J_0, J_0 \rangle_r}, \tag{8.21}$$

where $x = \xi_n r / a$. It can be shown (see Exercise D.6) that $\langle J_0, J_0 \rangle_r = \frac{1}{2} a^2 J_1^2(\xi_n)$ and so

$$A_n = \frac{2}{a^2 J_1^2(\xi_n)} \int_0^a r g(r) J_0\left(\frac{r\xi_n}{a}\right) dr. \tag{8.22}$$

The scope for pencil and paper calculations is limited by our ability to evaluate integrals involving Bessel functions. Perhaps the simplest scenario involves the initial data

$$g(r) = \begin{cases} 1 & \text{for } 0 \leq r < b, \\ 0 & \text{for } b < r \leq a, \end{cases} \tag{8.23}$$

for some $0 < b < a$. This corresponds to an initial condition in which a smaller concentric disc of radius b is heated to a uniform temperature. Then

$$\int_0^a r g(r) J_0\left(\frac{r\xi_n}{a}\right) dr = \int_0^b r J_0\left(\frac{r\xi_n}{a}\right) dr$$

and, using Exercise D.4, we find that

$$A_n = \left(\frac{2b}{a\xi_n}\right) \frac{J_1(b\xi_n/a)}{J_1^2(\xi_n)}. \tag{8.24}$$

The corresponding solution (8.21) is shown in Fig. 8.4 when $b = 1/4$, $a = 1/2$. To facilitate a comparison with Fig. 8.3 (left), where the inner quarter of a rod was uniformly heated (here a quarter of the area of the disc is initially heated) the solution that is shown has been extended to an even function of r in the interval $-a \leq r \leq a$ for each value of t. $\diamond$

Example 8.6 (The heat equation with spherical symmetry) Determine the solution of the heat equation $u_t = \nabla^2 u$ in a spherical ball of radius a when the temperature at the boundary satisfies $u = 0$ and the initial temperature distribution, $u = g$ at $t = 0$, has spherical symmetry.

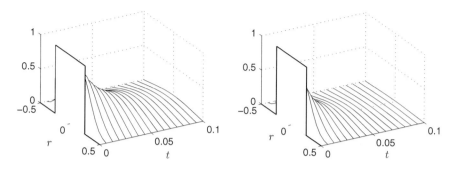

Fig. 8.4 The cross-section of the solution on any diameter of the initial boundary value problem for the heat equation in a disc (Example 8.5, *left*) and a sphere (Example 8.6, *right*) for $0 \le t \le 0.1$

In spherical polar coordinates $x = r \cos \theta \sin \phi$, $y = r \sin \theta \sin \phi$, the spherical symmetry of the initial data means that g depends only on the radial distance r from the centre of the sphere. Then u is a function of r and t only and satisfies

$$u_t = u_{rr} + \frac{2}{r} u_r, \tag{8.25}$$

which can be written in the form (8.13) with independent variables t and r and weight function $w(r) = r^2$. The appropriate boundary conditions are $u_r(0, t) = u(a, t) = 0$ (see Exercise 8.8). It can be shown (see Exercise 8.10) that the general solution may be written in the form

$$u(r, t) = \sum_{n=1}^{\infty} A_n \frac{\sin (n\pi r/a)}{r} e^{-(n\pi/a)^2 t}. \tag{8.26}$$

The coefficients A_n can be readily evaluated (see Exercise 8.11) in the specific case of an initial condition $u(r, 0) = g(r)$ where g is the step function defined by (8.23) with $a = 1/2$ and $b = 1/4$. They are given by

$$A_n = \begin{cases} \dfrac{1}{(2m-1)^2 \pi^2} (-1)^{m-1} & \text{for } n = 2m - 1, \\ \dfrac{1}{4\pi m} (-1)^{m-1} & \text{for } n = 2m. \end{cases} \tag{8.27}$$

The resulting temperature cross-section profiles are shown in Fig. 8.4. The rate of cooling in $\mathbb{R}^3$ is visibly more rapid than it is in $\mathbb{R}^2$. (Since the surface/volume ratio is larger!) $\diamond$

8.1.2 ... and to Inhomogeneous Data

Data that is associated with boundary value problems come in three flavours; (a) initial conditions, (b) source terms in the PDE, and (c) boundary values. A systematic way of solving general linear BVPs is to separate the three scenarios, so that only one of these three data sources is nonzero, and then to appeal to the principle of superposition to combine component solutions. All examples considered thus far have been of type (a). Thus, to solve general problems the question to be addressed is: can separation of variables be applied to type (b) and type (c) problems? This is the issue explored in this section.

We will consider type (b) problems first. Thus, instead of (8.13), we would like to find a solution to the following BVP,

$$
\left.
\begin{array}{ll}
u_t + \frac{1}{w(x)}\mathcal{L}u = f(x,t), & 0 < x < a \\
\mathcal{B}u(0,t) = \mathcal{B}u(a,t) = 0, & u(x,0) = 0,
\end{array}
\right\}
\tag{8.28}
$$

where, as previously, neither $\mathcal{L}$ nor $\mathcal{B}$ involves derivatives or coefficients depending on t, and where $w(x) > 0$ is a weight function. To make progress, we will need to insist that the source term $f(x,t)$ is separable, that is, it takes the form[3] $f(x,t) = F(x)G(t)$. Then, if $F(x)$ is expanded in terms of the eigenfunctions (8.14), so that

$$
F(x) = \sum_{n=1}^{\infty} B_n X_n(x), \quad B_n = \frac{\langle F, X_n \rangle_w}{\langle X_n, X_n \rangle_w},
\tag{8.29}
$$

then a solution u can be found which is itself an expansion in terms of these eigenfunctions, that is

$$
u(x,t) = \sum_{n=1}^{\infty} A_n(t) X_n(x)
\tag{8.30}
$$

but with coefficients that explicitly depend on t. Substituting these expansions into the PDE and using the relationship $\mathcal{L}X_n(x) = \lambda_n w(x) X_n(x)$ gives

$$
\sum_{n=1}^{\infty} \big(A_n'(t) + \lambda_n A_n(t) - B_n G(t) \big) X_n(x) = 0.
\tag{8.31}
$$

Moreover, the mutual orthogonality of the eigenfunctions X_n means that the coefficients A_n must satisfy the ODEs

$$
A_n'(t) + \lambda_n A_n(t) = B_n G(t), \quad n = 1, 2, \dots.
$$

Next, multiplying by the integrating factor $\exp(\lambda_n t)$ gives, on rearranging,

---

[3]The case where f is the sum of separable functions naturally arises when we consider inhomogeneous boundary conditions later in the section.

$$\left(e^{\lambda_n t} A_n(t)\right)' = B_n G(t) e^{\lambda_n t}$$

which can be integrated to give

$$A_n(t) = e^{-\lambda_n t} A_n(0) + B_n \int_0^t e^{-\lambda_n (t-s)} G(s)\, ds. \tag{8.32}$$

This, in turn, leads us to the general solution

$$u(x,t) = \sum_{n=1}^{\infty} e^{-\lambda_n t} A_n(0) X_n(x) + \sum_{n=1}^{\infty} B_n X_n(x) \int_0^t e^{-\lambda_n (t-s)} G(s)\, ds. \tag{8.33}$$

Note that the first sum on the right hand side is the general solution of the homogeneous PDE and the second sum represents a particular solution. The initial condition $u(x,0) = 0$ means that $A_n(0) = 0$ for $n = 1, 2, \ldots$ so, when solving (8.28), only the second sum remains, that is

$$u(x,t) = \sum_{n=1}^{\infty} B_n X_n(x) \int_0^t e^{-\lambda_n (t-s)} G(s)\, ds. \tag{8.34}$$

To complete the picture we need to consider type (c) problems. That is, we would like to find a solution to the following BVP,

$$\left. \begin{array}{l} u_t + \frac{1}{w(x)} \mathcal{L} u = 0, \quad 0 < x < a \\ \mathcal{B} u(0,t) = \alpha(t), \quad \mathcal{B} u(a,t) = \beta(t), \quad u(x,0) = 0, \end{array} \right\} \tag{8.35}$$

with $\mathcal{B}$ defined by (5.21), that is,

$$\left. \begin{array}{l} \mathcal{B} u(0,t) := a_0 u(0,t) - b_0 u_x(0,t), \\ \mathcal{B} u(a,t) := a_1 u(a,t) + b_1 u_x(a,t). \end{array} \right\} \tag{8.36}$$

At this point, we introduce polynomials $\phi_0(x)$ and $\phi_1(x)$ of degree ≤ 2, satisfying the following conditions

$$\mathcal{B}\phi_0(0) = 1, \quad \mathcal{B}\phi_0(a) = 0, \quad \mathcal{B}\phi_1(0) = 0, \quad \mathcal{B}\phi_1(a) = 1.$$

This is a clever move! The function $\Phi(x,t) = \phi_0(x)\alpha(t) + \phi_1(x)\beta(t)$ is forced to satisfy the same boundary conditions as the solution of (8.35); that is, $\mathcal{B}\Phi(0,t) = \alpha(t)$ and $\mathcal{B}\Phi(a,t) = \beta(t)$. Consequently the function $v(x,t) = \Phi(x,t) - u(x,t)$ satisfies the *homogeneous* conditions $\mathcal{B}v(x,t) = 0$ at $x = 0$ and $x = a$. Note that, if the boundary functions $\alpha(t)$ and $\beta(t)$ are consistent with the initial condition, then

$\alpha(0) = \beta(0) = 0$, from which we deduce that $v(x, 0) = 0$. The upshot is that $v(x, t)$ satisfies a BVP of the form (8.28), where the source term

$$f(x, t) = \alpha'(t)\phi_0(x) + \beta'(t)\phi_1(x) + \left(\frac{1}{w(x)}\mathcal{L}\phi_0(x)\right)\alpha(t) + \left(\frac{1}{w(x)}\mathcal{L}\phi_1(x)\right)\beta(t)$$

is the sum of four separable terms of the form $F(x)G(t)$. The solution can then be readily computed using (8.34). The complete solution strategy is best shown by looking at a specific example. This comes next.

Example 8.7 Solve the heat equation in the semi-infinite strip $\{(x, t) : 0 < x < a, t > 0\}$ with end conditions $u_x(0, t) = 0$, $u(a, t) = \sin 2\pi t$ and with a homogeneous initial condition $u(x, 0) = 0$.

First it may be verified that $\Phi(x, t) = \sin 2\pi t$ satisfies both boundary conditions and the initial condition, so $v(x, t) = \Phi(x, t) - u(x, t)$ satisfies

$$\left.\begin{array}{ll} v_t - v_{xx} = 2\pi\cos 2\pi t, & 0 < x < a \\ v_x(0, t) = 0, \quad v(a, t) = 0, \quad v(x, 0) = 0. \end{array}\right\} \tag{8.37}$$

The associated eigenvalue problem is

$$\left.\begin{array}{l} -X''(x) = \lambda X(x), \quad 0 < x < a, \\ X'(0) = X(a) = 0, \end{array}\right\} \tag{8.38}$$

which has eigenvalues $\lambda_n = (n - \frac{1}{2})^2\pi^2/a^2$ and corresponding eigenfunctions

$$X_n(x) = \cos\left((n - \tfrac{1}{2})\frac{\pi x}{a}\right), \quad n = 1, 2, \ldots.$$

The source term in (8.37) takes the form $F(x)G(t)$ with $F(x) = 2\pi$, $G(t) = \cos 2\pi t$. Next, since $w(x) = 1$ for the heat equation in Cartesian coordinates, we find that

$$\langle F, X_n\rangle = 2\pi\int_0^a \cos\left((n - \tfrac{1}{2})\frac{\pi x}{a}\right)dx = 2a\frac{(-1)^{n-1}}{(n - \frac{1}{2})}.$$

Thus, using (8.29),

$$F(x) = \sum_{n=1}^{\infty} B_n \cos\left((n - \tfrac{1}{2})\frac{\pi x}{a}\right), \quad B_n = 4\frac{(-1)^{n-1}}{(n - \frac{1}{2})}.$$

(The slow decay in the coefficients is because $F(x)$ does not satisfy the boundary conditions.) With $G(s) = \cos 2\pi s$, the integral in (8.34) can be evaluated to give

$$\int_0^t e^{-\lambda_n(t-s)}G(s)\,ds = \frac{1}{\lambda_n^2 + 4\pi^2}\left(\lambda_n\cos 2\pi t + 2\pi\sin 2\pi t - \lambda_n e^{-\lambda_n t}\right),$$

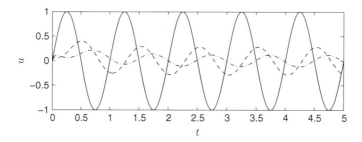

Fig. 8.5 The boundary condition $u(x, 1) = \sin 2\pi t$ (*solid line*) and the solution $u(0, t)$ of Example 8.7 for $0 \le t \le 5$ when $a = 1$ (*dashed line*) and $a = 4/3$ (*dot-dashed line*)

which completes the solution of the BVP (8.37). Since $u(x, t) = \Phi(x, t) - v(x, t)$ we deduce that

$$u(x, t) = P(x, t) + \sum_{n=1}^{\infty} \frac{\lambda_n B_n}{\lambda_n^2 + 4\pi^2} e^{-\lambda_n t} \cos\left((n - \tfrac{1}{2})\frac{\pi x}{a}\right), \qquad (8.39)$$

with

$$P(x, t) = \sin 2\pi t - \sum_{n=1}^{\infty} \frac{B_n}{\lambda_n^2 + 4\pi^2} \left(\lambda_n \cos 2\pi t + 2\pi \sin 2\pi t\right) \cos\left((n - \tfrac{1}{2})\frac{\pi x}{a}\right).$$

The exponential terms in (8.39) represent an initial transient that dies off relatively rapidly leaving $u(x, t) \approx P(x, t)$, a periodic function of t with unit period. Figure 8.5 shows the imposed boundary condition at $x = 1$ (solid line) together with the solutions at $x = 0$ for interval lengths $a = 1, 4/3$. This can be regarded as a grossly simplified model of the dissipation of heat through the wall of a building of thickness a. In a hot climate and with time measured in days, the exterior wall is heated by the sun during the day and is cooled at night. By adjusting the thickness of the wall (in this case $a = 4/3$) it can be arranged for the interior wall to be coolest when the exterior wall is at its hottest, and vice versa, thus providing a comfortable environment without the need for air-conditioning!   ◇

8.2 The Wave Equation Revisited

As discussed in Chap. 3, the wave equation is important in a variety of practical applications. It is the prototypical hyperbolic PDE. It can also be solved by separation of variables.

Example 8.8 Solve the wave equation (**pde.5**)

$$u_{tt} = c^2 u_{xx} \tag{8.40}$$

in the semi-infinite strip $S = \{(x, t) : 0 < x < 1, t > 0\}$, with initial conditions:

$$u(x, 0) = g_0(x), \quad u_t(x, 0) = g_1(x), \qquad 0 < x < 1,$$

and homogeneous Dirichlet boundary conditions, $u(0, t) = u(1, t) = 0, t > 0$.

The procedure mirrors that used to solve the heat equation in Example 8.1. Looking for a separated solution $u(x, t) = X(x)T(t)$, the boundary conditions $u(0, t) = u(1, t) = 0$ imply that $X(0) = X(1) = 0$. Substituting the ansatz for $u(x, t)$ into (8.40) gives the slightly different characterisation

$$\frac{1}{c^2} \frac{1}{T(t)} \frac{d^2 T(t)}{dt^2} = \frac{1}{X(x)} \frac{d^2 X(x)}{dx^2} = \text{constant}. \tag{8.41}$$

We follow Example 8.1 and set the separation constant to $-\lambda$. This means that X must satisfy the eigenvalue problem (8.4). Solving this problem gives

$$\lambda_n = n^2 \pi^2, \quad X_n(x) = \sin n\pi x, \qquad n = 1, 2, \dots.$$

Note that the ODE for T is second order (whereas for the heat equation it is only first order):

$$T''(t) + \lambda_n c^2 T(t) = 0.$$

With $\lambda_n = n^2 \pi^2$, the general solution to this ODE is given by

$$T(t) = A \cos cn\pi t + B \sin cn\pi t,$$

where A and B are arbitrary constants. As a consequence, there are two sequences of fundamental solutions of the wave equation (8.40), namely

$$\{\cos cn\pi t \sin n\pi x\}_{n=1}^{\infty} \text{ and } \{\sin cn\pi t \sin n\pi x\}_{n=1}^{\infty}.$$

These solutions are periodic in both space (having period 2) and time (having period $2/c$). Note the contrast with the fundamental solutions (8.5) of the heat equation, which decay exponentially with time. Any initial stimulus to the wave equation will persist for all time, whereas solutions to the heat equation invariably tend to a constant as $t \to \infty$.

A linear combination of these fundamental solutions,

$$u(x, t) = \sum_{n=1}^{\infty} (A_n \cos cn\pi t + B_n \sin cn\pi t) \sin n\pi x \tag{8.42}$$

provides a general solution of the PDE (8.40) that also satisfies homogeneous Dirichlet boundary conditions. The two sequences of arbitrary constants $\{A_n, B_n\}$ are again determined from the given initial conditions. First, the initial condition $u(x, 0) = g_0(x)$ gives the characterisation

$$g_0(x) = \sum_{n=1}^{\infty} A_n \sin n\pi x. \tag{8.43}$$

Second, differentiating (8.42) with respect to t, gives

$$\frac{\partial u}{\partial t} = \sum_{n=1}^{\infty} cn\pi \left(-A_n \sin cn\pi t + B_n \cos cn\pi t\right) \sin n\pi x,$$

so the initial condition $u_t(x, 0) = g_1(x)$ implies that

$$g_1(x) = \sum_{n=1}^{\infty} cn\pi B_n \sin n\pi x. \tag{8.44}$$

Finally, we can explicitly compute the coefficients using the inversion formula (8.10). Doing this gives

$$A_n = 2 \int_0^1 g_0(x) \sin n\pi x \, dx, \quad B_n = \frac{2}{cn\pi} \int_0^1 g_1(x) \sin n\pi x \, dx. \tag{8.45}$$

The combination (8.45) and (8.42) is the unique solution to the BVP.   ◊

Some specific problems that mirror ones solved earlier in the chapter are worked out in the following examples.

Example 8.9 Compare and contrast the solutions of the wave equation (8.40) with Dirichlet boundary conditions when the initial velocity is given by $u_t(x, 0) = 0$ and the initial displacement $u(x, 0) = g_0(x)$ is the same as that in Example 8.4; that is

(a) $g(x) = 1$ for $3/8 < x < 5/8$ and equal to zero otherwise,
(b) $g(x) = \max\{0, 1 - |8x - 4|\}$.

The coefficients $\{A_n\}$ are as given in Example 8.4, so the respective solutions to the wave equation are

$$u(x, t) = \sum_{\substack{n=1 \\ n \text{ odd}}}^{\infty} \sin \tfrac{1}{2} n\pi \left(\frac{4}{n\pi} \sin \tfrac{1}{8} \pi n\right) \cos cn\pi t \sin n\pi x, \tag{8.46a}$$

$$u(x, t) = \sum_{\substack{n=1 \\ n \text{ odd}}}^{\infty} \sin \tfrac{1}{2} n\pi \left(\frac{8}{n\pi}\right)^2 (\sin \tfrac{1}{16} \pi n)^2 \cos cn\pi t \sin n\pi x. \tag{8.46b}$$

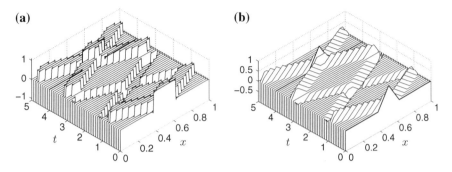

Fig. 8.6 Solutions of the wave equation for Example 8.9 for initial conditions **a** (*left*) and **b** (*right*) with $c = 1/2$. The solutions at time $t = 4$ (after one period) are highlighted by thicker lines

The two solutions are visualised in Fig. 8.6 for $0 \leq t \leq 5$ when $c = 1/2$. They both have the same geometric structure. The initial functions both have unit height and in both cases they immediately split into two smaller waves of amplitude $1/2$, one travelling to the right and the other to the left—the reason for this behaviour may be deduced from Exercise 8.15. When these waves hit the boundary, they are reflected as in a perfect elastic collision—their velocity changes from $\pm c$ to $\mp c$ while their amplitude changes from $+1/2$ to $-1/2$. The process repeats itself every time a pulse hits a boundary. The solutions in (8.46) are time periodic and both return to their initial state when t is an integer multiple of $2/c$.

Close inspection of Fig. 8.6 (right) for the initial condition (b) shows that the plotted solution remains piecewise linear as time evolves. In contrast there are small imperfections in the plotted solution corresponding to the initial condition (a). These imperfections are known as *Gibb's phenomenon* (see Example E.1). They are a direct consequenc of expressing a piecewise continuous function as a sum of trigonometric functions.[4] This severely limits the usefulness of separation of variables when solving hyperbolic equations with piecewise continuous data. The bad solution behaviour can also be identified from the series solution (8.46a) where the coefficients only decay like $1/n$. The initial condition (b) is continuous with a piecewise continuous derivative. In this case the coefficients in (8.46b) decay like $1/n^2$ which means that the series converges uniformly.   ◇

Following the procedure described in Sect. 8.1.1, separation of variables can also be used to construct series solutions to generalised hyperbolic BVPs given by

$$\left. \begin{array}{l} u_{tt} + \frac{1}{w(x)}\mathcal{L}u = 0, \quad 0 < x < a \\ \mathcal{B}u(0, t) = \mathcal{B}u(a, t) = 0, \quad u(x, 0) = g_0(x), \quad u_t(x, 0) = g_1(x). \end{array} \right\} \quad (8.47)$$

---

[4]In order to produce graphical solutions the sums in (8.46) need to be truncated to a finite number of terms. Although convergence of the series is assured—see the discussion following Definition 8.3—it is not uniform. As x approaches a discontinuity more and more terms are required in the summation (8.46a) in order to achieve any particular level of accuracy (in Fig. 8.6 we have used 100 terms).

More specifically, if the eigenvalues and corresponding eigenfunctions of (8.14) are denoted by λ_n and $X_n(x)$, respectively, then we can construct fundamental solutions $u(x, t) = X_n(x)T_n(t)$, where $T_n(t)$ satisfies the second-order ODE

$$T''(t) + \lambda_n T(t) = 0.$$

This immediately gives us the general solution to (8.47),

$$u(x, t) = \sum_{n=1}^{\infty} \left(A_n \sin \sqrt{\lambda_n}t + B_n \cos \sqrt{\lambda_n}t \right) X_n(x). \tag{8.48}$$

Two specific examples are worked out below.

Example 8.10 (*The wave equation with circular symmetry*) The wave equation with circular symmetry is given by

$$u_{tt} = c^2 \frac{1}{r} \partial_r (r u_r). \tag{8.49}$$

The general solution on the interval $0 < r < a$ with homogeneous boundary conditions $u_r(0, t) = 0$ and $u(a, t) = 0$ is given by (8.48), where the eigenvalues and eigenfunctions are those in Example 8.5. With the initial conditions $u(r, 0) = g(r)$ and $u_t(r, 0) = 0$, we have $B_n = 0$ with A_n given by (8.22). $\diamond$

Example 8.11 (*The wave equation with spherical symmetry*) The wave equation with spherical symmetry is given by

$$u_{tt} = \frac{c^2}{r^2} \partial_r (r^2 u_r). \tag{8.50}$$

The general solution on the interval $0 < r < a$ with homogeneous boundary conditions $u_r(0, t) = 0$ and $u(a, t) = 0$, is also given by (8.48) with eigenvalues and eigenfunctions as in Example 8.6. For initial conditions $u(r, 0) = g(r)$ and $u_t(r, 0) = 0$, we have $B_n = 0$, but A_n is given by (8.27) in this case. $\diamond$

Figure 8.7 gives a direct comparison of the wave equation solutions for the three different geometries. It compares solutions at time $t = 0.75$ for Example 8.10 (circular geometry), Example 8.11 (spherical geometry) as well as the solution in Cartesian geometry (see Exercise 8.17) all with the same initial conditions $g_0(x) = \max(0, 1 - (x/b)^2)$ and $g_1(x) = 0$. It can be observed that the initial condition (shown dotted) has been translated by the same amount ($ct = 0.375$) in each case, despite the marked differences in the form of the respective solutions!

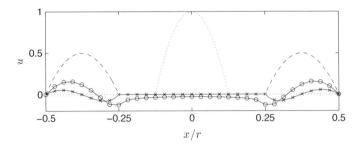

Fig. 8.7 Solutions of the wave equation on a line (- -), a disc with circular symmetry (o) and a sphere with spherical symmetry ($\times$) at $t = 0.75$ with $c = 1/2$. The initial condition is shown *dotted*

8.3 Laplace's Equation

Laplace's equation is the prototypical elliptic PDE. Since it only differs from the wave equation by changing a minus to a plus sign, the separation of variables procedure is going to be very similar to that in the previous section. The major difference lies in the imposition of boundary conditions. As discussed in Chap. 4, there is no time-like variable in Laplace's equation (both the independent variables are on an equal footing) so there are no initial conditions, only boundary conditions.

Example 8.12 (Laplace's equation) Solve Laplace's equation (**pde.3**)

$$u_{xx} + u_{yy} = 0 \tag{8.51}$$

in the rectangle $\Omega = \{(x, y) : 0 < x < a, 0 < y < b\}$ together with Dirichlet boundary conditions:

$$u(x, 0) = g_1(x), \quad u(x, b) = g_3(x), \qquad 0 < x < a,$$
$$u(a, y) = g_2(y), \quad u(0, y) = g_4(y), \qquad 0 < y < b,$$

for four given functions $\{g_i\}$. The situation is illustrated in Fig. 8.8 (left).

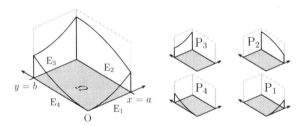

Fig. 8.8 *Left* A decomposition of Laplace's equation on a rectangular domain with Dirichlet boundary conditions. *Right* The four subproblems P_1, P_2, P_3, P_4, each having zero boundary conditions on three edges of the domain

In order to establish a relationship with an appropriate eigenvalue problem, at least two boundary conditions need to be homogeneous. If the four edges $y = 0$, $x = a$, $y = b$ and $x = 0$ of Ω are labelled E_j, $(j = 1, .., 4)$, respectively, then linearity of the BVP justifies its decomposition into four subproblems P_1, P_2, P_3, P_4. This is visualised in Fig. 8.8 (right). For subproblem P_j, the condition $u = g_j$ is imposed on edge E_j and $u = 0$ is imposed on the other three edges. The solution to the original BVP is simply the sum of the solutions of the four subproblems.

We will show how P_1 may be solved and leave the remaining subproblems to be dealt with in a similar fashion. Thus, looking at P_1, we require that the ansatz $u(x, y) = X(x)Y(y)$ satisfies the homogeneous boundary condition for this subproblem. That is, the BCs $u(0, y) = u(a, y) = 0$ for $0 < y < b$ require $X(0) = X(a) = 0$ and the BC $u(x, b) = 0$ for $0 < x < a$ requires $Y(b) = 0$.

Substituting $u(x, y) = X(x)Y(y)$ into (8.51) and rearranging gives

$$-\frac{1}{X(x)}\frac{d^2 X(x)}{dx^2} = \frac{1}{Y(y)}\frac{d^2 Y(y)}{dy^2} = \text{constant} = \lambda, \tag{8.52}$$

where the separation constant λ is selected so that X solves the eigenvalue problem

$$\left.\begin{aligned} -X''(x) = \lambda X(x), \quad 0 < x < a \\ X(0) = X(a) = 0. \end{aligned}\right\} \tag{8.53}$$

This is slightly more general than (8.4), since the domain length is a not 1. It can be immediately checked that the rescaled eigenfunctions and eigenvalues are

$$X_n(x) = \sin\left(\frac{n\pi x}{a}\right), \quad \lambda_n = \left(\frac{n\pi}{a}\right)^2. \tag{8.54}$$

It then follows from (8.52) that Y satisfies

$$Y''(y) = \lambda_n Y(y), \quad 0 < y < b,$$

with $Y(b) = 0$ (from above). This ODE has only one boundary condition, so has the family of solutions

$$Y_n(y) = C \sinh\left(\sqrt{\lambda_n}(b - y)\right),$$

where C is an arbitrary constant. Choosing C so that $Y_n(0) = 1$ leads to the infinite family of fundamental solutions

$$u_n(x, y) = \frac{\sinh\left(n\pi(1 - y/b)/a\right)}{\sinh\left(n\pi/a\right)} \sin\left(\frac{n\pi x}{a}\right), \quad n = 1, 2, \ldots \tag{8.55}$$

The first three solutions are illustrated in Fig. 8.9 for a rectangular domain. They resemble the fundamental solutions of the heat equation shown in Fig. 8.1 in that the higher the frequency, the less they penetrate the domain. This is an indication that

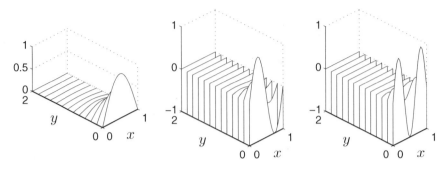

Fig. 8.9 The first three fundamental solutions of subproblem P_1 for Laplace's equation on the rectangle $0 \leq x \leq 2, 0 \leq y \leq 1$

fine detail in boundary values has minimal influence on the behaviour of solutions in the heart of the domain. By taking a linear combination of the solutions (8.55), we arrive at the general solution of P_1, namely

$$u(x, y) = \sum_{n=1}^{\infty} A_n \frac{\sinh\left(n\pi(1 - y/b)/a\right)}{\sinh\left(n\pi/a\right)} \sin\left(\frac{n\pi x}{a}\right). \qquad (8.56)$$

The coefficients $\{A_n\}$ are uniquely determined by the nonhomogeneous boundary condition $u(x, 0) = g_1(x)$ on the edge E_1. Thus, setting $y = 0$ in (8.56) gives

$$g_1(x) = \sum_{n=1}^{\infty} A_n \sin\left(\frac{n\pi x}{a}\right),$$

and using the orthogonality of the eigenfunctions in (8.54) (with respect to the standard inner product on $(0, a)$), gives the characterisation

$$A_n = \frac{2}{a} \int_0^a g_1(x) \sin\left(\frac{n\pi x}{a}\right) dx. \qquad (8.57)$$

Solutions to subproblems P_2, P_3 and P_4 can be constructed in exactly the same way (see also Exercise 8.19). $\qquad \qquad \Diamond$

Example 8.13 Solve the subproblem P_1 shown in Fig. 8.8 with $a = 1$ and $b = 1$ with the boundary condition $u(x) = g_1(x)$ on E_1, where, as in Example 8.4, we have

(a) $g_1(x) = 1$ for $3/8 < x < 5/8$ and equal to zero otherwise,
(b) $g_1(x) = \max\{0, 1 - |8x - 4|\}$.

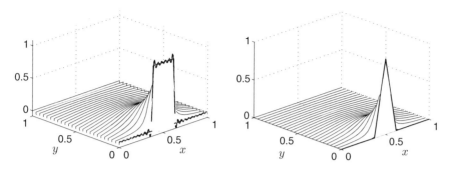

Fig. 8.10 The solution of subproblem P_1 for Laplace's equation in Example 8.13 with two choices of boundary condition on $y = 0$

When $a = 1$, the coefficients A_n may be deduced from Example 8.4. The resulting solutions, computed with the first 25 nonzero terms in their respective series are shown in Fig. 8.10. Gibb's phenomenon (see Example E.1) is clearly visible in the discontinuous boundary data on the left (where g_1 is represented by its series expansion), but no oscillations are evident in the interior of the domain showing that these imperfections decay very rapidly away from the boundary. (This contrasts with the corresponding solution to the wave equation in Fig. 8.6, where the boundary imperfections persist for all time.) Both solutions are smooth in the interior and decay rapidly with distance from the boundary spike. ◇

The discontinuity in the boundary condition in the previous example means that the presence of Gibb's phenomenon is to be expected. What is perhaps unexpected is that it can also occur when the boundary data is continuous. This is because the individual subproblems all have discontinuities at the corners of the domain unless the prescribed data is zero there (see Fig. 8.8 (right)). This is explored in the next example, along with a technique for its removal.

Example 8.14 Solve Laplace's equation in the square $\{(x, y), 0 < x, y < 1\}$ with boundary conditions $u(x, 0) = x^2$ on $y = 0$, $u(1, y) = 1 - y$ on $x = 1$ and $u = 0$ on the remaining two sides.

This is a particular instance of Example 8.12 with $a = b = 1$, $g_1(x) = x^2$, $g_2(y) = 1 - y$ with $g_3(x)=0$ and $g_4(y)=0$. Only subproblems P_1 and P_2 have nontrivial solutions.

The solution to subproblem P_1 is given by (8.56). Evaluating the integral in (8.57) (see Exercise 8.20) gives the coefficients

$$A_n = (-1)^{n-1}\frac{2}{n\pi} - (1 + (-1)^n)\frac{4}{(n\pi)^3}. \tag{8.58}$$

The solution to subproblem P_2 is worked out in Exercise 8.21. The combination of the two subproblem solutions (with $N = 40$ terms in each series expansion) is shown in

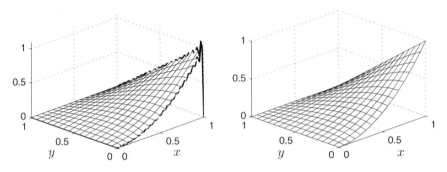

Fig. 8.11 The separation of variables solution of Example 8.14 (*left*) and the solution to the same problem after it is preprocessed to make the boundary data zero at the corners (*right*)

Fig. 8.11 (left). Spurious Gibbs oscillations are again evident. They arise because the boundary functions $g_1(x)$ and $g_2(y)$ do not satisfy the same boundary conditions as the eigenfunctions in the respective directions—the separation of variables solution is necessarily zero at the corner, whereas the boundary condition implies that $u = 1$ at the same point.

The discontinuity can be circumvented however. The idea is to subtract a suitably chosen polynomial from the solution so that $v = u - p$ is zero at the vertices of the domain. The following *bilinear* function $p(x, y)$ (also known as a *ruled surface*) does this perfectly

$$p(x, y) := \big((a - x)(b - y)u(0, 0) + x(b - y)u(a, 0)$$
$$+ xyu(a, b) + (a - x)yu(0, b)\big)/(ab), \tag{8.59}$$

since, by construction, $p(x, y)$ is equal to $u(x, y)$ at the four corners. Note that since $p(x, y)$ is bilinear it trivially satisfies Laplace's equation, which in turn means that the modified function $v = u - p$ satisfies Laplace's equation in the domain $\{(x, y), 0 < x < a, 0 < y < b\}$, with the new boundary conditions

$$v = g_k - p \text{ on edge } E_k, \quad k = 1, \dots, 4.$$

In our case, evaluating the boundary conditions at the four corner points gives $u(0, 0) = 0$, $u(1, 0) = 1$, $u(1, 1) = 0$ and $u(0, 1) = 0$, so the modification function is given by $p(x, y) = x(1 - y)$. This simple modification makes a big difference. In particular, the boundary conditions for v are much more straightforward: on the edge E_1 we have $v(x, 0) = x^2 - p(x, 0) = x(x - 1)$, but on edges E_2, E_3, E_4, we find that v is zero. Thus, to compute v, we simply have to solve subproblem P_1. Easy! (The solution is given by (8.56) with coefficients determined by (8.57).)

Having computed v, we can immediately write down the solution $(u = p + v)$:

$$u(x, y) = x(1 - y) - \sum_{\substack{n=1 \\ n \text{ odd}}}^{\infty} \frac{8}{(n\pi)^3} \frac{\sinh(n\pi(1 - y))}{\sinh n\pi} \sin n\pi x. \tag{8.60}$$

This is the solution that is plotted in Fig. 8.11 (right).[5]   ◊

Example 8.15 (*Eigenvalues of the Laplacian on a rectangle*) Determine the eigen-values of the Laplacian operator with Dirichlet boundary conditions on the rectangle $\Omega := \{(x, y) : 0 < x < a, 0 < y < b\}$. That is, find real numbers μ associated with nontrivial solutions of[6]

$$-\nabla^2 u = \mu u \quad \text{in } \Omega, \tag{8.61}$$

where $u = 0$ on the boundary of the rectangular domain Ω.

We look for eigenfunctions in the separated form $u(x, y) = X(x)Y(y)$. Since $u \neq 0$, the corresponding boundary conditions are $X(0) = X(a) = Y(0) = Y(b) = 0$. Then, substituting $u(x, y) = X(x)Y(y)$ into (8.61) and rearranging, we get

$$-\frac{X''(x)}{X(x)} = \frac{Y''(y)}{Y(y)} - \mu = \text{constant} = \lambda,$$

where the separation constant λ is chosen so that $X(x)$ satisfies the eigenvalue problem (8.53). With this choice, the function $Y(y)$ satisfies

$$\left. \begin{array}{l} -Y''(y) = (\mu - \lambda_n)Y(y), \quad 0 < y < b \\ Y(0) = Y(b) = 0, \end{array} \right\} \tag{8.62}$$

which must define another eigenvalue problem (with eigenvalue $\tilde{\mu} = \mu - \lambda_n$) if nontrivial solutions are to emerge. It has solutions given by

$$Y_m(y) = \sin\left(\frac{m\pi y}{b}\right), \quad \tilde{\mu}_m = \left(\frac{m\pi}{b}\right)^2.$$

Combining the components X_n and Y_m generates eigenfunctions of (8.61) that involve a pair of indices,

$$u_{m,n}(x, y) = X_n(x)Y_m(y) = \sin\left(\frac{n\pi x}{a}\right) \sin\left(\frac{m\pi y}{b}\right), \quad m, n = 1, 2, \ldots$$

with corresponding eigenvalues given by

---

[5]This series converges much more rapidly than that associated with (8.58). Taking only the first 5 terms in (8.60) gives an accuracy of three significant digits!

[6]We have denoted the general eigenvalue by μ rather than λ in order to avoid confusion with the solutions of the corresponding one-dimensional eigenvalue problem.

$$\mu_{m,n} = \lambda_n + \tilde{\mu}_m = \left(\frac{n\pi}{a}\right)^2 + \left(\frac{m\pi}{b}\right)^2. \tag{8.63}$$

The eigenvalues $\mu_{m,n}$ are all strictly positive but, unlike the underlying one-dimensional eigenvalue problems they do not have to be *simple*. For example, for a square domain with $a = b = 1$ we note that $\mu_{m,n} = \mu_{n,m}$ with eigenfunctions $X_n(x)Y_m(y)$ and $X_m(x)Y_n(y)$ related by symmetry (either eigenfunction may be obtained from the other by interchanging x and y).   ◇

   The eigenfunctions of the Laplacian are intrinsically useful. They are intimately connected to *resonance* in physics and they play a crucial role in numerical analysis. As shown next, they also provide a way of constructing a series solution to the Poisson equation.

Example 8.16 (Poisson's equation revisited) Solve the Poisson equation $-(u_{xx} + u_{yy}) = f(x, y)$ in a rectangle $0 < x < a, 0 < y < b$, with the homogeneous condition $u = 0$ at all points of the boundary.
   When dealing with the inhomogeneous heat equation in Sect. 8.1.2 it was assumed that f was a product of functions of one variable (or a sum of such functions). We can accomplish the same end here by expanding f in terms of the eigenfunctions of the Laplacian. That is, we determine coefficients such that

$$f(x, y) = \sum_{m=1}^{\infty} \sum_{n=1}^{\infty} A_{m,n} X_n(x)Y_m(y), \tag{8.64}$$

where $X_n(x)Y_m(y)$ are the eigenfunctions determined in the previous example. Multiplying (8.64) by the eigenfunction and integrating over the domain, and then using the orthogonality of functions $\{X_n(x)\}$ on $0 < x < a$ and of $\{Y_m(y)\}$ on $0 < y < b$ leads to the following characterisation of the generic coefficient $A_{k,\ell}$,

$$\int_0^b \int_0^a f(x, y)X_k(x)Y_\ell(y)\, dx\, dy$$

$$= \sum_{m=1}^{\infty} \sum_{n=1}^{\infty} A_{m,n} \int_0^a X_k(x)X_n(x)\, dx \int_0^b Y_m(y)Y_\ell(y)\, dy = \frac{1}{4}ab\, A_{k,\ell}. \tag{8.65}$$

Next, using the fact that $u_{m,n}(x, y) = X_n(x)Y_m(y)$ satisfies (8.61) with eigenvalues (8.63), we can rewrite the right hand side of (8.64) as follows,

$$f(x, y) = -\nabla^2 \left(\sum_{m=1}^{\infty} \sum_{n=1}^{\infty} \frac{A_{m,n}}{\mu_{m,n}} X_n(x)Y_m(y) \right).$$

Thus, since our solution $u(x, y)$ satisfies $-\nabla^2 u = f(x, y)$, we can construct a new function

$$v(x, y) = u(x, y) - \sum_{m=1}^{\infty} \sum_{n=1}^{\infty} \frac{A_{m,n}}{\mu_{m,n}} X_n(x) Y_m(y) \qquad (8.66)$$

which satisfies Laplace's equation $-\nabla^2 v = 0$ in the domain and also satisfies $v = 0$ on the boundary (since the eigenfunctions also satisfy the homogeneous boundary condition, see (8.61)). Clearly $v(x, y) = 0$ is a possible solution. At this juncture we can make use of the theory developed in Chap. 7. In particular, since we have an inverse monotone operator (see Corollary 7.7) the function $v(x, y) = 0$ must be unique (see Theorem 5.3). This, in turn, implies that the unique solution to Poisson's equation $-\nabla^2 u = f$ is given by the double sum

$$u(x, y) = \sum_{m=1}^{\infty} \sum_{n=1}^{\infty} \frac{A_{m,n}}{\mu_{m,n}} X_n(x) Y_m(y), \qquad (8.67)$$

where the coefficients $A_{m,n}$ are defined as in (8.65). Note that the coefficients $A_{m,n}/\mu_{m,n}$ tend to zero much faster that those in (8.64), signalling the fact that the solution u is much smoother than the data f.

$\Diamond$

Exercises

8.1 Work out the details that were omitted in Example 8.2.

8.2 Suppose that $g(x) = \max\{0, 1 - |8x - 4|\}$. By expressing $g(x)$ in the form $Ax + B$ on the separate subintervals $[0, 3/8), (3/8, 1/2), (1/2, 5/8)$ and $(5/8, 1]$, compute the inner products $\langle g, X_n \rangle$ when $X_n(x) = \sin n\pi x$. Hence verify that the series solution in Example 8.4 (b) has the coefficients (8.12).

8.3 Any function $g(x)$ that satisfies $g(1 - x) = g(x)$ for $x \in [0, 1]$ is symmetric about $x = 1/2$. Show that $g'(1/2) = 0$, provided that $g(x)$ is continuously differentiable in the neighbourhood of $x = 1/2$. Next, suppose that the initial condition $g(x)$ in Example 8.1 is symmetric about $x = 1/2$ and show that the function $v(s, t) = u(1 - s, t)$ satisfies the same PDE and boundary/initial conditions as $u(x, t)$. (This shows that the solution of the BVP must also be symmetric about the centre of the interval.)

8.4 Using separation of variables, construct the solution of the heat equation $u_t = \kappa u_{xx}$ for $0 < x < 1/2$, $t > 0$ with BCs $u(0, t) = 0$, $u_x(1/2, t) = 0$ for $t > 0$ and initial condition $u(x, 0) = g(x)$, where $g(x) = 0$ for $0 < x < 3/8$ and $g(x) = 1$ for $3/8 < x < 1/2$. Why is this solution identical to the series solution that is explicitly constructed in Example 8.4(a)?

8.5 This exercise builds on Exercise 8.3. Both problems in Example 8.4 have series solutions in which the coefficients of the even-numbered terms are equal to zero. Show that this will always be the case when homogeneous Dirichlet conditions $u = 0$ are imposed at the end points (still assuming that the initial condition $g(x)$ is symmetric about $x = 1/2$).

8.6 Show that the advection–diffusion equation $u_t - u_{xx} + 2u_x = 0$ with boundary conditions $u(0, t) = u(1, t) = 0$ for $t > 0$, has fundamental solutions of the form $u(x, t) = \exp(x - \alpha t) \sin \beta x$ for suitably chosen constants α and β. Hence, construct the solution when the initial condition is given by $u(x, 0) = 1$ for $0 < x < 1$.

8.7 Suppose that $u(x, y)$ is a continuously differentiable, circularly symmetric function, so that when expressed in polar coordinates $x = r \cos \theta$, $y = r \sin \theta$, it depends solely on the radius r; that is $u = f(r)$. Show that $u_x(x, y) = f'(r) \cos \theta$ and hence deduce that $f'(0) = 0$, which implies the Neumann boundary condition $u_r = 0$ when $r = 0$.

8.8 Suppose that $u(x, y, z)$ is a continuously differentiable, spherically symmetric function, so that when expressed in spherical polar coordinates $x = r \cos \theta \sin \phi$, $y = r \sin \theta \sin \phi$, $z = r \cos \phi$, it depends solely on the radius r; that is $u = f(r)$. Show that $u_r = 0$ at $r = 0$.

[Hint: use the chain rule to express the partial derivatives u_θ, u_ϕ and u_r in terms of u_x, u_y and u_z and then set $u_\theta = 0$ and $u_\phi = 0$.]

8.9 Show that the coefficients in the solution (8.21) of the heat equation with circular symmetry are given by

$$A_n = \frac{2}{\xi_n^2 J_1^2(\xi_n)} \left(\frac{4}{\omega} J_1(\omega) - 2 J_0(\omega) \right), \quad \omega = \frac{\xi_n b}{a},$$

when $g(r) = 1 - (r/b)^2$ for $0 \le r \le b$ and $g(r) = 0$ for $b < r \le a$.

[Hint: $\int x^3 J_0(x) \, dx = 2x^2 J_0(x) + x(x^2 - 4) J_1(x)$.]

8.10 By identifying the differential operator $\mathcal{L}$, show that the heat equation in a sphere (8.25) can be expressed in the generic form (8.13). Look for a separated solution $u(r, t) = R(r) T(t)$ for Example 8.6 and show that the eigenvalue problem for $R(r)$ can be written as $X''(r) + \lambda X(r) = 0$, $X(0) = X(a) = 0$, where $X(r) = r R(r)$. If $g(r)$ is the initial condition, show that the general solution to the problem is given by (8.26) with coefficients $\{A_n\}$ given by

$$A_n = \frac{2}{a} \int_0^a r g(r) \sin \frac{n \pi r}{a} \, dr, \quad n = 1, 2, \ldots$$

8.11 * This exercise builds on the previous exercise. Verify that the coefficients $\{A_n\}$ are given by (8.27) when $g(r)$ is the step function defined by (8.23).

8.12 Using the expression for the coefficients given in Exercise 8.10, show that the heat equation in a sphere with initial condition $g(r) = 1 - (r/b)^2$, $0 \le r \le b$ and $g(r) = 0$ for $b < r \le a$ has the series solution (8.26) with coefficients

$$A_n = \frac{4b^2}{a} \left(3 \frac{\sin \omega}{\omega^4} - 3 \frac{\cos \omega}{\omega^3} - \frac{\sin \omega}{\omega^2} \right), \quad \omega = \frac{n \pi b}{a}.$$

8.13 Suppose that a solid sphere of unit radius is held at a uniform temperature $u = 1$ when, at $t = 0$, it is plunged into a bath of freezing water. Use separation of variables to determine the solution u (assumed to be spherically symmetric for all time) that satisfies the heat equation (8.25), the initial condition $u(r, 0) = 1$ and the boundary condition $u(1, t) = 0$. Deduce from this solution that the temperature at the centre of the sphere is given by

$$u(0, t) = 2\sum_{n=1}^{\infty}(-1)^{n-1}e^{-n^2\pi^2 t}.$$

What is the dominating behaviour at large times?

8.14 Using separation of variables, construct the solution of the BVP

$$\left.\begin{array}{l} u_t = \kappa u_{xx} + u, \quad 0 < x < L, \, t > 0 \\ u(0, t) = u(L, t) = 0, \quad u(x, 0) = \sin(n\pi x/L). \end{array}\right\}$$

Show that there exists a value $L = L^*$, say, such that, $u(x, t) \to 0$ for $L < L^*$ as $t \to \infty$, whereas $u(x, t) \to \infty$ for $L > L^*$. This illustrates a general feature: physical diffusion becomes less and less effective in damping solutions of the heat equation as the size of the domain increases.

8.15☆ If the functions $g_0(x)$ and $g_1(x)$ are given by the series expansions (8.43) and (8.44), respectively, show that d'Alembert's solution (4.20) leads to the series solution (8.42).

8.16 Using separation of variables, construct the general solution of the wave equation $u_{tt} = u_{xx}$ for $0 < x < 1$, $t > 0$ with BCs $u(0, t) = 0$, $u_x(1, t) = 0$, for $t > 0$ and initial condition $u(x, 0) = 0$. What other condition might determine the remaining constants in this solution?

8.17 Determine the general solution of the wave equation $u_{tt} = u_{xx}$ for $-a < x < a$, $t > 0$ with BCs $u(\pm a, t) = 0$, for $t > 0$ by separation of variables. Find the specific solution satisfying the initial conditions $g_0(x) = \max(0, 1 - (x/b)^2)$ and $g_1(x) = 0$ $(b < a)$.

8.18 Use separation of variables to solve the PDE

$$u_{tt} + 2u_t = u_{xx}$$

in the strip $0 < x < 1$, $t > 0$ subject to the boundary conditions $u(0, t) = u(1, t) = 0$ for $t \geq 0$ and initial conditions $u(x, 0) = \sin \pi x$, $u_t(x, 0) = 0$ for $0 \leq x \leq 1$.

8.19 By applying interchanges such as $x \leftrightarrow a - x$, $x \leftrightarrow y$, $a \leftrightarrow b$ to the general solution (8.56) of subproblem P_1 in Example 8.12, write down the series solutions for subproblems P_2, P_3 and P_4.

8.20 * Evaluate the integral (8.57) when $a = 1$, $g_1(x) = x^2$ and show that it leads to (8.58).

8.21 * Show that the solution to Laplace's equation in the square domain $0 < x, y < 1$ with $u(1, y) = 1 - y$ and $u = 0$ on the remaining edges is given by the series

$$u(x, y) = \sum_{n=1}^{\infty} \left(\frac{2}{n\pi} \right) \frac{\sinh n\pi x}{\sinh n\pi} \sin n\pi y.$$

8.22 Using separation of variables, construct the solution of the elliptic PDE $u_{xx} + u_{yy} = cu$ (in which c is a constant) in the rectangle $0 < x < 2, 0 < y < 1$, with BCs $u(0, y) = u(2, y) = 0$ for $0 \leq y \leq 1$ and $u_y(x, 0) = u_y(x, 1) = 0$ for $0 \leq x \leq 2$. For what values of c are nontrivial fundamental solutions possible?

8.23 Using separation of variables, find fundamental solutions of the elliptic PDE $-(u_{xx} + u_{yy}) + 2u_x = 0$ in the square domain $0 < x, y < 1$ with homogeneous Dirichlet BCs $u = 0$ on the edges $y = 0$ and $y = 1$ and the Neumann BC $u_x = 0$ on $x = 1$. Can you construct an explicit solution that satisfies the additional BC $u(0, y) = y(1 - y)$?

8.24 Suppose that $u(r, \theta)$ satisfies Laplace's equation in the circular region defined by $0 \leq \theta < 2\pi, 0 \leq r \leq a$ in polar coordinates. Look for a separated solution $u(r, \theta) = R(r)\Theta(\theta)$ and construct the ODEs satisfied by $R(r)$ and $\Theta(\theta)$. Show that requiring $\Theta(\theta)$ to be a periodic function of θ with period 2π uniquely identifies the set of associated eigenvalues: $\lambda_n = n^2$, $n = 0, 1, 2, \ldots$.

Show that the ODE satisfied by $R(r)$ has solutions of the form Ar^{α} (for $n > 0$) and identify the two possible values of α. Explain why requiring solutions to be bounded means that one of the two solutions can be disregarded and write down the resulting series solution of the BVP.

Finally, if u is subject to the BC $u(a, \theta) = g(\theta)$, show that $u(0, 0) = \frac{1}{2\pi} \int_0^{2\pi} g(\theta) \, d\theta$: the mean value of a harmonic function over a circle is equal to its value at the centre.

8.25 Suppose that $u(r, \theta)$ satisfies Laplace's equation in polar coordinates in the annular region defined by $0 \leq \theta < \pi/4, 1 \leq r \leq 2$.

Look for a separated solution $u(r, \theta) = R(r)\Theta(\theta)$ and construct the ODEs satisfied by $R(r)$ and $\Theta(\theta)$ appropriate when $u = 0$ on the boundary except on $r = 2$ $(0 < \theta < \pi/4)$, where $u = g(\theta)$. Verify that the associated eigenvalues are given by $\lambda_n = 16n^2$ and show that the corresponding solutions for R are of the form Ar^{α}, where A is an arbitrary constant and α is related to n.

8.26 Repeat the previous exercise when the BCs are changed so that $u = g(r)$ on $\theta = 0$ $(1 < r < 2)$ and is otherwise equal to zero. Verify that the associated eigenvalues are given by $\lambda_n = (n\pi/\log 2)^2$. [Hint: apply the change of variable $s = \log r$ to simplify the differential equations for R.]

8.27 * Consider the eigenvalue problem $-\nabla^2 u = \lambda u$ for the negative Laplacian $-\nabla^2$ in a circle of radius a with $u = 0$ on the boundary. Using separation of variables, show that the eigenvalues are given by

$$\lambda_{m,n} := \left(\frac{\xi_{n,m}}{a}\right)^2, \quad n = 0, 1, 2, \ldots, \quad m = 1, 2, \ldots$$

with corresponding eigenfunctions given by

$$J_n\left(\frac{r}{a}\xi_{n,m}\right)\sin n\theta, \quad J_n\left(\frac{r}{a}\xi_{n,m}\right)\cos n\theta,$$

where $0 = \xi_{n,0} < \xi_{n,1} < \xi_{n,2} < \cdots$ denote the zeros of the Bessel function $J_n(x)$.

8.28 Use the solution of the previous exercise to determine the general solution of the wave equation $u_{tt} = c^2\nabla^2 u$, where c is a constant, in a circle of radius a with homogeneous Dirichlet boundary conditions.

Chapter 9
The Method of Characteristics

Abstract This chapter describes a classical technique for constructing solutions of hyperbolic PDEs. The method is applied to linear systems of PDEs and to nonlinear PDE problems. This naturally leads to a discussion of more advanced topics including shocks, Riemann problems and weak solutions.

The study of characteristics initiated in Chap. 4 is continued in this chapter. We begin by considering systems of *linear* first-order hyperbolic PDEs and we continue by looking at how the *method of characteristics* can be extended to second-order hyperbolic PDEs. The chapter concludes with a detailed study of semi-linear and quasi-linear PDEs of first order. The solutions to these problems will be shown to exhibit behaviour not found in linear problems; for example, shock waves—the spontaneous development of discontinuities—leading to a reassessment of what is meant by a solution of a PDE.

9.1 First-Order Systems of PDEs

As a starting point, we recall from Sect. 4.1 that the simplest linear PDE

$$pu_x + qu_y = f \tag{9.1}$$

in which p, q and f are functions of x, y and u, can be identified with characteristic equations

$$\frac{\mathrm{d}x}{p} = \frac{\mathrm{d}y}{q} = \frac{\mathrm{d}u}{f} \tag{9.2}$$

which have to be solved in order to determine the relationships between x, y and u. These relationships are often parameterized in terms of k, say, and the paths $(x(k), y(k))$ are referred to as the characteristics.

In this section, we shall consider *systems* of first-order PDEs with *constant coefficients*, $P\boldsymbol{u}_x + Q\boldsymbol{u}_y = \boldsymbol{0}$, where the solution $\boldsymbol{u}(x, y)$ represents a vector in $\mathbb{R}^d$ and P and Q are matrices in $\mathbb{R}^{d \times d}$. For simplicity, it is helpful to assume that at least

© Springer International Publishing Switzerland 2015
D.F. Griffiths et al., *Essential Partial Differential Equations*, Springer
Undergraduate Mathematics Series, DOI 10.1007/978-3-319-22569-2_9

one of these matrices is nonsingular (to avoid degenerate situations that are generally underdetermined) so let us suppose that Q is nonsingular. Then, premultiplying by Q^{-1} and writing $A = Q^{-1}P$ gives

$$A\boldsymbol{u}_x + \boldsymbol{u}_y = \boldsymbol{0}. \tag{9.3}$$

This generic PDE system can be *diagonalised* into d uncoupled scalar problems by making use of the eigenvalue decomposition—however, it turns out that it is the eigenvalue problem for A^T, rather than A, that is the key.[1] To see this, suppose that λ is an eigenvalue of A^T with corresponding eigenvector $\boldsymbol{v}$ so

$$A^\mathsf{T}\boldsymbol{v} = \lambda\boldsymbol{v}.$$

Since, $\boldsymbol{v}^\mathsf{T}A = \lambda\boldsymbol{v}^\mathsf{T}$, multiplying both sides of (9.3) by $\boldsymbol{v}^\mathsf{T}$ leads to

$$\boldsymbol{v}^\mathsf{T}A\boldsymbol{u}_x + \boldsymbol{v}^\mathsf{T}\boldsymbol{u}_y = \boldsymbol{0},$$

that is

$$\lambda(\boldsymbol{v}^\mathsf{T}\boldsymbol{u})_x + (\boldsymbol{v}^\mathsf{T}\boldsymbol{u})_y = \boldsymbol{0}. \tag{9.4}$$

When the eigenvalues of A^T are real and *distinct* this is a first-order hyperbolic equation having as dependent variable the scalar $\boldsymbol{v}^\mathsf{T}\boldsymbol{u}$ and the characteristic equations

$$\frac{\mathrm{d}x}{\lambda} = \frac{\mathrm{d}y}{1} = \frac{\mathrm{d}(\boldsymbol{v}^\mathsf{T}\boldsymbol{u})}{0}.$$

Thus, for each eigenvalue λ, the component of $\boldsymbol{u}$ in the direction of the corresponding eigenvector $\boldsymbol{v}$ is constant along the characteristic $x - \lambda y = \text{constant}$. For the remainder of this section we will assume that the system (9.3) is hyperbolic and we will replace y by t in order to emphasize the point that it represents a time-like variable.

Example 9.1 Solve the system $A\boldsymbol{u}_x + \boldsymbol{u}_t = \boldsymbol{0}$ for $t > 0$, $x \in \mathbb{R}$ when subject to the initial condition $\boldsymbol{u}(x, 0) = \boldsymbol{g}(x)$, $x \in \mathbb{R}$, given that A has real distinct eigenvalues, and show that the solution is bounded by the data.

In order to determine the solution at some point $\mathrm{P}(X, T)$ (with $T > 0$), the characteristics through P are followed backwards in time until they intersect the initial line $t = 0$. The process is illustrated in Fig. 9.1. When A^T has eigenvalues $\lambda_1 < \lambda_2 < \cdots < \lambda_d$ with corresponding eigenvectors $\boldsymbol{v}_1, \boldsymbol{v}_2, \ldots, \boldsymbol{v}_d$, the characteristics are solutions of the ODEs $\mathrm{d}x/\mathrm{d}t = \lambda_j$ with $x(T) = X$. They are given by

$$x - \lambda_j t = X - \lambda_j T, \quad j = 1, \ldots, d$$

---

[1]Note that matrices A and A^T have the same eigenvalues, but that the corresponding eigenvectors are different in general.

Fig. 9.1 The characteristics through a general point P(X, T) for a typical d-dimensional hyperbolic system

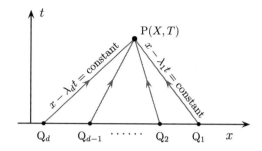

and intersect the x-axis at points $Q_j(x_j, 0)$, where $x_j = X - \lambda_j T$. Moreover, since $\boldsymbol{v}_j^T \boldsymbol{u}$ is constant along the jth characteristic, we can construct a system of d linear algebraic equations

$$\boldsymbol{v}_j^T \boldsymbol{u}(X, T) = \boldsymbol{v}_j^T \boldsymbol{g}(X - \lambda_j T), \quad j = 1, \ldots, d, \tag{9.5}$$

which can be solved to give $\boldsymbol{u}(X, T)$. This characteristic system can be written in matrix-vector form as

$$V^T \boldsymbol{u}(X, T) = \boldsymbol{f}, \tag{9.6}$$

where $V = [\boldsymbol{v}_1, \boldsymbol{v}_2, \ldots, \boldsymbol{v}_d]$ is the matrix of eigenvectors and

$$\boldsymbol{f} = [\boldsymbol{v}_1^T \boldsymbol{g}(X - \lambda_1 T), \boldsymbol{v}_2^T \boldsymbol{g}(X - \lambda_2 T), \ldots, \boldsymbol{v}_d^T \boldsymbol{g}(X - \lambda_d T)]^T.$$

Eigenvectors corresponding to distinct eigenvalues are linearly independent so the nonsingularity of V^T is assured. This, in turn, means that (9.6) has a unique solution.

In order to bound the solution in terms of the data we can use the vector and matrix norms discussed in Sect. B.1. Thus, taking the maximum vector-norm of both sides of $\boldsymbol{u} = V^{-T} \boldsymbol{f}$ leads to

$$\|\boldsymbol{u}(X, T)\|_\infty = \|V^{-T} \boldsymbol{f}\|_\infty \leq \|V^{-T}\|_\infty \|\boldsymbol{f}\|_\infty = \|V^{-1}\|_1 \|\boldsymbol{f}\|_\infty. \tag{9.7}$$

In order to bound $\|\boldsymbol{f}\|_\infty$ we assume that the initial data $\boldsymbol{g}(x)$ is bounded, that is $\|\boldsymbol{g}(x)\|_\infty \leq M_\infty$ for each $x \in \mathbb{R}$. This leads to the bound

$$|\boldsymbol{v}_j^T \boldsymbol{g}(X - \lambda_j T)| \leq \sum_{1 \leq i \leq d} |v_{ij}| \, |g_i(X - \lambda_j T)| \leq M_\infty \sum_{1 \leq i \leq d} |v_{ij}|$$

and thus

$$\|\boldsymbol{f}\|_\infty = \max_{1 \leq j \leq d} |\boldsymbol{v}_j^T \boldsymbol{g}(X - \lambda_j T)| \leq M_\infty \max_{1 \leq j \leq d} \sum_{1 \leq i \leq d} |v_{ij}| = M_\infty \|V\|_1.$$

Combining this result with (9.7) gives the final estimate

$$\|u(X, T)\|_\infty \leq \kappa_1(V)M_\infty \tag{9.8}$$

where $\kappa_1(V) = \|V\|_1\|V^{-1}\|_1$ is known as the 1-condition number. (See Exercise 9.3 for a related result in a different norm.)   ◇

To complete this section we look at an example that extends the use of characteristics to a system of three PDEs that involves a mixture of initial and boundary conditions.

Example 9.2 Suppose that the 3×3 matrix A has eigenvalues $\lambda_1 = -1$, $\lambda_2 = 1$ and $\lambda_3 = 2$ and that the corresponding eigenvectors of A^T are $v_1 = [1, 0, 1]^T$, $v_2 = [0, 1, 1]^T$ and $v_3 = [1, 1, 0]^T$. Consider the boundary value problem in which the PDE $Au_x + u_t = 0$ is to be solved in the first quadrant of the x-t plane with initial data that specifies the three components of $u = [u, v, w]^T$ on the line $t = 0$, and with two boundary conditions $u(0, t) = v(0, t)$ and $w(0, t) = 0$ specified on the positive t-axis. Use the method of characteristics to determine the solution at the point $P(X_2, T_2)$ with $T_2 < X_2 < 2T_2$.

There are three families of characteristics:

$$\Gamma_1 : \lambda_1 = -1, \quad x + t = \text{constant}, \quad v_1^T u = u + w = \text{constant},$$
$$\Gamma_2 : \lambda_2 = 1, \quad x - t = \text{constant}, \quad v_2^T u = v + w = \text{constant},$$
$$\Gamma_3 : \lambda_3 = 2, \quad x - 2t = \text{constant}, \quad v_3^T u = u + v = \text{constant}.$$

Note that, as discussed in Sect. 4.1, the fact that two of the three characteristic families (Γ_2 and Γ_3) are directed into the domain along the t-axis is the reason why *two* boundary conditions need to be specified there.

The first quadrant is divided into three regions by the two incoming characteristics ($x - t = 0$ and $x - 2t = 0$) that pass through the origin. Typical points P_1 ($X_1 > 2T_1$), P_2 ($T_2 < X_2 < 2T_2$) and P_3 ($X_3 < T_3$) in each of these regions are shown in Fig. 9.2.

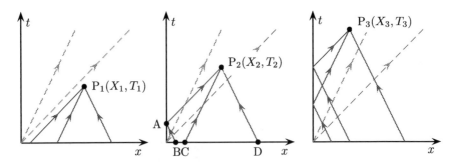

Fig. 9.2 The characteristics through points P_1, P_2, and P_3 for Example 9.2 drawn backwards in time, with reflections when they intersect the t-axis. The *dashed lines* show the characteristics $x - t = 0$ and $x - 2t = 0$ that pass through the origin

The three characteristics through P_1 intersect the x-axis directly while, for P_2, the Γ_3 characteristic $x - 2t = X_2 - 2T_2$ intersects the t-axis at $A(0, T_2 - \frac{1}{2}X_2)$ and we have to include the Γ_1 characteristic $x + t = T_2 - \frac{1}{2}X_2$ that passes through A and intersects the x-axis at B with abscissa $x = T_2 - \frac{1}{2}X_2$. The Γ_1 characteristic AP_2 is referred to as the reflection of BA in the t-axis: this makes sense because the characteristic is outgoing rather than incoming.

In order to illustrate the characteristic solution process, we shall focus exclusively on the point P_2. (For P_3, where $X_3 < T_3$, the two characteristics backwards in time both intersect the t-axis and their "reflections" have to be included.) We start where the characteristics backwards in time, or their reflections, intersect the x-axis and simply move forwards to the point where we want the solution. By inspection there are four sections:

BA: The Γ_1 characteristic along which $v_1^{\mathsf{T}} u$ is constant. Thus[2] $u(A) + w(A) = u(B) + w(B)$ and the boundary condition $w = 0$ on the t-axis gives

$$u(A) = u(B) + w(B). \qquad (9.9)$$

AP_2: The Γ_3 characteristic along which $v_3^{\mathsf{T}} u$ is constant. Thus $u(P_2) + v(P_2) = u(A) + v(A)$ and the boundary condition $u = v$ on the t-axis together with (9.9) leads to

$$u(P_2) + v(P_2) = 2\big(u(B) + w(B)\big). \qquad (9.10)$$

CP_2: The Γ_2 characteristic along which $v_2^{\mathsf{T}} u$ is constant. Thus

$$v(P_2) + w(P_2) = v(C) + w(C). \qquad (9.11)$$

DP_2: The Γ_1 characteristic along which $v_1^{\mathsf{T}} u$ is constant. Thus

$$u(P_2) + w(P_2) = u(D) + w(D). \qquad (9.12)$$

We now have three linear algebraic equations (9.10)–(9.12) with which to determine $u(P_2)$. Alternatively, the system can be written in the succinct form

$$V^{\mathsf{T}} u(P_2) = f, \qquad (9.13)$$

where $f = [u(D) + w(D), v(C) + w(C), 2(u(B) + w(B))]^{\mathsf{T}}$. Since the coefficient matrix in (9.13) is nonsingular, we deduce that the solution is uniquely specified by the given initial data on the line $t = 0$. ◊

---

[2]We use the informal notation that $u(A)$ denotes the value of u at the point A.

9.2 Second-Order Hyperbolic PDEs

Before proceeding to examine the solution of constant coefficient second-order PDEs, we show how such problems can arise from 2×2 systems of the form (9.3). First, supposing that

$$A = \begin{bmatrix} \alpha & \beta \\ \gamma & \delta \end{bmatrix}, \quad \boldsymbol{u} = \begin{bmatrix} u \\ v \end{bmatrix}, \tag{9.14}$$

the terms in (9.3) may be reorganised to read

$$(\alpha \partial_x + \partial_y) u + \beta \partial_x v = 0,$$
$$\gamma \partial_x u + (\delta \partial_x + \partial_y) v = 0.$$

Next, eliminating v by subtracting the multiple $\beta \partial_x$ of the second equation from the multiple $\delta \partial_x + \partial_y$ of the first equation gives the second-order PDE

$$\det(A)\, u_{xx} + \operatorname{tr}(A)\, u_{xy} + u_{yy} = 0, \tag{9.15}$$

where $\operatorname{tr}(A) = \alpha + \delta$ is the trace of A. Finally, using the matrix properties $\det(A) = \lambda_1 \lambda_2$ and $\operatorname{tr}(A) = \lambda_1 + \lambda_2$, where λ_1, λ_2 are the eigenvalues of A, this equation can be rewritten as

$$\lambda_1 \lambda_2\, u_{xx} + (\lambda_1 + \lambda_2)\, u_{xy} + u_{yy} = 0 \tag{9.16}$$

whose discriminant is $(\lambda_1 + \lambda_2)^2 - 4\lambda_1 \lambda_2 = (\lambda_1 - \lambda_2)^2 \geq 0$. Thus the PDE (9.16) is readily factorized[3]

$$(\lambda_1 \partial_x + \partial_y)(\lambda_2 \partial_x + \partial_y) u = 0,$$

and implies that the PDE can be directly integrated. The process is illustrated in the next example.

Example 9.3 (Quarter-plane problem) Use the method of characteristics to solve the wave equation $u_{tt} - c^2 u_{xx} = 0$ in the quarter plane $\{(x, t) : x > 0, t > 0\}$ with homogeneous initial conditions $u(x, 0) = u_t(x, 0) = 0$ $(0 \leq x < \infty)$ and with boundary condition $u(0, t) = g(t)$ for $t > 0$.

In view of the factorization

$$u_{tt} - c^2 u_{xx} = (\partial_t + c\, \partial_x)(\partial_t - c\, \partial_x) u$$

---

[3] Under the classification in Sect. 4.2 the system is hyperbolic as long as $\lambda_1 \neq \lambda_2$. This is why the eigenvalues of A were required to be distinct in Sect. 9.1.

the wave equation may be written as $(\partial_t + c\partial_x)U^+ = 0$, where $U^+ = (\partial_t - c\partial_x)u$, which has the characteristic equations

$$\frac{dt}{1} = \frac{dx}{c} = \frac{dU^+}{0},$$

or as $(\partial_t - c\partial_x)U^- = 0$, where $U^- = (\partial_t + c\partial_x)u$, which has the characteristic equations

$$\frac{dt}{1} = \frac{dx}{-c} = \frac{dU^-}{0}.$$

Together these give two families of distinct characteristics:

$$\left.\begin{matrix} \Gamma^+ : U^+ = u_t - cu_x = \text{constant} \\ \Gamma^- : U^- = u_t + cu_x = \text{constant} \end{matrix}\right\} \text{ along } \begin{cases} x - ct = \text{constant} \\ x + ct = \text{constant} \end{cases},$$

where the Γ^+ (Γ^-) characteristics transmit information in the positive (negative) x-direction.

In order to determine the solution at a point P, say, we trace the characteristics through P back in time until they meet either the x-axis, where the initial conditions can be applied, or else the boundary where the boundary condition is applied. When P has coordinates (x, t) where $x - ct < 0$, as shown in Fig. 9.3, there are three component sections:

CP: The Γ^- characteristic that intersects the x-axis at C, that is, $(x + ct, 0)$. Since $U^- = \text{constant}$ on Γ^- and $u_x = u_t = 0$ on $t = 0$, we have

$$u_t(P) + cu_x(P) = 0. \tag{9.17}$$

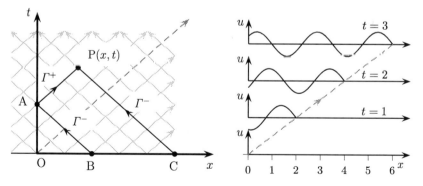

Fig. 9.3 The two families of characteristics for Example 9.3 are shown on the *left*. Characteristics through a point P(X, T) (with $X < cT$) are traced back to intersect with the positive x-axis. The solution at times $t = 1, 2, 3$, for $c = 2$ and boundary condition $u(0, t) = \sin 5t$ is shown on the *right*. The *dashed line indicates* the characteristic $x = ct$ through the origin

AP: The Γ^+ characteristic that intersects the t-axis at A, that is $(0, t - x/c)$. Since U^+ constant on Γ^+ and $u_t(0, t) = g'(t)$,

$$u_t(P) - cu_x(P) = g'(t - x/c) - cu_x(A). \tag{9.18}$$

BA: The Γ^- characteristic that intersects the x-axis at B, that is $(ct - x, 0)$. Thus $u_t(A) + cu_x(A) = u_t(B) + cu_x(B)$, that is,

$$g'(t - x/c) + cu_x(A) = 0.$$

This, combined with (9.18), gives

$$u_t(P) - cu_x(P) = 2g'(t - x/c). \tag{9.19}$$

Solving (9.17) and (9.19) (as algebraic rather than differential equations) for $u_t(P)$ and $u_x(P)$ gives

$$u_t(x, t) = g'(t - x/c), \qquad u_x(x, t) = -(1/c) g'(t - x/c).$$

Integrating the first of these gives $u(x, t) = g(t - x/c) + f(x)$, where f is an arbitrary function. When this expression for u is substituted into the second equation, we find that $f(x)$ must be constant. Finally, the boundary condition $u(0, t) = g(t)$ implies that $f = 0$.

This completes the solution in the region $x < ct$. A similar (though simpler) process for $x > ct$ leads to $u(x, t) = 0$ and so the overall solution is

$$u(x, t) = \begin{cases} g(t - x/c) & 0 \le x \le ct, \\ 0 & ct < x < \infty. \end{cases} \tag{9.20}$$

The solution at three different snapshot times is shown in Fig. 9.3 when $g(t) = \sin 5t$ with the wave speed $c = 2$. This problem models the motion of a thin inextensible string lying initially along the positive x-axis that has its left end vibrated vertically in a sinusoidal fashion.   ◊

Example 9.4 (Half-plane problem) Use the method of characteristics to solve the inhomogeneous (or *forced*) wave equation $u_{tt} - c^2 u_{xx} = f(x, t)$ in the upper half plane $\{(x, t) : x \in \mathbb{R}, t > 0\}$ with homogeneous initial conditions $u(x, 0) = u_t(x, 0) = 0$, $x \in \mathbb{R}$, for a time-independent forcing function $f(x, t) = 2(2x^2 - 1) \exp(-x^2)$.

Using the factorization and the notation in the previous example, we have the characteristic equations

$$\frac{dt}{1} = \frac{dx}{c} = \frac{dU^-}{f}.$$

We deduce that $\frac{dU^-}{dt} = f$ along the Γ^- characteristic, shown as CP in Fig. 9.3. A general point on CP has coordinates $(x + c(t-\tau), \tau)$ for $0 \le \tau \le t$. Moreover, since $U^- = u_t + cu_x = 0$ along the x-axis, we have that $U^-(C) = 0$ and then integrating along the characteristic gives

$$U^-(P) = u_t(P) + cu_x(P) = \int_0^t f(x + c(t-\tau), \tau)\, d\tau. \qquad (9.21)$$

Using exactly the same argument for the Γ^+ characteristic through P gives

$$U^+(P) = u_t(P) - cu_x(P) = \int_0^t f(x - c(t-\tau), \tau)\, d\tau. \qquad (9.22)$$

Thus subtracting (9.21) from (9.22) gives

$$u_x(x,t) = \frac{1}{2c} \int_0^t \Big(f(x + c(t-\tau), \tau) - f(x - c(t-\tau), \tau) \Big)\, d\tau. \qquad (9.23)$$

The next step is to recognize that, by Leibniz's rule for differentiating an integral with variable limits,

$$f(x + c(t-\tau), \tau) - f(x - c(t-\tau), \tau) = \partial_x \int_{x-c(t-\tau)}^{x+c(t-\tau)} f(\xi, \tau)\, d\xi.$$

This means that (9.23) can be integrated to give

$$u(x,t) = \frac{1}{2c} \int_0^t \int_{x-c(t-\tau)}^{x+c(t-\tau)} f(\xi, \tau)\, d\xi\, d\tau + A(t), \qquad (9.24)$$

where A is an arbitrary function. When this expression for u is substituted into (9.22), we find that $A(t)$ must be constant. Moreover, the initial condition $u(x,0) = 0$ implies that $A = 0$. Thus, in conclusion, the solution at a typical point $P(x,t)$ is a constant multiple $(1/2c)$ of the integral of the source term over the triangle formed by the two characteristics through P and the x-axis (shown as the shaded triangle APB in Fig. 9.4).

In the case that $f(x,t) = 2(1 - 2x^2)\exp(-x^2)$ the double integral in (9.24) can be readily evaluated (see Exercise 9.4) to give

$$u(x,t) = \frac{1}{2c^2}\Big(2e^{-x^2} - e^{-(x-ct)^2} - e^{-(x+ct)^2}\Big). \qquad (9.25)$$

The solution at three different snapshot times is shown in Fig. 9.4. It is recognisable as being a combination of two travelling waves and a standing wave.   ◊

In the preceding example the solution of the inhomogeneous wave equation at a specific point P was seen to depend only on values of the source term at points lying

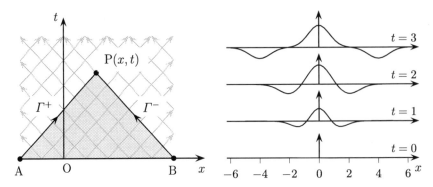

Fig. 9.4 The two families of characteristics for Example 9.4 are shown on the *left*. The solution to the forced wave equation at P is proportional to the integral of the source term $f(x, t)$ over the triangle APB. A specific solution with $c = 1$ and source term $f(x, t) = 2(1 - 2x^2)\exp(-x^2)$ at times $t = 1, 2, 3$, is shown on the *right*

in the shaded triangle APB in Fig. 9.4. D'Alembert's solution (4.20) for the unforced wave equation similarly shows that the solution at P depends only on the initial data along AB. These observations, along with the principle of superposition, allow us to conclude that the solution at P is unaffected by any data relating to points lying outside the triangle APB. This localisation property holds more widely for hyperbolic PDEs and leads to a set of generic definitions that are stated below which and which are illustrated in Fig. 9.5.

Definition 9.5 (*Causality for hyperbolic PDEs*) The *domain of dependence* of a point P is the region between the characteristics passing through P and the initial line. The intersection of the domain of dependence with the initial line is known as the *interval of dependence* of the point P. The *domain of influence* of a point C on the initial line is the set of all points that can be affected by initial data in a small interval containing C. When solving systems of first-order PDEs the triangle APB should be chosen so as to include all characteristics that pass through the point P.

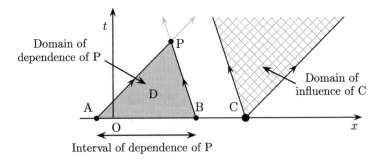

Fig. 9.5 Domains of dependence and influence for a hyperbolic PDE

9.3 First-Order Nonlinear PDES

The method of characteristics is also applicable to nonlinear PDEs. The usefulness of the method will be demonstrated by going through a sequence of examples that show increasingly complex solution behaviour.

Example 9.6 (A semi-linear PDE) Use the method of characteristics to find the general solution of the semi-linear PDE $u_t - xu_x = -u^2$ in the upper half plane $\{(x, t) : x \in \mathbb{R}, t > 0\}$. Compare and contrast the solutions satisfying the initial conditions $u(x, 0) = \sin^2 x$ and $u(x, 0) = \sin x$, $x \in \mathbb{R}$.

The characteristic equations are

$$\frac{dt}{1} = \frac{dx}{-x} = \frac{du}{-u^2},$$

and the associated ODEs are

$$\frac{dx}{dt} = -x, \quad \frac{du}{dt} = -u^2.$$

The general solution of the first of these is $x(t) = ke^{-t}$ in which different values of the constant of integration, k, distinguish different characteristics (see Fig. 9.6). The second ODE has the general solution $u(t) = 1/(A(k) + t)$ in which the constant of integration $A(k)$ depends on the particular characteristic, and therefore on k. Thus k plays the role of a parameter which, when eliminated, gives the general solution

$$u(x, t) = \frac{1}{A(xe^t) + t},$$

where the arbitrary function $A(\cdot)$ may be determined from the initial condition. The specific initial condition $u(x, 0) = g(x)$ leads to the relation $A(x) = 1/g(x)$, and so the generic solution is given by

$$u(x, t) = \frac{g(xe^t)}{1 + tg(xe^t)}. \tag{9.26}$$

The solutions associated with the specified initial conditions $g(x) = \sin^2 x$ and $g(x) = \sin x$ share the property that the distance between zeros of $g(xe^t)$ (and those of the solution u) is πe^{-t}. Thus, in either case, the frequency of zeros increases exponentially as time evolves. The character of these two solutions is fundamentally different however. For $g(x) - \sin^2 x$ we have $g(xe^t) \geq 0$, for all x and t, and consequently $u(x, t) \to 0$ for all x as $t \to \infty$. This contrasts with what happens for $g(x) = \sin x$, where $u(x, t)$ becomes infinite as $t \to 1$ for any value of x where $\sin(x/e) = -1$. This is an example of a *finite-time singularity*, which is a feature one needs to be aware of when solving semi-linear PDEs. The solution blow-up in finite

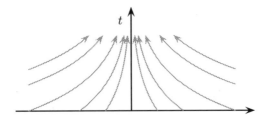

Fig. 9.6 A sketch of the characteristics associated with Example 9.6

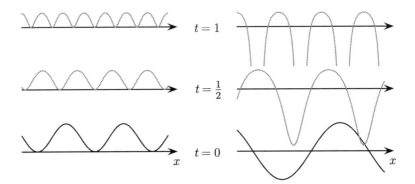

Fig. 9.7 Solutions to the semi-linear PDE in Example 9.6 at three snapshot times for initial conditions $u(x, 0) = \sin^2 x$ (*left*) and $u(x, 0) = \sin x$ (*right*). Note that the number of zeros of u increases as time evolves; in both cases

time is illustrated in Fig. 9.7 where the two solutions are plotted over the interval $-4 \le x \le 4$ for three different snapshot times.   ◇

In general, for a first-order semi-linear PDE the coefficients of the derivative terms do not involve u and so the ODEs that describe the characteristics may be solved, as in the previous example, without reference to u. This is not the case for quasi-linear PDEs, as we shall see in the next example.

Example 9.7 (A quasi-linear PDE) Use the method of characteristics to find the general solution of the quasi-linear PDE $u_t + uu_x = -u$ in the upper half plane $\{(x, t) : x \in \mathbb{R}, t > 0\}$ when subject to the general initial condition $u(x, 0) = g(x)$, $x \in \mathbb{R}$.

The characteristic equations in this case are

$$\frac{dt}{1} = \frac{dx}{u} = \frac{du}{-u}$$

and the associated ODEs are

$$\frac{dx}{dt} = u, \quad \frac{du}{dt} = -u.$$

The first of these ODEs, which describes the characteristic path $x(t)$, cannot be solved without knowledge of u. Fortunately, the second ODE can be solved explicitly in this case: $u = A(k)e^{-t}$ so that the equation for x becomes

$$\frac{dx}{dt} = A(k)e^{-t}.$$

The choice of parameterization has a significant impact on the complexity of the subsequent calculations and, with this in mind, we suppose that the characteristic path $x(t)$ intersects the x-axis at $x(0) = k$. It follows that

$$x(t) = k + A(k)(1 - e^{-t}) = k + A(k) - u.$$

This shows that values of u on the characteristic are needed in order to find the characteristic. To this end, using the initial condition $g(x) = u(x, 0) = A(k)$ and noting that $x = k$ when $t = 0$ gives $A(k) = g(k)$ leading to the solution

$$u = g(k)e^{-t}, \quad x = k + g(k)(1 - e^{-t}) \tag{9.27}$$

that is parameterized by $k \in (-\infty, \infty)$.

We now encounter the first difficulty. Only in exceptional cases (see Exercises 9.18 and 9.19) can the second of these equations be solved for k in terms of x and t and thereby allow one to express u as a function of x and t. Nevertheless, (9.27) is perfectly adequate for graphing the solution at any given time t. For example, the solutions corresponding to the initial function

$$g(x) = \frac{\sigma}{1 + \sigma^2 x^2} \tag{9.28}$$

are illustrated for two different values of σ in Fig. 9.8. But now we encounter a deeper difficulty: when $\sigma = 2$ we get a "breaking wave" and there are combinations of x and t where the solution u has three possible values. Indeed, what is shown in the figure cannot be a solution of the PDE since an intrinsic property of functions is that they be single valued. When the second of the two equations (9.27) is used to draw the corresponding characteristics (see Fig. 9.8, bottom right) it is seen that some intersect each other—this is the souce of the problem: two colliding characteristics carry different (and contradictory) information.   ◊

Our task in the remainder of the chapter is to explore how a single-valued solution may be recovered when solving PDE problems like the one above. We will make a detailed study of the inviscid Burgers' equation in order to achieve this goal.

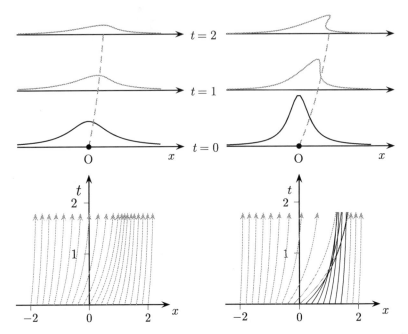

Fig. 9.8 The solution for Example 9.7 at $t = 0, 1, 2$ and the initial function $g(x) = \sigma/(1 + \sigma^2 x^2)$ with $\sigma = 1$ (*left*) and $\sigma = 2$ (*right*). The *dashed lines* show the characteristics through the origin ($k = 0$) in both cases. The lower figures show the corresponding characteristics

9.3.1 Characteristics of Burgers' Equation

The inviscid Burgers' equation (**pde.2**) is an important model in fluid dynamics. The physical basis of the model is discussed in detail in Sect. 3.2.5. The solution u typically represents the (scaled) depth of water in a simple model of river flow without rain or seepage.

The PDE problem is specified in the upper half plane $\{(x, t) : x \in \mathbb{R}, t > 0\}$,

$$u_t + u u_x = 0, \quad x \in \mathbb{R}, \ t > 0, \tag{9.29}$$

together with an initial condition $u(x, 0) = g(x)$. The solution process mirrors that in Example 9.7. The characteristic equations are

$$\frac{dt}{1} = \frac{dx}{u} = \frac{du}{0}$$

and the associated ODEs are

$$\frac{dx}{dt} = u, \quad \frac{du}{dt} = 0.$$

As in Example 9.7, the second ODE can be solved explicitly to give

$$u = A(k) \quad \text{and} \quad x = A(k)t + k.$$

We deduce that the characteristics are straight lines in the x-t plane and that the solution u is constant on a given characteristic. For the characteristic that intersects the x-axis at $x = k$ we have $u(0, x) = A(x) = g(x)$, so that the solution, in terms of k and t, becomes

$$u = g(k), \quad x = g(k)t + k. \tag{9.30}$$

It is also possible to eliminate k to obtain an implicit representation of the solution to the BVP in the form

$$u = g(x - ut). \tag{9.31}$$

Note that (9.31) is of limited utility since it usually requires root-finding software to evaluate u at any given values of t and x. The representation (9.30) that is parameterized by k and t is generally much more useful.

It is evident from (9.30) that characteristics are straight lines in the x-t plane with slope $1/g(k)$. When $g(k)$ is a decreasing function of k, faster characteristics start behind slower characteristics and they will inevitably intersect at some later time (see Fig. 9.9) after which time the solution will become multivalued because u has a different constant value on the two characteristics.

The next example is chosen to illustrate how easy it is to construct a dysfunctional solution to the inviscid Burgers' equation. It will show that it suffices to take monotonically decreasing initial data that is continuous and piecewise linear. To see why this inevitably leads to a breakdown, note that linear initial data $g(x) = \sigma(x - x_0)$, where σ is constant, leads, via (9.31) to the solution

$$u(x, t) = \sigma \frac{x - x_0}{1 + \sigma t}, \tag{9.32}$$

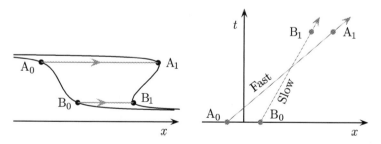

Fig. 9.9 Two points A_0 and B_0 on the initial function $g(x)$ evolve under the inviscid Burgers' equation to A_1 and B_1 at a later time (*left*). The movement is horizontal because u is constant along characteristics. The corresponding characteristics (*right*) intersect at some intermediate time

which is a linear function of x having a constant intercept on the x-axis for fixed $t > 0$. The slope of this solution at time t is $\sigma/(1 + \sigma t)$. For increasing data, $\sigma > 0$, there is no problem: the slope tends to zero as $t \to \infty$. For decreasing data the situation is different however: for $\sigma < 0$, the slope decreases without bound until $t = -1/\sigma$ when all the associated characteristics intersect simultaneously and the solution blows up! Note that beyond this critical time the slope is positive, so the dysfunctinal solution also tends to a constant in the limit $t \to \infty$.

Example 9.8 Determine the solution of Burgers' equation (9.29) when subject to the initial data

$$g(x) = \begin{cases} 3 & \text{for } x \leq -1 \\ 2 - x & \text{for } -1 \leq x \leq 1 \\ 1 & \text{for } x \geq 1. \end{cases}$$

The characteristics starting at $x \leq -1$ are all parallel and have speed 3 (and slope 1/3, see Fig. 9.10) while those starting at $x \geq 1$ are also parallel and have speed 1 (and slope 1). From (9.30) we see that characteristics starting at $x = k \, (-1 \leq k \leq 1)$ follow the paths $x = (2 - k)t + k$, for $-1 \leq k \leq 1$ (shown by the shaded region in Fig. 9.10). These all intersect simultaneously at $x = 2, t = 1$ when a discontinuity forms. The non-constant sections of the solution shown in Fig. 9.10 are given by (9.32) with $\sigma = -1$ and $x_0 = 2$. Solutions after $t = 1$ should be disregarded because they are multivalued.   ◊

The solution snapshots in Figs. 9.8 and 9.10 indicate that multivalued solutions are initiated by characteristics intersecting and by the slope of the solution becoming

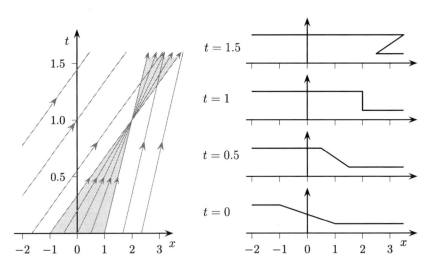

Fig. 9.10   The characteristics for Example 9.8 are shown on the *left*. The *shaded area highlights* the characteristics starting from the sloping part of the initial data on $(-1, 1)$. Snapshots of the solution are shown on the *right*

infinite at the same time. To investigate the connection between these two events, let us consider two neighbouring characteristics having paths $x = x(t, k)$ and $x = x(t, k+\delta k)$ in the x-t plane. The characteristics intersect when $x(t, k) = x(t, k+\delta k)$. If we assume that the partial derivative of x with respect to k is a continuous function then the *mean value theorem* implies that

$$x(t, k + \delta k) - x(t, k) = \delta k \cdot \partial_k x(t, k + \theta \delta k)$$

for some value $\theta \in (0, 1)$. This implies that at the specific time t (when neighbouring characteristics intersect) there is at least one point in space where $\partial_k x = 0$. Now, since the solution u at a fixed time depends on x and k, applying the chain rule $\partial_k u = \partial_x u \cdot \partial_k x$ gives $\partial_x u = \partial_k u / \partial_k x$. Thus the conclusion is that the intersection of neighbouring characteristics causes $\partial_k x$ to vanish at some point in space which, in turn, leads to the slope $\partial_x u$ becoming infinite.

Differentiating the solution (9.30) of Burgers' equation gives $\partial_k u = g'(k)$ and $\partial_k x = 1 + g'(k)t$ and so we deduce that

$$\partial_x u = \frac{g'(k)}{1 + g'(k)t}.$$

Crucially, if $g'(k)$ is negative then the slope u_x will become infinite at time t and position x satisfying

$$t = -\frac{1}{g'(k)}; \quad x = k + g(k)t = k - \frac{g(k)}{g'(k)}. \tag{9.33}$$

This represents a locus of *critical* points parameterised by k in (x, t) space.

For the piecewise linear initial data g given in Example 9.8 the solution to (9.33) is the single point $t = 1, x = 2$ (for $k \in (-1, 1)$). The locus of critical points is more interesting in our next example.

Example 9.9 Determine the solution of Burgers' equation (9.29) subject to the initial condition $u(x, 0) = g(x)$, where $g(x) = 1 - x/(1 + |x|)$, and investigate the locus of points where the slope of the solution becomes infinite.

We begin by noting that the nonlinear function $g(x)$ is continuously differentiable with $g'(k) = -1/(1 + |k|)^2 < 0$ (see Exercise 4.1). The characteristic solution is given by (9.30) and, using (9.33), we find that u_x becomes infinite when

$$t = (1 + |k|)^2 \quad \text{and} \quad x = (1 + |k|)^2 - k|k|. \tag{9.34}$$

The earliest time at which this occurs is when $k = 0$ so that $t = 1$ and $x = 1$. This point is shown by a solid dot in Fig. 9.11 (left). The two branches of the cusp emanating from this point are defined by (9.34). The rightmost branch is the envelope of characteristics corresponding to $k < 0$ while the leftmost branch is the envelope of characteristics corresponding to $k > 0$. This behaviour is fairly typical of problems in which characteristics collide, provided that they do not do so all at once.

The solution before, during and after the first collision of characteristics is shown on the right of Fig. 9.11 where the locus of points where u_x is infinite are shown by the dashed curve. ◊

The model in Example 9.9 breaks down when characteristics intersect each other, since a function cannot have three different values at the same place at the same time. One way of resolving this ambiguity is to introduce a discontinuity, or shock wave, into the solution at $x = s(t)$ so that, at any time t, $u(x, t)$ is continuous both to the left and right of this point. For a river, this would realistically approximate a flash flood—see Fig. 9.12 (right) where the shock connects the points P^+ and P^-. We investigate the principle governing the process of fitting such a shock to the solution in the next section.

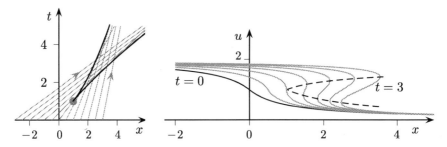

Fig. 9.11 Shown on the *left* are a selection of characteristics for Example 9.9. Those that intersect within the cusp defined by (9.34) are shown *highlighted*. Snapshots of the solutions are shown on the *right*. The *dashed curve indicates* the points where u_x is infinite

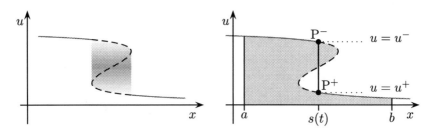

Fig. 9.12 The multivalued nature of the solution in the *shaded area* shown on the *left* is resolved by introducing a discontinuity (shock), shown on the *right*, that connects two points P^-, where $u = u^-$, to P^+, where $u = u^+$. The *shaded area* on the *right* represents the quantity $\int_a^b u \, dx$. If this quantity is to be conserved then the two lobes cut off by the shock must have equal area

9.3.2 Shock Waves

This section is where the discussion becomes more technical. To avoid difficulties associated with the interpretation of terms like u_x we need to start again from the fundamental conservation law presented in Sect. 3.1 Since we are in a one-dimensional setting the analogues of volume V and surface S are, respectively, an interval (a, b) and its endpoints, $x = a$ and $x = b$. If we suppose that the conserved quantity Q is the solution u itself then, for a given flux function $q(u)$ and source term F, the conservation law (3.2) states that

$$\frac{d}{dt} \int_a^b u(x, t) \, dx + q(u)\big|_{x=b} - q(u)\big|_{x=a} = \int_a^b F(u, x, t) \, dx. \qquad (9.35)$$

We shall use the notation $[u]$ to denote the jump in the value of u across the shock. That is, as illustrated in Fig. 9.12,

$$[u] = u^+ - u^- \quad \text{with} \quad \begin{cases} u^+ = u(s^+, t) = \lim_{\varepsilon \to 0^+} u(s(t) + \varepsilon, t) \\ u^- = u(s^-, t) = \lim_{\varepsilon \to 0^+} u(s(t) - \varepsilon, t). \end{cases}$$

On either side of the shock u varies continuously and satisfies the one-dimensional analogue of (3.3), that is

$$u_t + q_x(u) = F \iff u_t + q'(u)u_x = F. \qquad (9.36)$$

The first term on the left-hand side of (9.35) is treated by splitting the integral into two parts, one over $(a, s(t))$ and the other over $(s(t), b)$. Since u is a differentiable function on both intervals, we find, using Leibniz's rule for differentiating an integral with variable limits, that

$$\frac{d}{dt} \int_a^b u(x, t) \, dx = \frac{d}{dt} \int_a^s u(x, t) \, dx + \frac{d}{dt} \int_s^b u(x, t) \, dx$$

$$= \left(\int_a^s u_t(x, t) \, dx + u^- s'(t) \right) + \left(\int_s^b u_t(x, t) \, dx - u^+ s'(t) \right)$$

$$= \int_a^s u_t(x, t) \, dx + \int_s^b u_t(x, t) \, dx - [u]s'(t), \qquad (9.37)$$

where $s'(t) = \frac{ds(t)}{dt}$ is the speed of the shock. The flux terms in (9.35) can also be rearranged to give

$$q(u)\big|_{x=b} - q(u)\big|_{x=a} = \left(q(u)\big|_{x=b} - q(u^+) \right) + \left(q(u^-) - q(u)\big|_{x=a} \right) + [q(u)]$$

$$= \int_s^b q_x(u) \, dx + \int_a^s q_x(u) \, dx + [q(u)] \qquad (9.38)$$

which, when combined with (9.37) inside (9.35) gives

$$
0 = \frac{\mathrm{d}}{\mathrm{d}t} \int_a^b u(x,t) \, \mathrm{d}x + q(u)\big|_{x=b} - q(u)\big|_{x=a} - \int_a^b F(u,x,t) \, \mathrm{d}x
$$

$$
= \int_a^s (u_t + q_x(u) - F) \, \mathrm{d}x + \int_s^b (u_t + q_x(u) - F) \, \mathrm{d}x - [u]s'(t) + [q(u)].
$$

Since u satisfies (9.36) in both (a, s) and (s, b), it can be seen that (9.35) can only be satisfied if the shock speed is given by

$$
s'(t) = \frac{[q(u)]}{[u]}, \tag{9.39}
$$

which is known as the *Rankine–Hugoniot* condition.

An important observation is that the shock speed does not involve the source term (provided that F depends only on x, t, u and does not involve derivatives of u). The initial condition for the ODE (9.39) is provided by the point at which characteristics first intersect (as described in the previous section). Moreover, it must be solved in conjunction with the characteristic equations of (9.36), namely

$$
\frac{\mathrm{d}u}{\mathrm{d}t} = F, \qquad \frac{\mathrm{d}x}{\mathrm{d}t} = q'(u). \tag{9.40}
$$

As $u^{\pm} \to u$ the shock strength tends to zero and, in this limiting case, (9.39) gives $s'(t) \to q'(u)$ so the shock speed gets closer and closer to the characteristic speed. This has relevance in situations, such as that in Example 9.8 for times $t < 1$, where the solution is continuous but not differentiable at all points. It also confirms that such solutions satisfy the integral form (9.35) of the conservation law. This naturally leads us to a definition of admissible nonsmooth solutions.

Definition 9.10 (*Weak and classical solutions*) A piecewise continuous solution (such as one including shocks) or a solution having a continuous derivative is said to be a *weak solution* of a PDE if the underlying conservation law is satisfied at all points in time. Between successive discontinuities in the solution values (or discontinuities in derivative values) the solution satisfies the PDE itself. Solutions of nth-order PDEs that are n times continuously differentiable are called *classical* solutions.

The inherent lack of smoothness in solutions of nonlinear hyperbolic problems is one reason why the governing equations are always written as systems of first-order PDEs rather than a single higher-order PDE. Next we will see that the concept of weak solutions is what is needed to make sense of the dysfunctional solutions that were encountered in Sect. 9.3.1.

Example 9.11 (Example 9.8 revisited) Show that the introduction of a shock resolves the issue of solutions in Example 9.8 having multiple values after the intersection of the characteristics.

We can express Burgers' equation in conservation form by setting $C = 1/2$ in (3.12), this gives

$$u_t + \left(\frac{1}{2}u^2\right)_x = 0.$$

The Rankine–Hugoniot condition (9.39) then determines the shock speed:

$$s'(t) = \frac{[\frac{1}{2}u^2]}{[u]} = \frac{1}{2}\frac{(u^+)^2 - (u^-)^2}{(u^+ - u^-)} = \frac{1}{2}(u^+ + u^-),$$

which is simply the average of the solution immediately ahead and immediately behind the shock.

It was shown in Example 9.8 that characteristics first intersect at $x = 2, t = 1$. Moreover since $u^+ = 1$ and $u^- = 3$ the shock speed is constant $s'(t) = 2$. Thus, using the initial condition $s(1) = 2$, we deduce that the shock path is the straight line $s(t) = 2t$ passing through the origin. This defines a unique weak solution—the "corrected" version of Fig. 9.10 is that given Fig. 9.13. ◊

Example 9.12 (Example 9.9 revisited) Determine the shock speed for Example 9.9 and describe how a shock wave can be fitted so as to avoid multivalued solutions while conserving the mean value of the solution u at all points in time t.

The shock speed is, as in the previous example, given by $s'(t) = \frac{1}{2}(u^+ + u^-)$. Characteristics first intersect at $x = t = 1$ and so $s(1) = 1$ (the tip of the cusp in Fig. 9.11). In order to ascertain appropriate values of u^+ and u^-, we suppose that the characteristics with $k = k^-(t)$ and $k = k^+(t)$ intersect at $x = s(t)$ at time t.

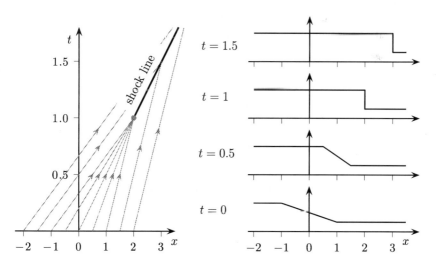

Fig. 9.13 The weak solution of the problem in Example 9.8 amended to include a shock wave at $s(t) = 2t$ for $t \geq 1$ as discussed in Example 9.11

Note that, since the characteristic through (1, 1) also passes through the origin, we know that $k^- < 0 < k^+$. The equation of the characteristics is given by (9.30) so the values of $k^\pm$ can be found in terms of t and s, by inverting the relation

$$s(t) = k^\pm + g(k^\pm)t \tag{9.41}$$

for the specific initial condition $g(x) = 1 - x/(1 + |x|)$. Moreover, since $u = g(k)$, the location of the shock can be determined by solving the initial value problem

$$s'(t) = \tfrac{1}{2}(g(k^+) + g(k^-)), \quad s(1) = 1, \tag{9.42}$$

using the values of $k^\pm$ obtained from (9.41). The details are worked out in Exercise 9.23.

We shall follow a different approach by exploiting the symmetry properties of the initial data. From (9.34), for a fixed time t, the parameter values at which u_x is infinite are given by $(1 + |k|)^2 = t$. Inverting this relation gives $k = \pm k^*$, say, where $k^* = \sqrt{t} - 1 > 0$ (recall that the shock is initiated at $t = 1$). Then, since $x = k + g(k)t$, we find that the locations $x(\pm k^*)$ at which these singularities occur satisfy

$$x(\pm k^*) - t = \pm \left(k^* - \frac{k^* t}{1 + k^*} \right),$$

so that the midpoint of these locations is simply given by

$$\tfrac{1}{2}(x(k^*) + x(-k^*)) = t.$$

This is the abscissa of a point on the $k = 0$ characteristic $x = g(0)t$ with associated solution value $u = g(0) = 1$. Next, given that the shock connecting P$^-$ to P$^+$ in Fig. 9.12 must be positioned so as to conserve the quantity $\int_a^b u \, dx$, it seems reasonable to suggest that the shock will follow the path taken by the midpoints (as visualised in Fig. 9.11). This suggests that $s(t) = t$. Moreover, since $g(k) - 1$ is an odd function of k, we note that (9.42) is immediately satisfied by setting $s(t) = t$ and choosing $k^- = -k^+$.

We can verify that the hypothesis is correct by computing the net area of the two lobes cut off by the shock (see Fig. 9.12). Switching the role of the dependent (u) and the independent (x) variables, the net area A is given by

$$A = \int_{u^+}^{u^-} (x - s) \, du.$$

Setting $x = k + ut$ and $q(u) = \frac{1}{2}u^2$ gives

$$A = \int_{u^+}^{u^-} k\, du - s \int_{u^+}^{u^-} du + t \int_{u^+}^{u^-} u\, du$$

$$= \int_{k^+}^{k^-} kg'(k)\, dk + s[u] - t[q(u)] = \int_{k^+}^{k^-} kg'(k)\, dk + [u]\left(s - ts'\right),$$

after making the change of variable $u = g(k)$ and enforcing the Rankine–Hugoniot condition. The integrand $kg'(k)$ is an odd function of k. Thus if $k^- = -k^+$ then the integral term vanishes and the net area will be zero provided

$$ts'(t) = s(t).$$

This ODE has general solution $s(t) = ct$, where c is a constant. Applying the initial condition $s(1) = 1$ then confirms that $s(t) = t$.

The associated weak solution is shown in Fig. 9.14 for selected values of t. Notice how the shock becomes more and more prominent as time evolves. ◇

Example 9.13 (Extension 3.2.4 revisited) Determine the shock speed for the traffic flow model described in Extension 3.2.4, that is the PDE

$$u_t - \frac{u_x}{u^2} = f.$$

Expressed in conservation form, the PDE is

$$u_t + (1/u)_x = f$$

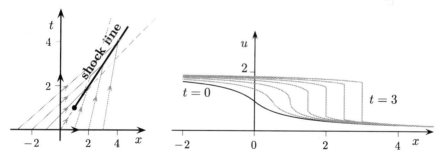

Fig. 9.14 The weak solution of the problem in Example 9.9 amended to include a shock wave at $s(t) = t$ for $t \geq 1$ as discussed in Example 9.12. The cusp (9.33) is shown by *dashed lines*

where the flux function is $q(u) = 1/u$. The condition (9.39) predicts that a shock (an abrupt change in traffic density) at $x = s(t)$ will travel at a speed

$$s'(t) = \frac{[1/u]}{[u]} = \frac{1/u^+ - 1/u^-}{u^+ - u^-} = -\frac{1}{u^+ u^-}.$$

Note that $s'(t) < 0$ since the traffic velocity u is proportional to the traffic density d (which is non-negative by definition).

As an illustration, one might visualize a high density convoy of vehicles travelling on the M6 towards Manchester at a uniform high speed of u^- when the lead vehicle meets a much lower speed restriction u^+ because of an accident. A short time later, vehicles at the head of the convoy are travelling with speed u^+, those much further back are travelling with speed u^-, while those in the vicinity of $x = s(t)$ will have to reduce their speed rapidly (due to the high density) from u^- to u^+. This region of discontent may be idealized as a shock that moves in the direction opposite to that of travel.   ◊

9.3.3 Riemann Problems and Expansion Fans

To conclude the chapter, we will broaden the discussion to include problems of the form

$$u_t + q'(u)u_x = 0 \quad \text{with} \quad u(x,0) = \begin{cases} u_L \text{ for } x < x_0 \\ u_R \text{ for } x > x_0, \end{cases} \tag{9.43}$$

where u_L and u_R are constants. Initial value problems of this type, with piecewise constant data and a single discontinuity, are known as *Riemann* problems. If there is a shock at $x = s(t)$, it will travel at the constant speed

$$s'(t) = \frac{q(u_L) - q(u_R)}{u_L - u_R}. \tag{9.44}$$

Shock waves will form when faster characteristics start behind slower characteristics. In this case characteristics must enter the shock from both sides at their natural speed, namely $\frac{dx}{dt} = q'(u)$, and this implies that the inequalities

$$q'(u_L) > \frac{q(u_L) - q(u_R)}{u_L - u_R} > q'(u_R)$$

must be satisfied for a shock to form. If $u_L > u_R$, then $q'(u)$ must be an increasing function of u so that $q''(u) > 0$ and therefore $q(u)$ must be a convex function of u (that is, a function that lies above any of its tangent lines). Alternatively, if $u_L < u_R$ then $q''(u) < 0$ and $q(u)$ must be a concave function. In the intermediate case $q''(u) = 0$ so that $q(u) = cu$, say, in which case the conservation law reduces to the (linear) one-way wave equation (**pde.1**). In this situation the Rankine–Hugoniot condition sets the shock speed to be equal to the characteristic speed: that is $s'(t) = c$. Discontinuities in this case are not strictly shocks since characteristics do not intersect and so they cannot enter a shock from both sides.

Example 9.14 Examine the Rankine–Hugoniot condition for the Riemann problem associated with the generic flux function $q(u) = u^\alpha$ $(u > 0)$.

Since $q''(u) = \alpha(\alpha - 1)u^{\alpha-2}$, the flux is concave for $0 < \alpha < 1$ and a shock will form if $u_L < u_R$ (a backward facing step). The flux is convex for $\alpha < 0$ or $\alpha > 1$ and in this case a shock will form if $u_L > u_R$ (a forward facing step). The expression (9.44) for the shock speed simplifies in some special cases:

$$
s'(t) = \begin{cases}
-1/(u_L u_R), & \alpha = -1, \ (u_L > u_R) \\
1/(u_L^{1/2} + u_R^{1/2}), & \alpha = 1/2, \ (u_L < u_R) \\
u_L + u_R, & \alpha = 2, \ (u_L > u_R).
\end{cases}
$$

The case $\alpha = 2$ demonstrates that shock waves can propagate positively with a decrease in u across the shock. The case $\alpha = -1$ also involves a decrease in u across the shock since the shock propagates negatively.   ◇

An important point is that different initial conditions can sometimes lead to the same solution at a later time, which implies that "information" is lost as characteristics enter a shock. This means that the reverse problem (equivalent to running time backwards starting with a shock in the initial data) does not have a unique solution and so is not well posed.

Let us suppose that the inequality $q'(u_L) > q'(u_R)$ is violated in the Riemann problem (9.43). In this situation a shock cannot be the correct form of solution because characteristics are leaving rather than entering the shock. The following example examines this issue within the context of Burgers' equation (the case $\alpha = 2$ in the previous example).

Example 9.15 Determine the solution of the Riemann problem (9.43) with $q(u) = \frac{1}{2}u^2$ and $u_L < u_R$ by solving a closely related problem having continuous initial data $u(x, 0) = g_\varepsilon(x)$, with a parameter $\varepsilon > 0$ so that

$$
g_\varepsilon(x) = \begin{cases}
u_L, & x < x_0, \\
u_L + [u](x - x_0)/\varepsilon, & x_0 \le x \le x_0 + \varepsilon, \\
u_R, & x_0 + \varepsilon < x,
\end{cases}
$$

where $[u] = u_R - u_L$, and then allowing $\varepsilon \to 0$.

For the original Riemann problem with $u_L < u_R$ there is a wedge-shaped region between $x = x_0 + u_L t$ and $x = x_0 + u_R t$ (shown shaded in Fig. 9.15) that is devoid of characteristics. For the modified function g_ε, where the jump from u_L to u_R is

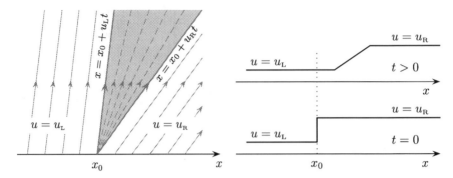

Fig. 9.15 Shown on the *left* are the characteristics for Example 9.15 with $0 < u_L < u_R$, where the expansion fan (9.47) is shown in the *shaded area* as a series of *dashed lines*. The associated weak solution at times $t = 0$ and $t > 0$ is shown on the *right*

replaced by a ramp of slope $1/\varepsilon$, there is a characteristic through each point of the plane and the solution, according to (9.30), is given by

$$u = g_\varepsilon(k), \quad x = g_\varepsilon(k)t + k.$$

Thus $u = u_L$ for $x < x_0 + u_L t$ (corresponding to $k < x_0$), $u = u_R$ for $x > x_0 + u_R t + \varepsilon$ (corresponding to $k > x_0 + \varepsilon$) and it remains to determine the solution corresponding to $x_0 < k < x_0 + \varepsilon$). Following Example 9.8 it may be shown (see Exercise 9.24) that this is given by

$$u = u_L + \frac{[u]}{\varepsilon + [u]t} (x - x_0 - u_L t). \tag{9.45}$$

Taking the limit of (9.45) when $\varepsilon \to 0$ we find that

$$u = \frac{x - x_0}{t}, \tag{9.46}$$

which is a special case of the solution (9.32) in the limit that σ, the initial slope, tends to infinity. The associated characteristics, which are given by

$$x = x_0 + ut, \quad u_L \le u \le u_R, \tag{9.47}$$

radiate outwards from the original point of discontinuity $(x_0, 0)$ and are shown as dashed lines in Fig. 9.15. This is called an *expansion fan*. The initial data and the solution at a later time are also shown in the figure.   ◇

In order to generalize this result to the Riemann problem (9.43) we note that shocks cannot be sustained if $q'(u_L) < q'(u_R)$ and we observe that the characteristic equations (9.40) with $F = 0$ imply that $u =$ constant along the characteristics

$$x = x_0 + q'(u)t, \quad q'(u_L) \le u \le q'(u_R).$$

This is a natural generalization of (9.47) and applies to a general Riemann problem where characteristics radiate away from the initial discontinuity rather than converge towards it. The solution at a point (x, t) in the associated expansion fan between $x = x_0 + q'(u_L)t$ and $x = x_0 + q'(u_R)t$ can be found by solving the equation

$$q'(u) = \frac{x - x_0}{t}. \tag{9.48}$$

In general, the underlying PDE is not satisfied at the extremities of the expansion fan (where $u = u_L$ and $u = u_R$) because the solution has discontinuous first derivatives at these points— the result is therefore an example of a weak solution. It is the only solution that is continuous for $t > 0$.

Example 9.16 Determine the form of the expansion fan for the traffic model in Example 9.13.

The flux function is $q(u) = u^{-1}$ so that $q'(u) = -u^{-2}$ and a shock cannot be formed with $0 < u_L < u_R$. For $u_L < u < u_R$ an expansion fan is obtained by solving $q'(u) = -u^{-2} = (x - x_0)/t$ giving the solution

$$u = \begin{cases} u_L, & x - x_0 < -t/u_L^2 \\ \sqrt{t/(x_0 - x)}, & -t/u_L^2 \le x - x_0 \le -t/u_R^2 \\ u_R, & -t/u_R^2 < x - x_0. \end{cases}$$

Our final example combines a shock wave and an expansion fan.

Example 9.17 Determine the solution of Burgers' equation (9.29) subject to the initial condition $u(x, 0) = g(x)$, where

$$g(x) = \begin{cases} 0, & x < 0 \text{ and } x > 1, \\ 1, & 0 \le x \le 1. \end{cases}$$

The solution development is in two phases. In the first phase the discontinuity at $x = 0$ (which has $u_L = 0$, $u_R = 1$) evolves into an expansion fan having characteristics (9.47) with $x_0 = 0$, that is $x = ut$, $0 \le u \le 1$. At the same time the discontinuity at $x = 1$ (which has $u_L = 1$, $u_R = 0$) proceeds as a shock travelling with a speed $s'(t) = \frac{1}{2}(u_L + u_R) = \frac{1}{2}$. Since the shock forms at $s(0) = 1$, the path is given by $s(t) = 1 + \frac{1}{2}t$.

The rightmost characteristic of the expansion fan ($x = t$) will catch up with the shock when $t = 2$ and $x = 2$. This triggers the second phase in which characteristics from the expansion fan meet up with the left side of the shock. Thus, at $x = s$ we have $u^- = s/t$ and $u^+ = 0$ and the Rankine–Hugoniot condition gives

$$s'(t) = \frac{s}{2t}, \quad s(2) = 2.$$

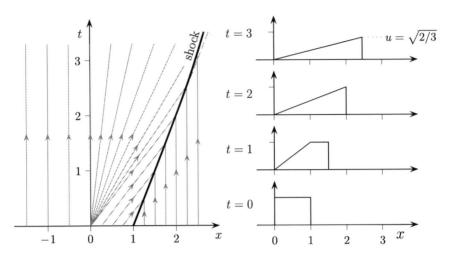

Fig. 9.16 The characteristics for Example 9.17 are shown on the *left*. A *dashed line* is included to emphasize the departure of the shock *line* from linearity. Snapshots of the associated weak solution are shown on the *right*

Solving this initial-value problem gives $s(t) = \sqrt{2t}$, and the associated PDE solution (valid for $t > 2$) is given by

$$u(x,t) = \begin{cases} 0, & x < 0 \text{ and } x > \sqrt{2t}, \\ x/t, & 0 \le x \le \sqrt{2t}. \end{cases}$$

The weak solution is shown in Fig. 9.16 for selected values of t. Note that, if t is fixed then the maximum value of u is given by $u(s,t) = \sqrt{2/t}$ which tends to zero as $t \to \infty$.   ◊

Exercises

9.1 Use the method of characteristics to find *general solutions* for the following PDEs for $u(x,t)$ both in terms of a characteristic variable and one of t or x, and in terms of t and x. In each case sketch the paths of the characteristics.

(a) $u_t + t u_x = u,$ (e) $t u_t + x u_x = x,$

(b) $t u_t - u_x = 1,$ (f) $t u_t - x u_x = t,$

(c) $u_t + x u_x = -u,$ (g) $x u_t - t u_x = xt,$

(d) $x u_t - u_x = t,$ (h) $x u_t + t u_x = -xu$

9.2 For the general solutions you have obtained from Exercise 9.1, apply the following boundary conditions, (a) to (a), (b) to (b), etc., and try to obtain unique solutions. For what values of t and x is each solution valid?

(a) $u(0, x) = \sin(x)$,   (e) $u(1, x) = x^3$

(b) $u(t, 0) = \exp(-t^2)$,   (f) $u(1, x) = 1/(1 + x^2)$

(c) $u(0, x) = x^2$,   (g) $u(0, x) = 1 + x$ for $x \geq 0$

(d) $u(t, 0) = \ln(1 + t^2)$,   (h) $u(0, x) = 1 - x$ for $x \geq 0$

9.3 Suppose that $u_1(X, T)$ satisfies the system $V^T u_1 = f_1$, which is a special case of (9.6) in which $f_1 = [v_1^T g(X - \lambda_1 T), 0, \ldots, 0]^T$. Show that $V^T u_1 = D_1 V^T g(X - \lambda_1 T)$, where D_1 is a certain $d \times d$ diagonal matrix. Deduce that $\|u_1(X, T)\|_2 \leq M_2 \kappa_2(V)$ if $\|g(x)\|_2 \leq M_2$ for $x \in \mathbb{R}$, where $\kappa_2 = \|V\|_2 \|V^{-1}\|_2$ is known as the 2-condition number of V.

Extend this result to show that $\|u(X, T)\|_2 \leq M_2 \kappa_2(V)$ when $u(X, T)$ is the solution of (9.6). Explain why this bound simplifies to $\|u(X, T)\|_2 \leq M_2$ when the matrix A of the system $A u_x + u_t = 0$ is symmetric.

9.4 ☆ Show that the eigenvalues λ of a general 2×2 matrix A satisfy the quadratic equation $\lambda^2 - \text{tr}(A)\lambda + \det(A) = 0$.

9.5 Verify that the solution of the boundary value problem in Example 9.2 at points $P(X_1, T_1)$ and $P(X_3, T_3)$ shown in Fig. 9.2 can also be written in the form (9.13), where the components of f contain linear combinations of values of $u(x, 0)$.

9.6 ☆ Show that the solution process described in Example 9.2 fails if the boundary conditions are changed to $u(0, t) = w(0, t) = 0$.

9.7 Suppose that Example 9.2 is modified so that the eigenvalues are $\lambda_1 = -2$, $\lambda_2 = -1$ and $\lambda_3 = 1$ (the corresponding eigenvectors remaining unchanged) and the boundary condition at $x = 0$ is given by $w(0, t) = 0$. Calculate the solution at the point $P(X, T)$, with $X < T$.

9.8 Consider the system of first-order PDEs (9.3) where

$$A = \begin{bmatrix} 0 & 1 & 0 \\ 0 & 0 & 1 \\ a & b & c \end{bmatrix}, \quad u = \begin{bmatrix} u \\ v \\ w \end{bmatrix}.$$

By generalizing the approach used in deriving (9.15), determine a third-order PDE satisfied by the first component u of u. How are the coefficients of this PDE related to the eigenvalues of A?

9.9 Write the first-order system $u_t + v_x = f(x, t)$ and $v_t + u_x = g(x, t)$ in matrix-vector form and hence determine the appropriate characteristic equations.

In the specific case $f(x, t) \equiv 0$ and $g(x, t) = \partial_t G(x, t)$, show that

$$u(x, t) = \tfrac{1}{2}(G(x + t, 0) - G(x - t, 0)),$$
$$v(x, t) = G(x, t) - \tfrac{1}{2}(G(x + t, 0) + G(x - t, 0))$$

is the solution satisfying the initial conditions $u(x, 0) = v(x, 0) = 0$.

9.10 Show that the PDE $2u_{xx} + 3u_{xy} + u_{yy} = 0$ is hyperbolic and determine its characteristics. Show that the general solution can be written as $u(x, y) = C(x - 2y) + D(x - y)$ for arbitrary functions C and D. Find the solution corresponding to the initial data $u(x, 0) = g_0(x)$, $u_y(x, 0) = g_1(x)$, $(x \in \mathbb{R})$.

9.11* This exercise builds on Exercise 9.10. Consider solving the PDE $2u_{xx} + 3u_{xy} + u_{yy} = 0$ in the first quadrant of the x-y plane with initial data $u(x, 0) = g_0(x)$, $u_y(x, 0) = g_1(x)$, $(x > 0)$ and boundary data $u(0, y) = f_0(y)$, $u_x(0, y) = f_1(y)$, $(y > 0)$.

Show that, for a point P having coordinates (X, Y), the solution $u(X, Y)$

(a) is identical to that in Exercise 9.10 for $X > 2Y > 0$,
(b) can be expressed as

$$u(X, Y) = 2 f_0(Y - \tfrac{1}{2}X) - f_0(Y - X) + 2 \int_{Y-X}^{Y - \frac{1}{2}X} f_1(s)\, ds$$

for $0 < X < Y$,
(c) can be expressed as
$$u(P) = u(Q) + u(R) - u(0, 0)$$

for $0 < Y < X < 2Y$, where Q, R are the points where the characteristics through P intersect the characteristics through the origin.

9.12 ☆ Verify that the solution to the problem in Example 9.4 with source term $f(x, t) = 2(1 - 2x^2) \exp(-x^2)$ is given by (9.25). [Hint: Differentiate $x \exp(-x^2)$.]

9.13 Show that the PDE $u_{tt} + u_{tx} - 2u_{xx} = t$, describing $u(x, t)$ for $x, t \in \mathbb{R} \times \mathbb{R}$, may be written as the pair of first-order PDEs $(\partial_t - \partial_x)v = t$ and $(\partial_t + 2\partial_x)u = v$ and hence determine its general solution.
   Find the solution that satisfies the initial conditions $u(0, x) \equiv 0$, and $u_t(0, x) \equiv 0$.

9.14 By using the factorization given in Example 9.3, or otherwise, show that the general solution of the PDE $u_{tt} - u_{xx} = 0$, which describes $u(x, t)$ for $x \in [0, \pi]$ and $t \in \mathbb{R}$, takes the form $u = A(t + x) + B(t - x)$.
   If we are given that $u(0, t) = 0$ for all t show that the solution takes the form $u = A(t + x) - A(t - x)$.

(a) If we are also given that $u(\pi, t) = 0$ for all t show that the function $A(\cdot)$ must be periodic with period 2π.

(b) If, instead, we were given that $u_x(\pi, t) = 0$, show that the function $A(\cdot)$ must satisfy $A'(z + 2\pi) = -A'(z)$ for any $z \in \mathbb{R}$. Deduce that the function $A(\cdot)$ must again be periodic and determine its period.

9.15 $\star$ Suppose that the variable coefficient differential operators $\mathcal{L}_1$ and $\mathcal{L}_2$ are defined by $\mathcal{L}_1 := a\partial_t + b\partial_x$ and $\mathcal{L}_2 := c\partial_t + d\partial_x$, where a, b, c and d are functions of x and t. Show that $\mathcal{L}_1\mathcal{L}_2 = \mathcal{L}$, where

$$\mathcal{L} = ac\partial_t^2 + (ad + bc)\partial_t\partial_x + bd\partial_x^2$$

if $ac_t + bc_x = 0$ and $ad_t + bd_x = 0$.

Illustrate this result by factorizing the operator associated with the PDE

$$u_{tt} + (t - 1)u_{tx} - tu_{xx} = 0.$$

If this equation holds in the upper half plane and initial conditions $u(x, 0) = g_0(x)$ and $u_t(x, 0) = g_1(x)$ are prescribed for $t = 0$, $x \in \mathbb{R}$ show, by following the characteristics of the operator $\mathcal{L}_1$, that

$$u_t - u_x = g_1(x - \tfrac{1}{2}t^2) - g_0'(x - \tfrac{1}{2}t^2).$$

Deduce that

$$u(x, t) = g_0(x + t) + \int_0^t \left(g_1(x + t - s - \tfrac{1}{2}s^2) - g_0'(x + t - s - \tfrac{1}{2}s^2) \right) ds.$$

9.16 $\star$ Use the factorization introduced in Exercise 9.15 to determine general solutions for the following PDEs:

(a) $u_{tt} + (1 + x)u_{tx} + xu_{xx} = 0$,

(b) $tu_{tt} + (x - t)u_{tx} - xu_{xx} = 0$,

(c) $xu_{tt} + (x - t)u_{tx} - tu_{xx} - 0$,

(d) $xu_{tt} + (1 + xt)u_{tx} + tu_{xx} = 0$,

(e) $xtu_{tt} + (x^2 - t^2)u_{xt} - xtu_{xx} = 0$

Show that characteristics are parallel at any points where the PDEs fail to be hyperbolic.

9.17 Show that the quasi-linear PDE $u_t + uu_x = -2u$ has the general solution $u = A(k)e^{-2t}$ and $x = k - \tfrac{1}{2}A(k)e^{-2t}$, where k and $A(k)$ are constant along any characteristic. If u satisfies the boundary condition $u(0, t) = e^{-t}$ show that

$kA(k) = 1/2$ and hence deduce that the solution for $x \geq 0$ is given by

$$u(x,t) = \frac{e^{-2t}}{x + \sqrt{x^2 + e^{-2t}}}.$$

9.18 Express the solution u in (9.27) as a function of x and t when $g(x) = \sigma x$, where σ is a constant. Deduce that (a) u remains a linear function of x for each t, (b) if $\sigma > -1$ then $u(x,t) \to 0$ for each value of x as $t \to \infty$, (c) if $\sigma < -1$, then $u(x,t) \to \infty$ for each value of x as $t \to \ln a/(1+a)$.

9.19 Suppose that $u(x,t)$ is a solution of $u_t + uu_x = -u$ as in Example 9.7. Show, by means of the change of variables $s = -x$, $v = -u$ that $u(x,t)$ is an odd function of x for each time t whenever the initial condition $u(x,0) = g(x)$ is an odd function.
Express the solution u in (9.27) as a function of x and t when $g(x) = x/(1+|x|)$.

9.20 ☆ Use (9.30) to show that the characteristics of Burgers' equation (9.29) with initial condition $g(x) = \sigma(x - x_0)$, where $\sigma < 0$ is a constant, all intersect at the same point and determine its coordinates.

9.21 ★ Consider Burgers' equation (9.29) with initial condition $u(x,0) = \max\{0, 1 - |x|\}$. Show that a shock forms at $t = 1$ and that the solution retains a triangular profile throughout its evolution. Verify that the area of the triangular profile remains constant in time.

9.22 ★ Consider the PDE $u_t + u^{1/2}u_x = 0$ with initial condition $u(0,x) = g(x)$, where

$$g(x) = \begin{cases} 4, & x \leq -1, \\ (1-x)^2, & -1 \leq x \leq 0, \\ 1, & x \geq 0. \end{cases}$$

Verify that the solution may be written as $u = g(k)$, with $x = k + \sqrt{g(k)}t$ and show that all characteristics for $-1 < k < 0$ collide at the point $x = t = 1$. Determine the shock speed and show that it follows the path $x = s(t) = (14t - 5)/9$ for $t \geq 1$.

9.23 Show that the roots of the equation (9.41) in Example 9.12 are given by

$$k^- = \tfrac{1}{2}\left(1 + s - 2t + \sqrt{(1 + s - 2t)^2 - 4(s - t)}\right),$$

$$k^+ = \tfrac{1}{2}\left(s - 1 + \sqrt{(s - 1)^2 + 4(s - t)}\right).$$

Hence verify that $s(t) = t$ is a solution of the initial value problem (9.42).

9.24 ☆ Show that the solution (9.45) satisfies Burgers' equation for $k \in (x_0, x_0 + \varepsilon)$ for the ramped initial data g_ε in Example 9.15.

9.25 ☆ Suppose that u is defined implicitly by (9.48). Show that u satisfies the conservation law $u_t + q'(u)u_x = 0$ provided that $q''(u) \neq 0$.

9.26 ☆ Show that the Riemann problem (9.43) cannot sustain a shock for the flux function $q(u) = 2u^{1/2}$ with $u_L > u_R > 0$. Determine the associated expansion fan and sketch the resulting solution at a time $t > 0$.

9.27 ☆ Determine the nature of the expansion fan for the Riemann problem (9.43) for the flux function $q(u) = \ln u$ when $u_L > u_R > 0$.

9.28 Determine the precise form of the solution to the problem in Example 9.17 for $0 < t < 2$.

9.29 Determine the solution to the problem in Example 9.17 when the initial value $g(x)$ is replaced by $1 + g(x)$.

9.30 Consider the second-order PDE

$$3u_{tt} + 10u_{xt} - 8u_{xx} = 0. \qquad (\star)$$

(a) By making an appropriate change of variables from $(x, t) \mapsto (y, s)$, show that the general solution of $(\star)$ can be written as $u(x, t) = A(3x + 2t) + B(x - 4t)$, where A and B are arbitrary functions.

(b) Determine the explicit form of the solution to $(\star)$ with the initial conditions $u(x, 0) = g_0(x)$, $u_t(x, 0) = g_1(x)$ where g_0 and g_1 are known functions, and show that

$$u(1, 3) = \tfrac{1}{14}\left(12g_0(3) + 2g_0(-11)\right) + \tfrac{3}{14}\int_{-11}^{3} g_1(s)\,\mathrm{d}s.$$

(c) Consider the pair of coupled PDEs

$$3u_t + 5u_x + 7v_x = 0, \quad 3v_t + 7u_x + 5v_x = 0. \qquad (\ddagger)$$

By expressing v_x and v_t in terms of u_x and u_t and imposing the condition $\partial_x v_t = \partial_t v_x$, eliminate v to show that u satisfying $(\ddagger)$ also satisfies $(\star)$. A suitable initial condition for u is given in part (b): can you determine the initial condition for v that ensures that $(\ddagger)$ is equivalent to $(\star)$?

(d) Express $(\ddagger)$ as a first-order system

$$\boldsymbol{u}_t + A\boldsymbol{u}_x = 0,$$

where $\boldsymbol{u} = (u, v)^{\mathrm{T}}$ and A is a 2×2 matrix and hence show that it has a (travelling wave) solution $\boldsymbol{u} = \boldsymbol{c}\phi(x - \lambda t)$, for any (differentiable) function ϕ provided that λ is an eigenvalue of A and $\boldsymbol{c}$ is the corresponding eigenvector.

(e) Show that the general solution of the first-order system is given by

$$\boldsymbol{u} = \boldsymbol{c}_1\phi(x - \lambda_1 t) + \boldsymbol{c}_2\psi(x - \lambda_2 t)$$

where ϕ and ψ are arbitrary scalar functions and determine c_i and λ_i for the specific problem ($\ddagger$). How does the solution for component u compare with the general solution given in part (a)?

(f) It was shown how ($\star$) could be deduced from ($\ddagger$) in part (c). We now consider the inverse operation. Verify that ($\star$) can be written as $\mathcal{L}^2 u = \mathcal{M}^2 u$ where $\mathcal{L} = 3\partial_t + 5\partial_x$ and $\mathcal{M} = -7\partial_x$. If the function v is related to u by the relationship $\mathcal{L}u = \mathcal{M}v$, show that $\mathcal{L}v = \mathcal{M}u$. Rearrange these relationships to give ($\ddagger$).

(g) Suppose that the substitution $\boldsymbol{u} = V\boldsymbol{v}$ is made to the first-order system in part (d) where $V = [\boldsymbol{c}_1, \boldsymbol{c}_2]$ denotes the matrix of eigenvectors of A. If we write $\boldsymbol{v} = [w, z]^{\mathsf{T}}$, use the fact that V diagonalizes A to show that

$$w_t + \lambda_1 w_x = 0 \quad \text{and} \quad z_t + \lambda_2 z_x = 0.$$

Write down the general solutions of these equations and deduce the general solution for $\boldsymbol{u}$ from $\boldsymbol{u} = V\boldsymbol{v}$. Compare the result with the explicit solutions found in parts (b) and (e). We would hope that no stone has been left unturned in this exercise!

Chapter 10
Finite Difference Methods for Elliptic PDEs

Abstract This self-contained chapter focuses on finite difference approximation of elliptic boundary value problems. A complete convergence theory is presented and two classical methods for improving the accuracy of computed solutions are described. Advanced topics include the extension to polar coordinates and a discussion of solution regularity when solving elliptic problems posed on nonconvex domains.

In this chapter the approximation methods developed in Chap. 6 will be extended to second-order linear elliptic PDEs. The basic finite difference schemes are natural extensions of the one-dimensional analogues as are the concepts of consistency, stability and convergence.

Apart from the increase in dimensionality, the main novelty is the need to deal with the shape of the domain.

An elliptic PDE takes the form $\mathcal{L}u = f$, where, in the case of two space dimensions, the differential operator is typically of the form

$$\mathcal{L}u := -(au_{xx} + 2bu_{xy} + cu_{yy}) + pu_x + qu_y + ru \qquad (10.1)$$

with coefficients a, b, c, p, q, and r representing functions of x and y (in practice they are often constant[1]). Such a PDE might hold on a domain $\Omega \subset \mathbb{R}^2$ like the one shown schematically in Fig. 10.1. Note that (10.1) is the natural generalization of the ODE operator in (5.2). The extension to three space dimensions will turn out to be perfectly straightforward.

An appropriate condition must be specified at each point of the boundary $\partial\Omega$ of the domain Ω if we are to have a uniquely defined solution. If we assume that the boundary condition is written as $\mathcal{B}u = g$, then the associated boundary value problem takes the form $\mathscr{L}u = \mathscr{F}$ and comprises

$$\mathscr{L}u = \begin{cases} \mathcal{L}u & \text{in } \Omega \\ \mathcal{B}u & \text{on } \partial\Omega \end{cases}, \quad \mathscr{F}u = \begin{cases} f & \text{in } \Omega \\ g & \text{on } \partial\Omega. \end{cases} \qquad (10.2)$$

---

[1] For example, setting $a = 1 = c$, $b = p = q = r = 0$ gives the *Poisson equation*.

© Springer International Publishing Switzerland 2015
D.F. Griffiths et al., *Essential Partial Differential Equations*, Springer
Undergraduate Mathematics Series, DOI 10.1007/978-3-319-22569-2_10

Fig. 10.1 An illustrative
domain Ω in $\mathbb{R}^2$

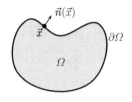

More specifically, if the boundary is decomposed into three disjoint sets $\Gamma_i, i = 1, 2, 3$ with $\partial\Omega = \Gamma_1 \cup \Gamma_2 \cup \Gamma_3$, then general boundary conditions associated with the differential operator (10.1) are given by

$$
\begin{array}{lrl}
\text{Dirichlet:} & u(\vec{x}) = g(\vec{x}), & \vec{x} \in \Gamma_1, \\
\text{Neumann:} & \vec{n}(x) \cdot \nabla u(\vec{x}) = g(\vec{x}), & \vec{x} \in \Gamma_2, \\
\text{Robin:} & \alpha(\vec{x})\, u(\vec{x}) + \beta(\vec{x})\, \vec{n}(\vec{x}) \cdot \nabla u(\vec{x}) = g(\vec{x}), & \vec{x} \in \Gamma_3,
\end{array}
$$

where α, β, g are given functions, ∇ is the standard gradient operator and $\vec{n}(x)$ denotes the unit outward normal vector to $\partial\Omega$ at the point $\vec{x}$.

10.1 A Dirichlet Problem in a Square Domain

Rather than treat the general case given by (10.1) the approximation process will be illustrated through the Dirichlet problem for the Poisson equation

$$
-\nabla^2 u = f \quad \text{in } \Omega, \tag{10.3}
$$

in which $\mathcal{L} := -\nabla^2 = -(u_{xx} + u_{yy})$ denotes the negative Laplacian in $\mathbb{R}^2$ and the boundary operator is simply $\mathcal{B}u := u$ on $\partial\Omega$. if the domain is the unit square $\Omega = (0, 1) \times (0, 1)$ then a finite difference approximation may be constructed using a grid of size $h \times h$, where $h = 1/M$, as illustrated in Fig. 10.2 (left). The internal grid points (shown as solid dots) are denoted by

Fig. 10.2 The grid Ω_h when
$M = 4$ (*left*) and the 5-point
finite difference stencil for
the Poisson equation (*right*)
applied at the grid point
indicated by o involves the
unknowns at the grid points
marked by •

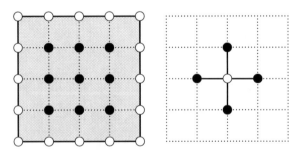

$$\Omega_h = \{(x_\ell, y_m) : x_\ell = \ell h, \ y_m = mh; \ \ell, m = 1, 2, \ldots, M - 1\}.$$

The grid points on the boundary (shown by circles) are likewise denoted by $\partial\Omega_h$, and the entire grid by $\overline{\Omega}_h = \Omega_h \cup \partial\Omega_h$. The finite difference solution can then be identified with the grid function U, whose value $U_{\ell,m}$ at a typical point $(\ell h, mh) \in \Omega_h$ approximates the exact solution $u_{\ell,m} := u(\ell h, mh)$ at that point. The values of U at the boundary nodes are known from the prescribed boundary condition whereas the values of U at the internal grid points are found by solving a system of finite difference equations.

To construct such a system, the partial derivatives of u at a generic internal grid point $(\ell h, mh)$ are usually approximated by second-order centered differences (see Sect. 6.1), so that, looking in the x-direction,

$$u_{xx}|_{\ell,m} = h^{-2}\left(u_{\ell-1,m} - 2u_{\ell,m} + u_{\ell+1,m}\right) + \mathcal{O}(h^2).$$

Introducing the second-order difference operator δ_x^2, defined by

$$\delta_x^2 u_{\ell,m} := u_{\ell-1,m} - 2u_{\ell,m} + u_{\ell+1,m}, \tag{10.4}$$

we have the more compact expression

$$u_{xx}|_{\ell,m} = h^{-2}\delta_x^2 u_{\ell,m} + \mathcal{O}(h^2).$$

Likewise, looking in the y-direction and introducing

$$\delta_y^2 u_{\ell,m} := u_{\ell,m-1} - 2u_{\ell,m} + u_{\ell,m+1} \tag{10.5}$$

leads to the overall approximation

$$-\nabla^2 u\Big|_{\ell,m} = -h^{-2}\left(\delta_x^2 u_{\ell,m} + \delta_y^2 u_{\ell,m}\right) + \mathcal{O}(h^2).$$

The finite difference equations that approximate the PDE are generated by discarding the remainder terms and replacing the exact solution $u_{\ell,m}$ by the grid approximation $U_{\ell,m}$ at each grid point. This gives the algebraic equation

$$-h^{-2}(\delta_x^2 + \delta_y^2)U_{\ell,m} = f_{\ell,m},$$

which can also be written as

$$4U_{\ell,m} - U_{\ell+1,m} - U_{\ell,m+1} - U_{\ell-1,m} - U_{\ell,m-1} = h^2 f_{\ell,m} \tag{10.6}$$

for each $(x_\ell, y_m) \in \Omega_h$. The values of the grid function are known on the boundary $\partial\Omega_h$, so (10.6) gives $(M - 1)^2$ linear equations to determine the unknown grid function values on Ω_h. The 5-point stencil in Fig. 10.2 (right) is used to indicate the values of U connected by equation (10.6); the grid point (x_ℓ, y_m) (the target point) is shown by an open circle (∘). In the special case of Laplace's equation ($f \equiv 0$) we have

$$U_{\ell,m} = \tfrac{1}{4}\left(U_{\ell+1,m} + U_{\ell,m+1} + U_{\ell-1,m} + U_{\ell,m-1}\right),$$

and the numerical solution at any internal grid point is simply the mean of its values at its four nearest neighbours taken horizontally and vertically.

10.1.1 Linear Algebra Aspects

By defining the discrete/grid operators so that

$$\left.\begin{aligned}
\mathcal{L}_h U_{\ell,m} &:= -h^{-2}\left(\delta_x^2 + \delta_y^2\right) U_{\ell,m} \\
&= h^{-2}\left(4U_{\ell,m} - U_{\ell+1,m} - U_{\ell,m+1} - U_{\ell-1,m} - U_{\ell,m-1}\right) \\
\mathcal{B}_h U_{\ell,m} &:= U_{\ell,m},
\end{aligned}\right\} \tag{10.7}$$

the $(M+1)^2$ finite difference equations approximating the problem (10.2) may be written in the compact form $\mathscr{L}_h U = \mathscr{F}_h$ with

$$\mathscr{L}_h U = \begin{cases} \mathcal{L}_h U & \text{in } \Omega_h \\ \mathcal{B}_h U & \text{on } \partial\Omega_h \end{cases}, \qquad \mathscr{F}_h = \begin{cases} f & \text{in } \Omega_h \\ g & \text{on } \partial\Omega_h \end{cases}. \tag{10.8}$$

A few items need to be addressed if these finite difference equations are to be expressed as a standard matrix–vector equation.

(i) The unknowns $\{U_{\ell,m}\}$ are indexed by two subscripts, so a mechanism is required to organize these into a column vector whose components have a single index. A natural approach is to lay out the interior grid values as a square matrix and then to collect the entries columnwise so that

$$U = [\,u_1\ u_2\ \dots\ u_{M-1}],$$

where $u_\ell \in \mathbb{R}^{M-1}$ is the vector that indexes the unknowns in the ℓth column of the grid

$$u_\ell = [U_{\ell,1}, U_{\ell,2}, \dots, U_{\ell,M-1}]^\mathsf{T}.$$

A single vector containing the $(M-1)^2$ unknowns can then be constructed by stacking the column vectors on top of each other[2]

---

[2] In linear algebra this operation is commonly written as $u = \mathrm{vec}(U)$.

$$u = \begin{bmatrix} u_1 \\ u_2 \\ \vdots \\ u_{M-1} \end{bmatrix}.$$

(ii) The matrix of coefficients representing the left side of (10.6) will contain at most five nonzero entries in each row. Moreover, if the equations are ordered in the same way as the unknowns then the coefficient 4 will be on the diagonal. When (10.6) is applied at a grid point adjacent to the boundary, one or more of the neighbouring values of $U_{\ell,m}$ will be known from the Dirichlet boundary condition and the corresponding term(s) moved to the right side of the relevant equation. For instance, when $\ell = m = 1$, we have

$$4U_{1,1} - U_{1,2} - U_{2,1} = h^2 f_{1,1} + g_{1,0} + g_{0,1},$$

so that the first row of the matrix will contain only three nonzero entries.

(iii) The focus now switches to the equations on the ℓth column of the grid. The unknowns in this column are linked only to unknowns on the two neighbouring columns. Thus the difference equations can be expressed as

$$-u_{\ell-1} + Du_\ell - u_{\ell+1} = h^2 f_\ell + g_\ell,$$

where D is the $(M-1) \times (M-1)$ tridiagonal matrix

$$D = \begin{bmatrix} 4 & -1 & & & \\ -1 & 4 & -1 & & \\ & \ddots & \ddots & \ddots & \\ & & -1 & 4 & -1 \\ & & & -1 & 4 \end{bmatrix} \quad \text{and } g_\ell = \begin{bmatrix} g_{\ell,0} \\ 0 \\ \vdots \\ 0 \\ g_{\ell,M} \end{bmatrix}$$

arises from the condition on the top and bottom boundaries. Also, when $\ell = 1$ or $M - 1$, boundary conditions from the vertical edges are applied so that $u_0 = g_0$ and $u_M = g_M$, where

$$g_0 = \begin{bmatrix} g_{0,1} \\ g_{0,2} \\ \vdots \\ g_{0,M-1} \end{bmatrix}, \quad g_M = \begin{bmatrix} g_{M,1} \\ g_{M,2} \\ \vdots \\ g_{M,M-1} \end{bmatrix}.$$

The difference equations can now be expressed as a system of the form

$$Au = f, \tag{10.9}$$

where $\boldsymbol{u}, \boldsymbol{f} \in \mathbb{R}^{(M-1)^2}$. The matrix A is a $(M-1)^2 \times (M-1)^2$ matrix with a characteristic *block tridiagonal* structure

$$A = \frac{1}{h^2} \begin{bmatrix} D & -I & & & \\ -I & D & -I & & \\ & \ddots & \ddots & \ddots & \\ & & -I & D & -I \\ & & & -I & D \end{bmatrix}, \tag{10.10}$$

where I is the $(M-1) \times (M-1)$ identity matrix. The associated right-hand side vector is given by

$$\boldsymbol{f} = \begin{bmatrix} \boldsymbol{f}_1 \\ \boldsymbol{f}_2 \\ \vdots \\ \boldsymbol{f}_{M-2} \\ \boldsymbol{f}_{M-1} \end{bmatrix} + \frac{1}{h^2} \begin{bmatrix} \boldsymbol{g}_0 + \boldsymbol{g}_1 \\ \boldsymbol{g}_2 \\ \vdots \\ \boldsymbol{g}_{M-2} \\ \boldsymbol{g}_{M-1} + \boldsymbol{g}_M \end{bmatrix}.$$

The coefficient matrix A has around M^4 entries but fewer than $5M^2$ nonzeros: these have a very regular structure, as may be seen in Fig. 10.3 when $M = 8$. This means that the matrix A should be stored as a *sparse* matrix, where only the nonzero entries are stored, along with the indices that specify their location.

A common method of solving the system $A\boldsymbol{u} = \boldsymbol{f}$ is to first factorise A into the product LR, where L is a unit lower triangular matrix (that is $L_{ij} = 0$ for $j > i$ and $L_{ii} = 1$) and R is an upper triangular matrix (that is $R_{ij} = 0$ for $j < i$) (see, for example, Golub and Van Loan [6, Sect. 4.3] where it is known as the L–U decomposition). The system $A\boldsymbol{u} = \boldsymbol{f}$ is then replaced by the pair of equations

$$L\boldsymbol{v} = \boldsymbol{f}, \qquad R\boldsymbol{u} = \boldsymbol{v}, \tag{10.11}$$

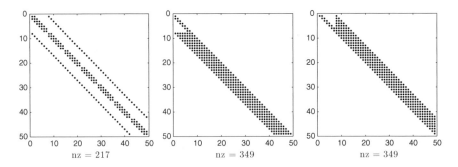

Fig. 10.3 Pattern of the nonzero entries in the 49×49 matrix A in (10.10) (*left*) and the LR factors of A (*centre* and *right*). The quantity nz gives a count of the nonzero entries

in which v is an intermediate vector, and are solved sequentially—processes that are referred to as forward and backward substitution, respectively. The matrix A is a sparse matrix with *band width M*—this is the maximum horizontal distance of nonzero elements from the main diagonal—so that $A_{ij} = 0$ for $|i - j| > M$. As illustrated in Fig. 10.3, this band width is also inherited by the triangular factors L and R. When M is large, the number of arithmetic operations (flops) needed to compute the LR factorization of a banded matrix is asymptotically proportional to the product of its dimension and the square of its bandwidth (see, for example, Golub and Van Loan [6, Sect. 5.3]). Thus the time needed to factorize our matrix A on a computer will typically increase as M^4 when the grid in Fig. 10.2 is refined. The number of operations needed for the forward/backward substitutions in (10.11) is proportional to the product of the dimension and the band width, that is M^3. This means that the solution of the linear system $Au = f$ is dominated by the factorization phase. This is illustrated in the following example.

Example 10.1 Apply the finite difference method (10.7)–(10.8) to the Poisson equation in the unit square with a Dirichlet condition $u = g$ on $\partial\Omega$, where

$$g(x, y) = -xy \log((x - a)^2 + (y - b)^2)$$

for fixed constants a and b,[3] together with a right hand side function given by

$$f(x, y) = 4\frac{2xy - bx - ay}{(x - a)^2 + (y - b)^2}.$$

Note that $f = -\nabla^2 g$ which means that the exact solution to the boundary value problem $\mathcal{L}u = \mathcal{F}$ is simply $u = g$.

The numerical solution with $M = 16$, $a = b = 5/4$ is visualised in Fig. 10.4 (left) together with the global error $E = u - U$. The error has a maximum of about 10^{-3} which suggests that the finite difference method successfully generates an accurate solution. The behaviour of the maximum norm of the global error $\|E\|_{h,\infty}$ (defined by a natural generalization of (6.28)),

$$\|u - U\|_{h,\infty} := \max_{\ell,m} |u_{\ell,m} - U_{\ell,m}|, \tag{10.12}$$

is assessed in Table 10.1. The measure of global error can be seen to reduce by a factor of four whenever M is increased by a factor of two, and this gives a strong indication that the global error is $\mathcal{O}(h^2)$. The table also shows the overall time taken to compute the numerical solution for each M. The discussion earlier in this section suggested that the time taken should be proportional to M^4 and therefore increase by a factor of 16 each time M is doubled. We see that this is a realistic estimate[4] for

---

[3]We will take $a = b > 1$ to ensure that $(a, b) \notin \Omega$.
[4]Timing of computational algorithms is notoriously difficult since it depends on the number of processes that are running concurrently.

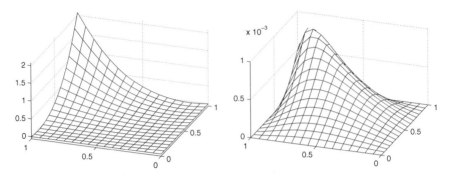

Fig. 10.4 Numerical solution for Example 10.1 with $M = 16$, $a = b = 5/4$ (*left*) and the corresponding global error (*right*)

Table 10.1 The maximum global error as a function of $h = 1/M$ for Example 10.1

h	$\|E\|_h$	Ratio	cpu time	%
1/8	0.0034	—	0.0001	39
1/16	0.00093	3.61	0.0008	5
1/32	0.00024	3.95	0.008	2
1/64	0.000060	3.98	0.09	3
1/128	0.000015	4.00	1.	1
1/256	0.0000037	4.00	14.	0.8

The third column shows the ratio of the global error with grid size $2h$ to the global error obtained using a grid size h. The final column gives the percentage of total cpu time needed to compute the forward/backward solves in (10.11)

the larger values of M. The final column supports our assertion that the time taken for the forward and back substitutions in (10.11) is negligible compared to the total time taken.

Perhaps the most important (not to say sobering) conclusion to be drawn from the statistics is that, if the method is indeed second-order convergent, then M^2 has to be increased by a factor of 10 in order to gain one extra digit of accuracy and the cost of achieving this could be a factor of 100 in cpu time. $\Diamond$

10.1.2 Convergence Theory

The foundation of the theory needed to establish convergence is described in Sect. 6.3 and can be summarised by the dictum "*consistency and stability imply convergence* ". We shall only deal with situations where each of these quantities is measured in the ℓ_∞-norm (10.12).

First, since the boundary value problem that we are dealing with is a Dirichlet problem (so that the boundary condition is replicated exactly) it is only necessary to check that the finite difference method is consistent with the PDE. That is, it has to be shown that the local truncation error $\mathcal{R}_h := \mathcal{L}_h u - \mathcal{F}_h = \mathcal{O}(h^p)$ with $p > 0$. Now, since $\mathcal{L}_h u|_{\ell,m} = -h^{-2}(\delta_x^2 + \delta_y^2)\, u_{\ell,m}$ and $\mathcal{F}_h|_{\ell,m} = f_{\ell,m}$, we can simply use (6.14) to get

$$
\begin{aligned}
\mathcal{R}_h|_{\ell,m} &= -h^{-2}(\delta_x^2 + \delta_y^2)\, u_{\ell,m} - f_{\ell,m} \\
&= -(u_{xx} + u_{yy})\big|_{\ell,m} - f_{\ell,m} - \tfrac{1}{12} h^2 (u_{xxxx} + u_{yyyy})\big|_{\ell,m} + \mathcal{O}(h^4) \\
&= -\tfrac{1}{12} h^2 (u_{xxxx} + u_{yyyy})\big|_{\ell,m} + \mathcal{O}(h^4),
\end{aligned}
\tag{10.13}
$$

since $-(u_{xx} + u_{yy}) = f$. Thus the method is consistent of order two if u and its partial derivatives up to order four are continuous on the domain Ω.[5]

Second, according to Lemma 6.8 the finite difference operator $\mathcal{L}_h$ will be a *stable* operator provided that it can be shown to be inverse monotone and have a comparison function Φ (bounded independently of h) satisfying $\mathcal{L}_h \Phi \geq 1$. We shall first prove that $\mathcal{L}_h$ satisfies a *maximum principle* so that it mimics the behaviour of the continuous problem (as discussed in Theorem 7.5).

Theorem 10.2 (Discrete maximum principle) *Suppose that $\mathcal{L}_h U_{\ell,m}$ is defined by (10.7) and that $\mathcal{L}_h U \leq 0$ at all grid points in Ω_h. Then either U is constant in Ω_h or else it achieves its maximum on the boundary $\partial \Omega_h$.*

Proof We sketch a proof that mirrors that for Theorem 6.10. Thus, by contradiction, we suppose that U attains its maximum value at a point (x_ℓ, y_m) in the interior of the domain, so that $0 < \ell, m < M$. The inequality $\mathcal{L}_h U_{\ell,m} \leq 0$ may be rewritten as

$$
U_{\ell,m} \leq \tfrac{1}{4}\left(U_{\ell+1,m} + U_{\ell,m+1} + U_{\ell-1,m} + U_{\ell,m-1} \right),
\tag{10.14}
$$

from which it follows that either $U_{\ell,m} < \max\{U_{\ell+1,m}, U_{\ell,m+1}, U_{\ell-1,m}, U_{\ell,m-1}\}$, which would immediately lead to a contradiction, or else the five values $U_{\ell,m}, U_{\ell+1,m}, U_{\ell,m+1}, U_{\ell-1,m}, U_{\ell,m-1}$ must be equal to each other, in which case the grid function U will attain its maximum on the set of five points $(x_\ell, y_m), (x_{\ell\pm1}, y_m), (x_\ell, y_{m\pm1})$. The same argument can therefore be applied to all the nearest neighbours of the original point (unless they lie of the boundary) which will either generate a contradiction, or else will further increase the number of grid points at which U attains its maximum. Eventually all the interior grid points will be exhausted and U is a constant grid function (a contradiction). □

Corollary 10.3 *The difference operator $\mathcal{L}_h$ defined by (10.7) and (10.8) is stable.*

Proof Establishing the inverse monotonicity is left to Exercise 10.6. All we need to do here is to find a suitable comparison function. The quadratic function

$$
\Phi(x, y) = 1 + \tfrac{1}{2}x(1 - x)
$$

---

[5]This is the reason why we need to insist that $(a, b) \notin \Omega$ in Example 10.1.

used in the proof of Theorem 7.5 is a likely candidate. Indeed, since the local trunca-tion error is identically zero for this choice (since it depends on the fourth partial deriv-atives of Φ), we immediately arrive at the desired result $\mathcal{L}_h \Phi_{\ell,m} = -\nabla^2 \Phi \big|_{\ell,m} = 1$ by invoking Lemma 6.8. $\square$

Uniqueness of the numerical solution follows directly from Theorem 5.3. Finally, to conclude the section, we reiterate the classical convergence result.

Corollary 10.4 *The 5–point approximation to the Poisson equation with a Dirichlet boundary condition, described by (10.7)–(10.8) on the unit square Ω is second-order convergent if the fourth derivatives of the exact solution are bounded on Ω.*

Proof This follows from Theorem 6.7 and the results established earlier in this section. $\square$

10.1.3 Improving the Solution Accuracy

There are two clever strategies for monitoring the accuracy of a numerical solution in cases where the exact solution is not known. The first of these is named after Richardson[6] and is a generic process that may be applied to a broad range of numerical approximation methods.

Richardson's idea is quite simple. It extrapolates results from a number of dif-ferent grids to both estimate the rate of convergence and predict the exact solution. An estimate of the global error follows by comparing the difference between this prediction and any of the numerically computed values. The process is illustrated in the following example.

Example 10.5 (Richardson extrapolation) Suppose that Q is a quantity of specific interest, for example, the value of u at a specific point in the domain. To give an explicit example, the following data represents the numerical solution to the problem in Example 10.1 at the particular point P=($3/4$, $3/4$) for five different grid sizes h:

h	1/4	1/8	1/16	1/32	1/64
$Q(h)$	0.37771749	0.38651913	0.38902712	0.38967667	0.38984053

Our task is to estimate the rate of convergence of the numerical solution at P and then to determine an improved estimate of $u(3/4, 3/4)$.

We start by assuming that $Q(h) = Q(0) + Ch^p$. Some simple algebra gives $2^p = (Q(h) - Q(h/2))(Q(h/2) - Q(h/4))$. Thus

$$p = \frac{1}{\log 2} \log \left| \frac{Q(h) - Q(h/2)}{Q(h/2) - Q(h/4)} \right|, \quad Q(0) = \frac{2^p Q(h/4) - Q(h/2)}{2^p - 1}. \quad (10.15)$$

---

[6]Lewis Fry Richardson (1881–1953) was one of the pioneers of numerical PDEs. He developed a life-long interest in weather forecasting when living in Manchester.

Table 10.2 The actual error $E(h) = u(3/4, 3/4) - Q(h)$, the exponent p (EOC) and solution estimate $Q(0)$ computed via (10.15) when solving Example 10.1 for different values of h

h	$Q(h)$	$E(h)$	EOC	$Q(0)$	$Q(0) - Q(h)$
1/4	0.37772	0.01218	—	—	—
1/8	0.38652	0.00338	—	—	—
1/16	0.38903	0.00087	1.811	0.39003	0.00100
1/32	0.38968	0.00022	1.949	0.38990	0.00024
1/64	0.38984	0.00005	1.987	0.38990	0.00006

To apply this construction to the tabulated data, we suppose that the global error converges as $\mathcal{O}(h^p)$ so that

$$\underbrace{U(3/4, 3/4)}_{Q(h)} \approx \underbrace{u(3/4, 3/4)}_{Q(0)} + Ch^p$$

in which C is a constant independent of h and the exponent p is the *experimental order of convergence* or EOC (cf. the discussion in Example 6.2). The results given in Table 10.2 show that the point values $Q(h) = U(3/4, 3/4)$ have an EOC of 1.811 on the coarsest grids, rising to 1.987, as finer grids are used. These results are compatible with a second-order rate of convergence. The extrapolated values of $U(3/4, 3/4)$ in the limit $h \to 0$ are denoted by $Q(0)$ and are also shown in Table 10.2. The final estimate $Q(0) = 0.38990$ differs from the exact solution $u(3/4, 3/4)$ by only one unit the the last decimal place. It is no surprise, therefore, that the quantity $Q(0) - Q(h)$ shown in the final column gives an increasingly accurate estimate of the actual error $E(h)$. ◇

The above example demonstrates that Richardson extrapolation is capable of estimating and increasing the accuracy of numerical results with only a small amount of arithmetic. There are a couple of issues that undermine the process however. These are (a) theoretical justification for the ansatz $Q(h) = Q(0) + Ch^p$ is not always available (note that all higher-order terms have been neglected) and (b) the numerical solution has to be computed on three (nested) grids but the error estimate (or extrapolated value) is only available at points on the coarsest grid.

A second way of estimating the global error from the finite difference solution U satisfying $\mathcal{L}_h U = \mathcal{F}_h$ is to *post-process* an estimate, $\widehat{\mathcal{R}}_h$ say, of the local truncation error. (This approach is called *iterative refinement* in a numerical linear algebra setting.) If we substitute this estimate into the right hand side of global error equation $\mathcal{L}_h E = \mathcal{R}_h$, see (6.31), an estimate $\widehat{E}$ of the global error $u - U$ may be found by solving

$$\mathcal{L}_h \widehat{E} = \widehat{\mathcal{R}}_h. \tag{10.16}$$

The grid function $\widehat{U} = U + \widehat{E}$ is the enhanced approximation to the solution u. It is important to observe that it is the original finite difference operator $\mathcal{L}_h$ that appears on the left hand side. One such strategy is illustrated in the following example.

Example 10.6 (A post-processed error estimate) Determine a finite difference approximation $\widehat{\mathscr{R}}_h$ of the leading term in the local truncation error and thereby find an equation of the form (10.16) that may be solved to give an estimate of the global error $u - U$ for the problem in Example 10.1.

The boundary conditions in Example 10.1 are incorporated without approximation so the focus is simply on the differential equation. The finite difference equations at internal grid points are represented in operator form by the equation $\mathcal{L}_h U = \mathcal{F}_h$, where $\mathcal{L}_h$ is defined by (10.7). The local truncation error for this method is given by (10.13),

$$\mathcal{R}_h|_{\ell,m} = -\tfrac{1}{12}\, h^2 (u_{xxxx} + u_{yyyy})\Big|_{\ell,m} + \mathcal{O}(h^4)$$

A direct approximation of the leading term on the right hand side is possible but complicated, see Exercise 6.32. An alternative strategy is to observe that

$$u_{xxxx} + u_{yyyy} = (\partial_x^2 + \partial_y^2)^2 u - 2\partial_x^2 \partial_y^2 u$$
$$= -(f_{xx} + f_{yy}) - 2\partial_x^2 \partial_y^2 u,$$

and then to construct a finite difference approximation of the right hand side.

This gives the following "simple" estimate of the local truncation error

$$\widehat{\mathcal{R}}_h\Big|_{\ell,m} = \tfrac{1}{12}(\delta_x^2 + \delta_y^2) f_{\ell,m} + \tfrac{1}{6} h^{-2} \delta_x^2 \delta_y^2 U_{\ell,m} \tag{10.17}$$

which involves only the nine grid values $U_{\ell+j,m+k}$ with $j, k = 0, \pm 1$.

Next, we note that the first term on the right hand side of (10.17) can be expressed as $-\tfrac{1}{12} h^2 \mathcal{L}_h f|_{\ell,m}$ so that, rather than solving (10.16) directly to get the estimate $\widehat{E}$ we can introduce the intermediate quantity $\widetilde{E} = \widehat{E} + \tfrac{1}{12} h^2 f$ that satisfies the boundary value problem

$$\left.\begin{array}{ll} \mathcal{L}_h \widetilde{E}_{\ell,m} = \tfrac{1}{6} h^{-2} \delta_x^2 \delta_y^2 U_{\ell,m} & \text{for } (x_\ell, y_m) \in \Omega_h, \\ \widetilde{E}_{\ell,m} = \tfrac{1}{12} h^2 f_{\ell,m} & (x_\ell, y_m) \in \partial\Omega_h. \end{array}\right\} \tag{10.18}$$

In this way we avoid having to evaluate $\mathcal{L}_h f$. Applying this post-processing strategy to the computed finite difference solutions in Example 10.1 gives results shown in Table 10.3. The ratios of errors for successive grids (listed in the last column) suggest that the rate of convergence of the post-processed difference solutions is significantly faster than $\mathcal{O}(h^2)$.

It might appear that computing the post-processed solution doubles the cost of solving the BVP. However, this is not the case because the Eqs. (10.18) can be assembled into matrix-vector form $A\widetilde{e} = \widetilde{f}$ in which the coefficient matrix is exactly the same as that used in (10.9) to determine U. Thus, provided that the previously computed matrices L and R of A have been saved, $\widetilde{e}$ may be computed by a forward and backward solve, that is

$$L\widetilde{d} = \widetilde{f}, \qquad R\widetilde{e} = \widetilde{d}.$$

Table 10.3 Numerical results for Example 10.6 showing norms of the global error $\|E\|_{h,\infty}$ (from Table 10.1), the post-processed error estimates $\|\widehat{E}\|_{h,\infty}$ and the errors in the post-processed solution $U + \widehat{E}$

h	$\|E\|_h$	$\|\widehat{E}\|_{h,\infty}$	$\|u - U - \widehat{E}\|_{h,\infty}$	Ratio
1/8	0.0038	0.0034	0.00015	—
1/16	0.00093	0.00093	0.000028	5.49
1/32	0.00024	0.00024	0.0000027	10.03
1/64	0.000060	0.000060	0.00000021	13.19

As discussed earlier (see the final column of Table 10.1) the cost of this process—and therefore the cost of computing the error estimate—is negligible compared to the cost of computing the original solution U. ◇

The derivation of (10.16) is closely allied to that of higher-order methods, such as Numerov's method described in Sect. 6.4.2. A 9-point generalization of Numerov's method to the solution of PDEs is pursued in Exercise 10.4.

10.2 Advanced Topics and Extensions

The material in Sects. 10.1 and 10.2.1 gives a flavour of the issues involved in solving elliptic PDEs by finite difference methods. This may be deemed sufficient for a first foray into this area. A more ambitious reader might also want to look at some of the topics in the rest of the chapter.

10.2.1 Neumann and Robin Boundary Conditions

Suppose that the Poisson equation (10.3) is to be solved on the unit square, $\Omega = (0, 1) \times (0, 1)$, but this time subject to the Neumann boundary condition

$$\partial_n u = g(x, y), \quad \text{for } (x, y) \in \partial\Omega, \tag{10.19}$$

where $\partial_n u$ denotes differentiation in the direction of the outward normal to the boundary. The normal is not well defined at corners of the domain and, indeed, g need not be continuous there. Note that it was shown in Example 7.9 that this boundary value problem is not well posed unless the data f and g satisfies the compatability condition (7.8).

Finite difference approximations of the condition (10.19) can be constructed by generalising the process described in Sect. 6.4.1 for generating second-order approximations to derivative boundary conditions. To illustrate this process, we consider a grid point $(0, mh)$ (with $0 < m < M$ to avoid the vertices) on the boundary $x = 0$ where (10.19) gives $-u_x = g(0, y)$ $(0 < y < 1)$. Taking the forward difference operator $\triangle_x^+$ (see Table 6.1) gives

$$\triangle_x^+ u_{0,m} = u_{1,m} - u_{0,m} = h u_x|_{0,m} + \tfrac{1}{2} h^2 u_{xx}\big|_{0,m} + \mathcal{O}(h^3) \qquad (10.20)$$

and rearranging this we get

$$- u_x(0, mh) = -h^{-1} \triangle_x^+ u_{0,m} - \tfrac{1}{2} h u_{xx}\big|_{0,m} + \mathcal{O}(h^2). \qquad (10.21)$$

In order to remove the first order term in h, we use the PDE $u_{xx} = -u_{yy} - f$ so that, at the point $(0, mh)$

$$-u_{xx}|_{0,m} = f_{0,m} + u_{yy}\big|_{0,m} = f_{0,m} - h^{-2} \delta_y^2 u_{0,m} + \mathcal{O}(h^2).$$

Combining this with (10.21) gives

$$-u_x(0, mh) = -h^{-1} \triangle_x^+ u_{0,m} - \tfrac{1}{2} h^{-1} \delta_y^2 u_{0,m} - \tfrac{1}{2} h f_{0,m} + \mathcal{O}(h^2)$$

which leads to the numerical boundary condition

$$- h^{-1} \triangle_x^+ U_{0,m} - \tfrac{1}{2} h^{-1} \delta_y^2 U_{0,m} = g_{0,m} + \tfrac{1}{2} h f_{0,m}, \qquad (10.22a)$$

$(m = 1, 2, \ldots, M - 1)$. It can also be written in the explicit form

$$4U_{0,m} - 2U_{1,m} - U_{0,m-1} - U_{0,m+1} = 2h g_{0,m} + h^2 f_{0,m}, \qquad (10.22b)$$

with a stencil that is depicted on the left of Fig. 10.5. An alternative derivation is described in Exercise 10.9.

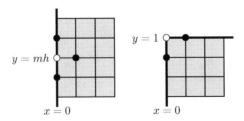

Fig. 10.5 The grid in the vicinity of the boundary point $(0, mh)$ and the corner point $(0, Mh)$. Approximations of the Neumann boundary condition at the target points (marked o) also involve the neighbouring grid points (marked •)

The form (10.22a) is most convenient for checking the consistency error of the boundary condition. The relevant component of the local truncation error is given by

$$\mathscr{R}_h|_{0,m} := -h^{-1}\Delta_x^+ u_{0,m} - \tfrac{1}{2}h^{-1}\delta_y^2 u_{0,m} - g_{0,m} - \tfrac{1}{2}hf_{0,m},$$

which, using the expansions listed in Table 6.1, leads to

$$\begin{aligned}
\mathscr{R}_h|_{0,m} &= -\left(u_x + \tfrac{1}{2}hu_{xx} + \mathcal{O}(h^2)\right)\big|_{0,m} - \tfrac{1}{2}h\left(u_{yy} + \mathcal{O}(h^2)\right)\big|_{0,m} - g_{0,m} - \tfrac{1}{2}hf_{0,m}, \\
&= (-u_x - g)\big|_{0,m} + \tfrac{1}{2}h(-u_{xx} - u_{yy} - f)\big|_{0,m} + \mathcal{O}(h^2) \\
&= (\mathcal{B}u - g)\big|_{0,m} + \tfrac{1}{2}h(\mathcal{L}u - f)\big|_{0,m} + \mathcal{O}(h^2).
\end{aligned} \tag{10.23}$$

Since $\mathcal{B}u = g$ and $\mathcal{L}u = f$ at $(0, mh)$, we conclude that the consistency error of the boundary condition is second order. Checking this more carefully (see Exercise 10.10) reveals that that second-order consistency is dependent on the derivative u_{xxx} being bounded in the interior of the domain, with u_{yyyy} being bounded on the edge $x = 0, 0 < y < 1$.

The derivation of a numerical boundary condition at a corner, $(0, 1)$, say, requires extra care in interpreting the boundary condition for the continuous problem. More specifically, along $x = 0$ the Neumann condition is $-u_x = g$ while along $y = 1$ it is $u_y = g$, but the value of g need not be the same as we approach the corner along the two edges. Taking the limits

$$\lim_{y \to 1^-} g(0, y) = g(0^-, 1) \quad \text{and} \quad \lim_{x \to 0^+} g(x, 1) = g(1^+, 0),$$

the two boundary conditions at the corner become $-u_x = g(0^-, 1)$ and $u_y = g(1^+, 0)$. By adding the backward difference approximation (10.20) for u_x to the backward difference approximation for u_y, that is,

$$h^{-1}\Delta_y^- u_{0,M} = h^{-1}(u_{0,M} - u_{0,M-1}) = u_y\big|_{0,M} - \tfrac{1}{2}hu_{yy}\big|_{0,M} + \mathcal{O}(h^2),$$

we arrive at the numerical boundary condition

$$-h^{-1}\Delta_x^+ U_{0,m} + h^{-1}\Delta_y^- U_{0,M} = \tfrac{1}{2}hf_{0,M} + g(1^+, 0) + g(0^-, 1). \tag{10.24a}$$

This may be explicitly written as

$$2U_{0,M} - U_{1,M} - U_{0,M-1} = \tfrac{1}{2}h^2 f_{0,M} + hg(1^+, 0) + hg(0^-, 1), \tag{10.24b}$$

and is depicted on the right of Fig. 10.5. The details are left to Exercise 10.11. Proceeding in this manner for each of the boundary segments and each corner, we arrive at $(M + 1)^2$ linear equations for the values of U on $\overline{\Omega}_h$.

We conclude this section with an illustration of how a homogeneous Neumann condition naturally arises when exploiting the inherent symmetry in a boundary value problem. Suppose that both the domain and the solution u are symmetric with respect to the y-axis—for the Poisson equation this requires that the source term f and boundary values g are also symmetric with respect to the y-axis—then

$$u(-x, y) = u(x, y) \quad \text{for all } x \in \Omega. \tag{10.25}$$

By expanding both sides of this equation in a Taylor series about $x = 0$, we find that $u_x(0, y) = 0$. In fact all the odd-order partial derivatives of u must vanish along $x = 0$. The original problem can then be reduced to the half of the domain lying in $x \geq 0$, with the Neumann condition $u_x = 0$ along $x = 0$ together with the original boundary conditions on the remainder of the boundary.

Suppose now that a finite difference method with a 5-point stencil is used to solve such a problem. When this method is applied at a point on the axis of symmetry (as shown by the symbol o in Fig. 10.6, left) the grid values $U_{1,m}$ (at the point labelled A) and $U_{-1,m}$ (at the point labelled B) must be equal. In particular, the five point finite difference approximation (10.6) applied at $m = 0$ immediately reduces to the numerical boundary condition (10.22b) with $g = 0$. The upshot is that the numerical solution need only be computed for one half of the original domain with a significant saving in computational cost.

By a similar argument, if the domain and solution are symmetric with respect to both coordinate axes (see Fig. 10.6, right) then $u_x(0, y) = 0$ and $u_y(x, 0) = 0$. Furthermore, when the five point finite difference approximation (10.6) is applied at the origin $l = m = 0$ we find that it reduces to

$$2U_{0,0} - U_{1,0} - U_{0,-1} = \tfrac{1}{2} h^2 f_{0,0},$$

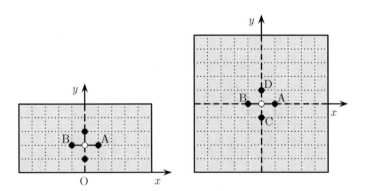

Fig. 10.6 An example of a domain that is symmetric about the y-axis *left* and, a domain that is symmetric with respect both x- and y-axes (*right*)

since symmetry implies that $U_{-1,0} = U_{1,0}$ and $U_{0,1} = U_{0,-1}$. This is identical to the numerical boundary condition (10.24b) with $g = 0$ once the difference in indexing in the two situations is taken into account. The numerical solution only needs to be computed over one quarter of the domain in this case.

10.2.2 Non-Rectangular Domains

The techniques described in earlier sections readily deal with domains whose boundaries consist of grid lines $x = \ell h$ or $y = mh$, for integers ℓ and m or even boundaries that pass diagonally through the grid. Although the range of problems can be further extended by allowing different grid sizes in the x and y directions (as illustrated by Exercise 10.1), a new procedure is clearly required to deal with domains having curved boundaries.

A possible process will be described through its application to solving the Dirichlet problem for the Poisson equation on a non-rectangular domain such as that illustrated in Fig. 10.1. When the domain is covered by a rectangular grid of size $h \times h$ (such as that shown in Fig. 10.7), there are difficulties in imposing the boundary conditions because the boundary does not, in general, pass through grid points. For the Dirichlet problem, the boundary condition takes the form $u(x, y) = g(x, y)$ for all points (x, y) lying on the boundary $\partial \Omega$. This data is available at those points where the boundary intersects the grid lines $x = \ell h$ and $y = mh$. These are marked with a circle (o) in Fig. 10.7 and constitute the boundary $\partial \Omega_h$ of our grid.

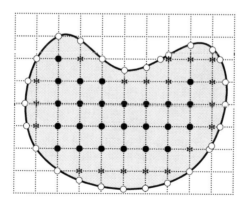

Fig. 10.7 A grid Ω_h on the domain Ω having a smoothly curved boundary. Points marked o show the intersection of grid lines with the boundary, grid points marked ● are those whose four nearest neighbours are each a distance h away and those marked ∗ have at least one of their nearest neighbours on the boundary and less that a distance h away

Fig. 10.8 The general form
of a 5–point finite difference
stencil at a grid point P
involving its four nearest
neighbours Q_j, $j = 1, 2, 3, 4$

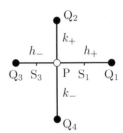

The set of grid points marked $*$ in Fig. 10.7 have the property that at least one of their
four nearest neighbours lies on the boundary $\partial \Omega_h$. We shall denote the set of these
grid points by Ω_h^*. The remaining points of the grid will be denoted by $\Omega_h^\bullet$ and are
marked $\bullet$ in Fig. 10.7. The entire grid $\overline{\Omega}_h$ is therefore made up of three categories of
grid points: $\partial \Omega_h$, Ω_h^* and $\Omega_h^\bullet$. The solution is known for grid points in $\partial \Omega_h$ from
the Dirichlet condition $U = g$ and the differential equation may be approximated
by the standard 5–point formula (10.6) for points in $\Omega_h^\bullet$. It therefore only remains to
construct an approximation for the differential equation at points of Ω_h^*.

Suppose that P denotes a general grid point in Ω_h^* then, in order to cope with all
possibilities, we assume that its nearest neighbours Q_j are at vertical or horizontal
distances $h_\pm$ and $k_\pm$ away, as shown in Fig. 10.8. To approximate u_{xx} at P we use
the Taylor expansions

$$u(x + h_+, y) = u + h_+ u_x + \tfrac{1}{2} h_+^2 u_{xx} + \mathcal{O}(h^3)$$
$$u(x - h_-, y) = u - h_- u_x + \tfrac{1}{2} h_-^2 u_{xx} + \mathcal{O}(h^3),$$

where h is the largest grid spacing in the x–direction and all the terms on the right
are evaluated at P. Eliminating u_x leads to

$$u_{xx}|_P = \frac{2}{h_+ + h_-} \left[\frac{u|_{Q_1} - u|_P}{h_+} - \frac{u|_P - u|_{Q_3}}{h_-} \right] + \mathcal{O}(h), \qquad (10.26)$$

where the truncation error is (formally) of first order. If we now suppose that S_1 and
S_3 denote the midpoints of the line segments PQ_1 and Q_3P, respectively, then (10.26)
can be interpreted as the approximation

$$u_{xx}|_P \approx \frac{u_x|_{S_1} - u_x|_{S_3}}{|S_1 S_3|} \quad \text{with } u_x|_{S_1} \approx \frac{u|_{Q_1} - u|_P}{|PQ_1|} \text{ and } u_x|_{S_3} \approx \frac{u|_P - u|_{Q_3}}{|PQ_3|}.$$

Applying a similar process in the y-direction leads to

$$u_{yy}|_P = \frac{2}{k_+ + k_-} \left[\frac{u|_{Q_2} - u|_P}{k_+} - \frac{u|_P - u|_{Q_4}}{k_-} \right] + \mathcal{O}(h), \qquad (10.27)$$

so the PDE $-u_{xx} - u_{yy} = f$ can be approximated at a typical point P by the difference formula

$$-\frac{2}{h_+ + h_-}\left[\frac{U_1 - U_0}{h_+} - \frac{U_0 - U_3}{h_-}\right] - \frac{2}{k_+ + k_-}\left[\frac{U_2 - U_0}{k_+} - \frac{U_0 - U_4}{k_-}\right] = f_0 \tag{10.28}$$

in which $U_0 \approx u|_{\text{P}}$, $U_1 \approx u|_{Q_1}$, etc. Note that when P is the node $(\ell h, mh)$ and $h_\pm = k_\pm = h$, this formula simplifies to that given by (10.6). An equation of the type (10.28) holds at each point in Ω_h^* while the 5–point replacement (10.6) holds at points in $\Omega_h^\bullet$. We have, therefore, a finite difference equation at each internal grid point. Once an ordering of the grid points has been decided upon, these equations may be written as a linear algebraic system $A\boldsymbol{u} = \boldsymbol{f}$.

Example 10.7 Construct a finite difference approximation to solve Laplace's equation in the quarter circle $\{(x, y) : x^2 + y^2 < 1\}$ with a symmetry boundary condition $u_y(x, 0) = 0$ and a Dirichlet condition $u = g$ on the remainder of the boundary—where g is the harmonic function $g(x, y) = x^5 - 10x^3y^2 + 5xy^4$ so that $u = g$ is the exact solution of the boundary value problem.

A subdivision with $M = 6$ is illustrated in Fig. 10.9. It is a straightforward exercise to apply (10.6) at points in $\Omega_h^\bullet$ and (10.28) at points in Ω_h^* to generate the set of finite difference equations. By following the procedure described in the previous section, a second-order difference approximation at the point $(x_\ell, 0)$ that incorporates an approximation of the Neumann condition can be shown (see Exercise 10.18) to be

$$2U_{\ell,0} - \tfrac{1}{2}U_{\ell-1,0} - \tfrac{1}{2}U_{\ell+1,0} - U_{\ell,1} = 0. \tag{10.29}$$

The corresponding stencil is depicted by open circles ($\circ$) in Fig. 10.9. The solution obtained with $M = 16$ and the associated global error $u - U$ is shown in Fig. 10.10. It can be seen that the error is largest error along $x = 0$ (a line of symmetry) and that it is relatively small near the curved boundary.

The maximum global error as a function of M is shown in Table 10.4. The final column shows the ratio by which the error is reduced when M is doubled (see Table 6.2) and the results are consistent with the method being convergent of second

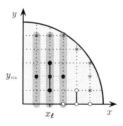

Fig. 10.9 A coarse grid for the boundary value problem in Example 10.7. The columns of grid values associated with the 5-point stencil at the point (x_ℓ, y_m) (marked $\bullet$) are shown highlighted

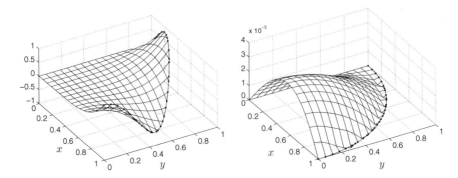

Fig. 10.10 Numerical solution for Example 10.7 with $h = 1/16$ (*left*) and the corresponding global error (*right*)

Table 10.4 The maximum norm of the global error as a function of $h = 1/M$ for Example 10.7

h	$\|u - U\|_{h,\infty}$	Ratio
1/16	0.0035	—
1/32	0.00091	3.85
1/64	0.00023	3.93
1/128	0.000058	3.96

order. This might appear surprising given that the consistency error in the boundary approximation (10.28) is only first order in h.   ◇

To establish the stability of finite difference approximations on a nonuniformly spaced grid we will need to generalise the notion of a positive operator that was given in Definition 6.9. The following construction will give us exactly what we need to handle such cases.

Definition 10.8 (*General positive type operator*) Suppose that P is a grid point and that $Q_1, Q_2, \ldots, Q_\nu$ are neighbouring grid points. The finite difference operator defined by[7]

$$\mathcal{L}_h U\big|_{\mathrm{P}} = \alpha_0\, U\big|_{\mathrm{P}} - \sum_{j=1}^{\nu} \alpha_j U\big|_{Q_j} \tag{10.30}$$

is said to be of *positive type* if the coefficients satisfy the inequalities

$$\alpha_j \geq 0, \quad j = 0, 1, \ldots, \nu; \qquad \alpha_0 \geq \sum_{j=1}^{\nu} \alpha_j. \tag{10.31}$$

Note that, for any point $\mathrm{P} \in \Omega_h^*$, the difference equation (10.28) can be written in the form (10.30) so it can be immediately checked that the overall finite difference

---

[7]Here ν and the coefficients $\{\alpha_j\}$ are generally different for different points P but the notation does not reflect this.

operator is of positive type. This, in turn, means that the finite difference approximation in the above example is *inverse monotone*, and following the exact same argument as in Sect. 6.3 we deduce that the discrete problem $\mathcal{L}_h U = \mathcal{R}_h$ associated with Example 10.7 has a unique solution. The somewhat surprising second-order convergence rate is studied next.

Theorem 10.9 (General discrete maximum principle) *Suppose that $\mathcal{L}_h U_{\ell,m}$ is defined by (10.7) at a general grid point of $\Omega_h^\bullet$ and by the left hand side of (10.28) at a general grid point of Ω_h^*. If $\mathcal{L}_h U \leq 0$ at all grid points in Ω_h, then either U is constant in Ω_h or it achieves its maximum on the boundary $\partial\Omega_h$.*

Proof Theorem 10.2 implies that U cannot have a maximum at a point in $\Omega_h^\bullet$, so it remains to prove that a maximum cannot occur in Ω_h^*. For any point $P \in \Omega_h^*$, the finite difference equation (10.28) can be written in the form (10.30) and is of positive type. It then follows from $\mathcal{L}_h U\big|_P \leq 0$ that

$$\alpha_0 \, U\big|_P \leq \left(\sum_{j=1}^4 \alpha_j\right) \max_j U\big|_{Q_j}$$

and so $U\big|_P \leq U\big|_{Q_j}$. Thus, either strict inequality holds giving a contradiction, or else U is constant on the set of points $\{P, Q_1, Q_2, Q_3, Q_4\}$. Applying the same argument to all points in Ω_h^* gives the desired contradiction. $\qquad\square$

Corollary 10.10 *The operator $\mathscr{L}_h$ comprising $\mathcal{L}_h$ at points of Ω_h together with a Dirichlet boundary operator $\mathcal{B}_h$ is stable.*

As in the proof of Corollary 10.3 we simply need to exhibit a suitable comparison function. If the domain Ω lies in the strip $0 \leq x \leq a$, then it is readily verified that

$$\Phi(x, y) = 1 + \tfrac{1}{2}x(a - x)$$

is a non-negative function on Ω satisfying $\mathscr{L}_h \Phi \geq 1$. $\qquad\sqcap$

Corollary 10.11 *The 5–point approximation to the Poisson equation with a Dirichlet boundary condition, described by (10.6) and (10.28) in a general, simply connected, domain Ω is second-order convergent if the fourth derivatives of the exact solution are bounded on Ω.*

Proof Convergence follows immediately from Theorem 6.7: establishing the second-order rate requires one more step. The key is to separate the errors on $\Omega_h^\bullet$ and Ω_h^*. Thus, suppose that $E^\bullet$ and E^* satisfy the boundary value problems

$$\mathscr{L}_h E^\bullet = \begin{cases} 0 & \text{on } \partial\Omega_h \\ 0 & \text{on } \Omega_h^* \\ \mathcal{R}_h^\bullet & \text{on } \Omega_h^\bullet \end{cases}, \quad \mathscr{L}_h E^* = \begin{cases} 0 & \text{on } \partial\Omega_h \\ \mathcal{R}_h^* & \text{on } \Omega_h^* \\ 0 & \text{on } \Omega_h^\bullet \end{cases}, \tag{10.32}$$

where $\mathcal{R}_h^\bullet := \mathcal{L}_h u - f$ on $\Omega_h^\bullet$ and $\mathcal{R}_h^* := \mathcal{L}_h u - f$ on Ω_h^*, so that $\mathcal{R}_h^\bullet = \mathcal{O}(h^2)$ and $\mathcal{R}_h^* = \mathcal{O}(h)$.

The standard argument in Theorem 6.7 is sufficient to prove that $E^\bullet = \mathcal{O}(h^2)$. To analyse the error function E^* we introduce a non-negative discrete function Ψ such that

$$\Psi\big|_P = \begin{cases} 0 & \text{on } \partial\Omega_h \\ h^2 & \text{on } \Omega_h \end{cases}.$$

This has the property that $\mathcal{L}_h\Psi\big|_P = 0$ for $P \in \Omega_h^\bullet$. For a point $P \in \Omega_h^*$, at least one of its nearest neighbours lies on the boundary. Suppose that this point is Q_1 then, using (10.31) gives

$$\mathcal{L}_h\Psi\big|_P = \alpha_0 \Psi\big|_P - \sum_{j=1}^4 \alpha_j \Psi\big|_{Q_j}$$

$$\geq h^2\left(\alpha_0 - \sum_{j=2}^4 \alpha_j\right) \text{ since } \Psi\big|_{Q_1} = 0,$$

$$\geq h^2\left(\alpha_0 - \sum_{j=1}^4 \alpha_j\right) + \alpha_1 h^2 \geq \alpha_1 h^2 \geq 1.$$

Note that if P has more than one nearest neighbour on the boundary, then the lower bound is increased. Thus, in all cases, $\mathcal{L}_h\Psi\big|_P \geq 1$ for all $P \in \Omega_h^*$.

Next, using (10.32), gives

$$\mathcal{L}_h(E^* - \|\mathcal{R}_h^*\|_{h,\infty}\Psi) = \begin{cases} 0 \\ \mathcal{L}_h E^* - \|\mathcal{R}_h^*\|_{h,\infty}\mathcal{L}_h\Psi \\ 0 \end{cases} \leq \begin{cases} 0 & \text{on } \partial\Omega_h \\ \mathcal{R}_h^* - \|\mathcal{R}_h^*\|_{h,\infty} & \text{on } \Omega_h^* \\ 0 & \text{on } \Omega_h^\bullet \end{cases}.$$

This means that $\mathcal{L}_h(E^* - \|\mathcal{R}_h^*\|_{h,\infty}\Psi) \leq 0$ and inverse monotonicity then implies that $E^* \leq \|\mathcal{R}_h^*\|_{h,\infty}\Psi$. An identical argument shows that $E^* + \|\mathcal{R}_h^*\|_{h,\infty}\Psi \geq 0$ and hence

$$-\|\mathcal{R}_h^*\|_{h,\infty}h^2 \leq E^* \leq \|\mathcal{R}_h^*\|_{h,\infty}h^2$$

so that the total contribution to the global error from the $\mathcal{O}(h)$ local error in Ω_h^* is $\mathcal{O}(h^3)$. We conclude that $E = E^* + E^\bullet = \mathcal{O}(h^2)$ with the usual proviso that the partial derivatives of u up to order four are bounded on Ω. $\qquad\square$

10.2.3 Polar Coordinates

A "difficult" geometry can sometimes be accommodated naturally by making a simple change of variables. The use of polar coordinates will be used to illustrate how this can be done.

Example 10.12 (Example 10.7 in polar coordinates) Construct a finite difference approximation to solve Laplace's equation in the quarter circle $\{(x, y) : x^2+y^2 < 1\}$, by transforming the problem into polar coordinates $(x, y) \mapsto (r, \theta)$ so that the boundary conditions in Example 10.7 are given by $u_\theta(r, 0) = 0$, $u(r, \pi/2) = 0$ and $u(1, \theta) = \cos 5\theta$.

Making the change of variable $x = r \cos\theta$, $y = r \sin\theta$ transforms Laplace's equation to the following form (see Example 5.14)

$$\frac{1}{r}\partial_r(r\partial_r)u + \frac{1}{r^2}\partial_\theta^2 u = 0.$$

The domain Ω is mapped to the rectangle $\{(r, \theta) : 0 < r < 1, 0 \leq \theta < \pi/2\}$ shown in Fig. 10.11, with the grid lines $r_\ell = \ell h$ (with $h = 1/M$) and $\theta_m = m\Delta\theta$ (with $\Delta\theta = \pi/(2M)$). We have chosen an equal number of grid lines in the two directions simply to avoid a profusion of symbols. The boundary $r = 0$ is shown as a dashed line as a reminder that the solution is the same at all points along it since these are all mapped to the origin $x = y = 0$. The internal grid points are denoted by Ω_h, where $\Omega_h = \{(r_\ell, \theta_m) : 0 < \ell < M, 0 < m < M\}$. The Neumann condition is satisfied at grid points $\{(r_\ell, \theta_0) : 0 < \ell < M\}$ and a Dirichlet condition is imposed at all other points on the boundary, that is, for the set of points $\{(r_\ell, \theta_M) : 0 < \ell < M\} \cup \{(1, \theta_m) : 0 < m \leq M\}$.

The PDE is approximated at a generic grid point $(r_\ell, \theta_m) \in \Omega_h$ by replacing the partial derivatives in r and θ by the first-order central difference operators $h^{-1}\delta_r$ and $\Delta\theta^{-1}\delta_\theta$ that were introduced in Sect. 6.1. This process leads to a system of finite difference equations $\mathcal{L}_h U = 0$, where

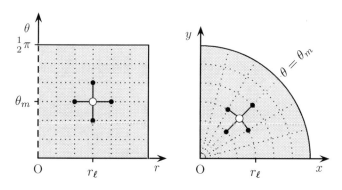

Fig. 10.11 A typical grid for Example 10.12 in the $r\theta$-plane (*left*) and the xy-plane (*right*)

$$\mathcal{L}_h U_{\ell,m} := \frac{1}{h^2 r_\ell} \delta_r(r_\ell \delta_r U_{\ell,m}) + \frac{1}{r_\ell^2 \Delta\theta^2} \delta_\theta^2 U_{\ell,m}. \tag{10.33}$$

Writing (10.33) explicitly gives

$$\mathcal{L}_h U_{\ell,m} := \frac{1}{h^2 r_\ell} \left(r_{\ell+1/2}(U_{\ell+1,m} - U_{\ell,m}) - r_{\ell-1/2}(U_{\ell,m} - U_{\ell-1,m}) \right)$$
$$+ \frac{1}{r_\ell^2 \Delta\theta^2} \left(U_{\ell,m+1} - 2U_{\ell,m} + U_{\ell,m-1} \right), \tag{10.33a}$$

where $r_{\ell\pm1/2} = (\ell \pm 1/2)h$. The associated stencil is illustrated in Fig. 10.11. The homogeneous Neumann condition reflects the symmetry about the x-axis, that is, $u(r, -\theta) = u(r, \theta)$. To enforce this condition we set $m = 0$ and $U_{\ell,-1} = U_{\ell,1}$ in (10.33a) to give the boundary operator

$$\mathcal{B}_h U_{\ell,0} := -\frac{1}{\Delta\theta} \left(U_{\ell,1} - U_{\ell,0} \right)$$
$$-\frac{\Delta\theta r_\ell}{2h^2} \left(r_{\ell+1/2}(U_{\ell+1,0} - U_{\ell,0}) - r_{\ell-1/2}(U_{\ell,0} - U_{\ell-1,0}) \right) \tag{10.33b}$$

having been rescaled so that $\mathcal{B}_h U_{\ell,0} = -\Delta\theta^{-1}\Delta_\theta^+ U_{\ell,0} + $ a correction term. This treatment is in line with the discussion at the end of Sect. 10.2.1. If these difference equations are assembled into a matrix-vector system using a column-ordering of unknowns (by which we mean columns in the $r\theta$-plane), then the structure of nonzeros in the coefficient matrix is the same as in the Cartesian case. (Note that each equation has to be multiplied by r_ℓ in order to get a symmetric matrix).

The solution obtained with $M = 16$ and the associated global error $u - U$ is shown in Fig. 10.12. This can be directly compared with the solution in Fig. 10.10. The behaviour of the maximum global error is compared in Table 10.5 with the analogous result when the same BVP is solved in xy-coordinates. The global error

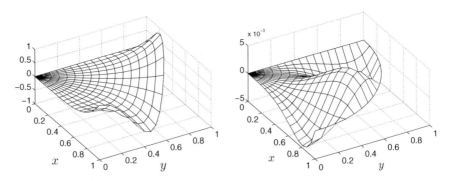

Fig. 10.12 Numerical solution for Example 10.12 computed with $M = 16$ (*left*) and the corresponding global error (*right*)

Table 10.5 The maximum norm of the global error as a function of $h = 1/M$ for Example 10.7 (using Cartesian coordinates) and for Example 10.12 (using polar coordinates)

h	Square grid		Polar grid	
	$\|u - U\|_{h,\infty}$	Ratio	$\|u - U\|_{h,\infty}$	Ratio
1/16	0.0035	—	0.0046	—
1/32	0.00091	3.85	0.0011	3.99
1/64	0.00023	3.93	0.00029	3.99
1/128	0.000058	3.96	0.000072	4.00

shows second-order convergence in both cases but is, for fixed M, about 25 % larger on the polar grid. This may be attributable to the Cartesian grid having a higher density of grid points on the curved boundary while the polar grid has a high density in the neighbourhood of the origin where there is very little variation in the solution.

$\Diamond$

The theory underpinning polar coordinate approximation is summarised within the following theorem.

Theorem 10.13 (Finite difference convergence in polar coordinates) *Suppose that $\mathcal{L}_h$ and $\mathcal{B}_h$ denote, respectively, the finite difference operators defined by (10.33a) in Ω_h and (10.33b) at points on the symmetry boundary. Then*

(a) *$\mathcal{L}_h$ is consistent of order two with the Laplacian operator written in polar coordinates in Ω_h and $\mathcal{B}_h$ is consistent of order two on the Neumann boundary.*
(b) *$\mathcal{L}_h$ satisfies a discrete maximum principle.*
(c) *The discrete operator $\mathscr{L}_h$, which comprises $-\mathcal{L}_h$ in the interior, $\mathcal{B}_h$ on the Neumann boundary and satisfies $\mathscr{L}_h U = U$ at grid points on the Dirichlet boundary, is stable with regards to the comparison function $\Phi(r, \theta) = 1 - \frac{1}{4}r^2$.*
(d) *The approximation method converges to the solution of Laplace's equation at a second-order rate.*

Proof We will simply outline the ingredients here. In order to check the consistency error of (10.33) we use the expansion (see Table 6.1)

$$h^{-1}\delta_r v_{\ell,m} = \partial_r v\big|_{\ell,m} + \tfrac{1}{24}h^2\partial_r^3 v\big|_{\ell,m} + \mathcal{O}(h^4)$$

for any smooth function $v(r, \theta)$. Thus, for the approximation of the r-derivatives,

$$h^{-2}\delta_r(r_\ell\delta_r u_{\ell,m}) = \left(\partial_r + \tfrac{1}{24}h^2\partial_r^3 + \mathcal{O}(h^4)\right) r \left(u_r + \tfrac{1}{24}h^2 u_{rrr} + \mathcal{O}(h^4)\right)$$

$$= \left(\partial_r + \tfrac{1}{24}h^2\partial_r^3 + \mathcal{O}(h^4)\right) \left(r u_r + \tfrac{1}{24}h^2 r u_{rrr} + \mathcal{O}(h^4)\right)$$

$$= \partial_r(r u_r) + \tfrac{1}{24}h^2\left((r u)_{rrr} + (r u_{rrr})_r\right) + \mathcal{O}(h^4),$$

where all the terms on the right hand side are evaluated at (r_ℓ, θ_m). The consistency error of the θ-derivatives is second order (exactly as in xy coordinates) so the overall truncation error is given by $\mathcal{R}_h := \mathcal{L}_h u - \mathcal{L} u = \mathcal{O}(h^2) + \mathcal{O}(\Delta\theta^2)$ whenever the fourth derivatives of u are bounded. Consistency of the approximation on the Neumann boundary is covered in Exercise 10.18.

The proof of the maximum principle follows by writing the discrete operator $\mathcal{L}_h$ in the form (10.30) and verifying that it is of positive type. Part (c) involves verification that the given function Φ is, indeed, a suitable comparison function. The last part follows immediately from Theorem 6.7. The details are left to Exercise 10.20.   □

Our final example involving polar coordinates addresses the degeneracy that occurs at the origin.

Example 10.14   Determine a finite difference approximation of the Poisson equation in polar coordinates that is valid at the origin.

The function $u(0, \theta)$ must be independent of θ if $u(r, \theta)$ is to be single-valued as $r \to 0$. Likewise, the numerical solution $U_{0,m}$ at $r = 0$ must be independent of m and we shall write $U_0 = U_{0,m}$ for all m to denote this common value. In this case, the finite difference operator defined by (10.33) becomes degenerate and a different approach is required on the line $r = 0$.

We will suppose that $\Delta\theta = 2\pi/M$, where M is a multiple of 4. Our approach is therefore not completely general but it turns out that the end result does not depend on this specific choice of M. A consequence of Exercise 4.13 is that the Laplacian is invariant under rotation of the coordinate axes. Thus, under rotation by the angle θ_m, the Laplacian can be approximated by the standard (Cartesian) 5–point formula (10.6) to give (see Fig. 10.13, right)

$$4U_0 = U_{1,m} + U_{1,m+M/4} + U_{1,m+M/2} + U_{1,m+3M/4} + h^2 f_0, \qquad (10.34)$$

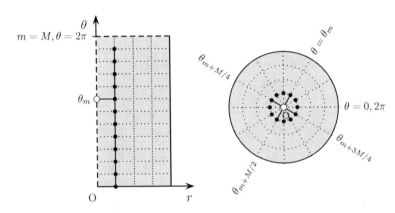

Fig. 10.13 A grid of dimensions h, $\Delta\theta$ for the rectangle $0 \le r \le 1$, $0 \le \theta < 2\pi$ in the $r\theta$-plane (*left*) is mapped by polar coordinates to a grid covering the unit circle (*right*). The finite difference approximation (10.35) at the origin involves M neighbouring points at $r = h$

where f_0 is the value of the source term at the origin. The nearest neighbours to the origin lie on the polar grid because M is a multiple of four. This formula is periodic in m of period $M/4$ and so provides $M/4$ independent formulae. Summing these and dividing by M then leads to the expression

$$U_0 = \frac{1}{M} \sum_{m=0}^{M-1} U_{0,m} + \frac{1}{4} h^2 f_0. \qquad (10.35)$$

Note that the expression (10.35) is well defined for any positive integer M (not just multiples of 4). Moreover, since it is the arithmetic mean of difference approximations that are consistent of order two, the discrete equation (10.35) must also be second-order accurate. $\diamond$

10.2.4 Regularity of Solutions

It has been tacitly assumed up to this point that, when assessing truncation errors, the solution u and its partial derivatives (usually up to order four) are well defined (bounded) functions on the domain Ω. We shall now investigate this assumption and give some indications of the consequences when it is not valid. This is a very delicate area of analysis and our aim is to give a flavour of what occurs in particular examples rather than a comprehensive overview.

We begin by considering the simplest case of Laplace's equation[8]

$$u_{xx} + u_{yy} = 0$$

in a simply connected domain Ω, when subject to a Dirichlet boundary condition $u = g$. Solutions of Laplace's equation possess derivatives of all orders on any open set in the plane so our attention focusses on the behaviour of a generic solution as the boundary is approached.

Let us suppose that the domain has a smooth boundary in the sense that it has a continuously turning tangent (as in an elliptical domain, or one similar to that depicted in Fig. 10.1, say). Any lack of smoothness in a solution must then be a consequence of the boundary values. A common situation occurs when the boundary data g is piecewise continuous. To take a concrete example, suppose that part of the boundary lies along the x-axis where $g(x) = 1$ for $x < a$ and is equal to zero otherwise (see Examples 4.8 and 8.4). In terms of polar coordinates $x = a + r \cos\theta$, $y = r \sin\theta$ centred at the discontinuity, Laplace's equation has the solution $u = A + B\theta$, for constants A and B. Setting $u = g$ when $y = 0$ then gives an analytic solution.

---

[8]Our conclusions will apply to general second-order elliptic PDEs with constant coefficients.

Fig. 10.14 A sector of a
circle having a re-entrant
corner with angle π/α and
an L-shaped domain with a
re-entrant angle $3\pi/2$

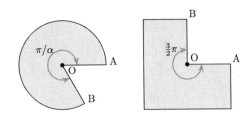

$$u(x, y) = \frac{1}{2} - \frac{1}{\pi} \tan^{-1}\left(\frac{x - a}{y}\right) + s(x, y),$$

where $s(x, y)$ is a smooth function with $s(x, 0) = 0$. This solution is singular when
we take the limit $r \to 0$. Indeed, it is easily shown that partial derivatives of order
n in x and y behave in proportion to $1/r^n$, so all derivatives become unbounded as
$r \to 0$.

The convergence theory that we have developed fails for problems such as this and,
indeed, numerical experiments confirm that the maximum global error $\|E\|_{h,\infty} \to c$
as $h \to 0$, where c is a positive nonzero constant. This would indicate that finite
difference methods cannot be used in the presence of discontinuous boundary data.
However, the use of Richardson extrapolation (see Example 10.5) suggests that the
global error E converges at a second-order rate.[9] Although seeming to contradict
each other, both results are correct. The maximum norm $\|E\|_{h,\infty}$ does not converge
to zero because, with successively finer grids, it measures errors at points that are
progressively closer to the discontinuity. Richardson extrapolation uses results on
three grids, of sizes h, $h/2$ and $h/4$ and provides an experimental order of convergence for points on the coarsest grid. Thus, during this process, results are compared
at points that stay a fixed distance from the discontinuity. These observations are
consistent with the global error being proportional to $(h/r)^2$ so, when $r = \mathcal{O}(h)$ we
find $\|E\|_{h,\infty} \to$ constant but, if P is any fixed point in the domain then $E|_P = \mathcal{O}(h^2)$.

When the boundary function is a little smoother so that it is continuous with a
piecewise continuous derivative we find that $\|E\|_{h,\infty} = \mathcal{O}(h)$ while, with a contin-
uous derivative and piecewise continuous second derivative, the standard converge
rate $\|E\|_{h,\infty} = \mathcal{O}(h^2)$ is regained.

Solutions of Laplace's equation may be *singular*, even when the boundary data
g is smooth, in any case where the boundary has one or more re-entrant corners as
depicted in Fig. 10.14. To illustrate this, let us suppose that $u = 0$ on the rays OA and
OB. Then, in terms of polar coordinates $x = r\cos\theta$, $y = r\sin\theta$ centred on O, the
general solution may be determined by separation of variables and may be shown to
be of the form

$$u = \sum_{n=1}^{\infty} A_n r^{n\alpha} \sin(n\alpha\theta),$$

[9]The observed quadratic rate of convergence is dependent on the point $x = a$ being a grid point,
with $g(a) = 1/2$ (see Definition 8.3 and the subsequent discussion).

where the interior angle at the corner is π/α and $\{A_n\}$ are arbitrary constants. When $1/2 < \alpha < 1$ the angle is reflex and all x and y partial derivatives of this solution are unbounded as $r \to 0$ (unless, by good fortune, $A_1 = 0$). The strength of the singularity is not as strong as in the case of discontinuous boundary data and numerical experiments[10] typically show that $\|E\|_{h,\infty} = \mathcal{O}(h^\alpha)$, whereas for a fixed point P a distance r from the origin, $E|_P = \mathcal{O}(h^{2\alpha})$, consistent with E being proportional to $(h^2/r)^\alpha$.

To complete the section, we briefly extend the discussion to cover the Poisson equation.

The next example shows that derivatives of a solution can be discontinuous even when the data appears to be smooth.

Example 10.15 Consider the boundary value problem $-(u_{xx} + u_{yy}) = 1$ in the quarter circle $\{(x, y) : x > 0, y > 0, x^2 + y^2 < 1\}$ with a homogeneous Dirichlet boundary condition, so that $u = 0$ on $x = 0$ and $y = 0$. Show that the partial derivatives u_{xx} and u_{yy} are not continuous at the origin.

On the x-axis we have $u(x, 0) = 0$ and so $u_{xx}(x, 0) = 0$. Whereas, on the y-axis, $u(0, y) = 0$ so $u_{yy}(0, y) = 0$ and using the PDE we find that $u_{xx}(0, y) = -1$. The result follows by taking the limits $x \to 0$ and $y \to 0$.   ◇

The crucial aspect of the previous example is that $-(u_{xx} + u_{yy}) = f(x, y)$ with $f(0, 0) \neq 0$. An explicit solution is given in Exercise 10.21 with a nonhomogeneous condition specified on the circular arc. Fortunately, the weak singularities in the corner have no discernible effect on the rate of convergence of finite difference solutions!

Solving the Poisson equation when the source term has jump discontinuities is usually problematic. In such cases, the standard 5–point approximation usually converges (despite the fact that the second partial derivatives are discontinuous and the third partial derivatives are unbounded) although the typical convergence rate will only be first order. Fornberg [4] has shown that second-order accuracy can often be restored by adjusting the values of f in a simple manner at grid points next to the discontinuity.

10.2.5 Anisotropic Diffusion

A feature of the (negative) Laplacian $-(u_{xx}+u_{yy})$ is its invariance under rotation (see Exercise 4.13). This makes it a faithful model of a diffusion process that acts equally in all directions (*isotropic* diffusion). To model an anisotropic diffusion process in two dimensions we will consider the general elliptic equation $\mathcal{L}u = f$, that was discussed in Sect. 4.2.3,

---

[10]Theory predicts that $\|E\|_{h,\infty} = \mathcal{O}(h^{\alpha-\varepsilon})$ for any positive number ε, no matter how small. We ignore ε in the discussion since it is not detectable in numerical experiments.

$$\mathcal{L}u = -\left(au_{xx} + 2bu_{xy} + cu_{yy}\right), \quad \text{with } b^2 < ac. \tag{10.36}$$

Note that we can assume, without loss of generality, that $a > 0$ from which it follows that $c > 0$. By interchanging a, x with c, y, if necessary, we can also assume that $a \geq c$.

Our first goal is to construct a 9–point finite difference approximation that is consistent of order two with the operator in (10.36) using only the grid points $(x_{\ell+j}, y_{m+k})$ for $j, k = 0, \pm 1$. If we add the proviso that the coefficients in the approximation must be inversely proportional to h^2, then there is a one-parameter family of methods (see Exercise 10.22), given by

$$\mathcal{L}_h U_{\ell,m} = -ah^{-2}\delta_x^2 U_{\ell,m} - 2bh^{-2}\Delta_x\Delta_y U_{\ell,m} - ch^{-2}\delta_y^2 U_{\ell,m} + \tfrac{1}{2}\gamma h^{-2}\delta_x^2\delta_y^2 U_{\ell,m}, \tag{10.37}$$

where γ is the parameter. The structure of this finite difference operator is shown in Fig. 10.15 using a sequence of four stencils, one for each term.

The inverse monotonicity of $\mathcal{L}_h$ is most conveniently investigated by expressing (10.37) in the form

$$\mathcal{L}_h U\big|_{\mathrm{P}} := \alpha_0 \, U\big|_{\mathrm{P}} - \sum_{k=1}^{8} \alpha_k U\big|_{Q_k}, \tag{10.38}$$

where $U\big|_{\mathrm{P}} := U_{\ell,m}$ and $\{U\big|_{Q_j}\}_{k=1}^8$ denote the values of U at the eight neighbouring grid points $Q_1, Q_2, \dots, Q_8$ listed counterclockwise, beginning at $(x_{\ell+1}, y_m)$. The idea here is to choose the parameter γ so that $\mathcal{L}_h$ is of *positive type* (as in Definition 10.8). In which case the argument followed in Sect. 6.3 can again be used to establish a discrete maximum principle.

To this end, making use of Fig. 10.15 gives

$$\left.\begin{aligned}
\alpha_0 &= 2h^{-2}(a + c - \gamma) \\
\alpha_1 = \alpha_5 &= \ h^{-2}(a - \gamma) \\
\alpha_2 = \alpha_6 &= \tfrac{1}{2}h^{-2}(\gamma - b) \\
\alpha_3 = \alpha_7 &= \ h^{-2}(c - \gamma) \\
\alpha_4 = \alpha_8 &= \tfrac{1}{2}h^{-2}(\gamma + b)
\end{aligned}\right\} \tag{10.39}$$

which will be non-negative (and satisfy the conditions of Definition 10.8) if γ is chosen so that

$$|b| \leq \gamma \leq \min(a, c). \tag{10.40}$$

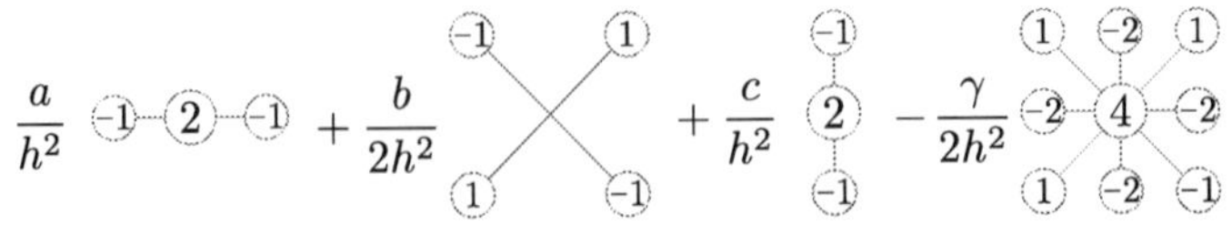

$$\frac{a}{h^2} \; \text{(-1)(-2)(-1)} \; + \; \frac{b}{2h^2} \; \boldsymbol{\times} \; + \; \frac{c}{h^2} \; \begin{matrix}\text{(-1)}\\\text{(2)}\\\text{(-1)}\end{matrix} \; - \; \frac{\gamma}{2h^2} \; \boldsymbol{\ast}$$

Fig. 10.15 A stencil representation of the operator $\mathcal{L}_h$ given in (10.37)

Fig. 10.16 The operator (10.36) is elliptic in the *shaded* area $1 \geq c/a > (b/a)^2$. The coefficients of the difference approximation (10.37) only satisfy the condition (10.31) in the cross-hatched region $1 \geq c/a \geq |b|/a$

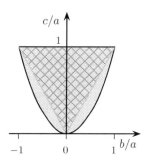

Since we arranged that $a \geq c$, this is possible if, and only if $c/a \geq |b|/a$. This region is shown cross-hatched in Fig. 10.16 and is a subset of the region $c/a \geq (b/a)^2$ (shown shaded in the same figure) where the operator (10.36) is elliptic. The upshot is that it is not possible to construct a second-order approximation of positive type—although it should be borne in mind that this is a sufficient, but not necessary, condition for inverse monotonicity.

This loss of positivity in the difference approximation has an elegant geometric interpretation. We know (from Sect. 4.2.3) that the effective diffusivity of the general second-order elliptic operator (10.36) is reflected in the level curves of the quadratic form $Q(x) = x^\mathsf{T} A x$, where $x = (x, y)$ and A is the matrix of PDE coefficients. Here, the major axis of any of the ellipses $Q(x) = $ constant makes an angle ϕ with the y-axis such that $|\phi| \leq \pi/4$ because of the condition $a \geq c$. The failed cases occur close to the bounding parabola $c/a = (b/a)^2$ (where the level curves degenerate to parallel lines) and so they become elongated—in fact, positivity is lost whenever the ratio of the lengths of the major to minor axes exceeds $1 + \sqrt{2}$, see Exercise B.8.

The problem above is easily fixed! What is needed is a different mesh size in the x and y directions. Setting these to h_x and h_y, respectively, (10.37) is replaced by the scheme

$$\mathcal{L}_h U_{\ell,m} := a h_x^{-2} \delta_x^2 U_{\ell,m} + 2 b h_x^{-1} h_y^{-1} \Delta_x \Delta_y U_{\ell,m} + c h_y^{-2} \delta_y^2 U_{\ell,m}$$
$$- \tfrac{1}{2} \gamma h_x^{-1} h_y^{-1} \delta_x^2 \delta_y^2 U_{\ell,m}. \quad (10.41)$$

Note that since h_x and h_y must both tend to zero simultaneously, it is natural to fix the mesh ratio h_x/h_y so that it is equal to ρ, say. As shown in Exercise 10.23, setting $\rho = \sqrt{a/c}$ and then selecting γ so that

$$|b| \leq \gamma \leq \sqrt{ac}, \quad (10.42)$$

ensures that the coefficients in (10.41) satisfy the required condition (10.31) for a positive type difference operator. Standard arguments can then be used to show that $\mathcal{L}_h$ in (10.41) is inverse monotone, so it is a stable approximation scheme for (10.36) in combination with a Dirichlet boundary condition. $\diamond$

10.2.6 Advection–Diffusion

To conclude our discussion of elliptic PDEs, we briefly consider finite difference approximation of the *steady* advection–diffusion equation, see (3.5), given in two dimensions by

$$- \varepsilon(u_{xx} + u_{yy}) + p u_x + q u_y = 0, \tag{10.43}$$

in situations where the advection terms $(p u_x + q u_y)$ dominate the diffusion term $-\varepsilon \nabla^2 u$. More specifically, if L is a length scale associated with the domain, we introduce the *Peclet* number, so that

$$\text{Pe} = \frac{L}{\varepsilon} \max\{|p|, |q|\} \tag{10.44}$$

and note that situations of interest in this section are characterised by $\text{Pe} \gg 1$. (Think of taking $\varepsilon = 10^{-6}$ with $\max\{|p|, |q|\} = 1$ as representative values in a unit square domain, so that $L = 1$). A one-dimensional example is considered in Example 6.14.

To avoid undue complexity, we shall suppose that the coefficients p and q are constant, and that a Dirichlet condition is imposed everywhere on the boundary of the unit square. As we have been at pains to emphasise throughout, the character of a PDE is normally dictated by the terms involving the highest derivatives in any coordinate direction. However, things are less clear cut when the coefficients of the highest derivative terms are particularly small relative to other coefficients. Taking the limit $\varepsilon \to 0$ in (10.43) gives the first-order PDE

$$p u_x + q u_y = 0 \tag{10.45}$$

which is sometimes referred to as the *reduced equation*. As discussed in Sect. 4.1 the general solution of (10.45) is constant along the characteristics

$$q x - p y = \text{constant}.$$

Each characteristic cuts the boundary twice so it is clear that a constant solution cannot satisfy the boundary condition at both end points. This over-specification can be resolved by recalling from Sect. 4.1 that boundary conditions on first-order PDEs should only be imposed on boundaries along which characteristics are directed into the domain. A discontinuity is created at points where characteristics leave the domain which, for the full advection–diffusion equation (10.43), is replaced by a layer of width $\mathcal{O}(\varepsilon)$, known as a *boundary layer*, over which the value of the solution changes by $\mathcal{O}(1)$. A specific instance of this abrupt solution behaviour is illustrated by the following example.

Example 10.16 Use separation of variables to determine the solution of the PDE

$$-\varepsilon(u_{xx} + u_{yy}) + u_x = 0$$

in the unit square, with boundary condition $u(0, y) = \sin \pi y$ $(0 < y < 1)$ and $u(x, y) = 0$ on the remainder of the boundary. Discuss the character of the solution when $0 < \varepsilon \ll 1$.

Following the construction used in Example 8.12 gives a simple solution, involving only the first term in the Fourier series,

$$u(x, y) = e^{x/(2\varepsilon)} \sin \pi y \frac{\sinh \gamma (1 - x)}{\sinh \gamma}, \quad \gamma = \frac{\sqrt{1 + 4\pi^2 \varepsilon^2}}{2\varepsilon}. \tag{10.46}$$

Unfortunately, this expression is not suitable for accurate computation of the solution when ε is small because the factors involving x and γ become extremely large. For instance, $\sinh \gamma > 10^{200}$ when $\varepsilon = 10^{-3}$. To make progress we need to express the hyperbolic sine terms as exponentials so that, after elementary algebraic manipulation, we arrive at the alternative representation,

$$u(x, y) = e^{x(1/(2\varepsilon) - \gamma)} \frac{1 - e^{-2\gamma(1-x)}}{1 - e^{-2\gamma}} \sin \pi y.$$

Note that, since $\gamma = \frac{1}{2\varepsilon} + \varepsilon \pi^2 + \mathcal{O}(\varepsilon^2)$ the term $e^{-2\gamma}$ in the denominator is exponentially small. Neglecting this term and truncating γ in the leading factor gives

$$u(x, y) \approx e^{-\varepsilon \pi^2 x} (1 - e^{-2\gamma(1-x)}) \sin \pi y, \tag{10.47}$$

which gives an excellent approximation to the solution for small values of ε. Note that the factor $1 - e^{-2\gamma(1-x)} \approx 1$ except when $1 - x = \mathcal{O}(\varepsilon)$, that is, when x lies in a thin boundary layer near $x = 1$. Moreover, outside this layer we have

$$u(x, y) \approx e^{-\varepsilon \pi^2 x} \sin \pi y \approx (1 - \varepsilon \pi^2 x) \sin \pi y + \mathcal{O}(\varepsilon^2), \tag{10.48}$$

the leading term of which is the solution of the reduced equation (10.45) with $p = 1$, $q = 0$ together with the condition $u(0, y) = \sin \pi y$ on the inflow boundary $x = 0$. The exact solution is shown in Fig. 10.17 (left) when $\varepsilon = 0.02$. ◇

The natural way to construct a finite difference approximation of (10.43) is to combine the standard 5–point stencil for the diffusion operator with standard central difference approximations

$$\triangle_x u_{\ell,m} = \tfrac{1}{2} h^{-1} (u_{\ell+1,m} - u_{\ell-1,m}), \quad \triangle_y u_{\ell,m} = \tfrac{1}{2} h^{-1} (u_{\ell,m+1} - u_{\ell+1,m}),$$

for the advection terms. For this specific choice, making use of the results in Table 6.1, the operator $\mathcal{L}_h$ defined by

$$\mathcal{L}_h U_{\ell,m} := -\varepsilon h^{-2} (\delta_x^2 + \delta_y^2) U_{\ell,m} + p h^{-1} \triangle_x U_{\ell,m} + q h^{-1} \triangle_y U_{\ell,m} \tag{10.49}$$

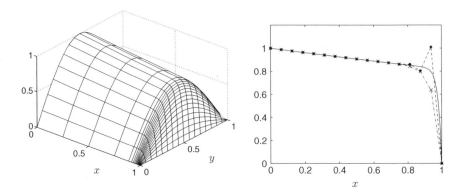

Fig. 10.17 *Left* The exact solution of Example 10.16 when $\varepsilon = 0.02$. *Right* The numerical solution along the line $y = 1/2$ with $M = 16$ using the centred difference approximation (10.49) ($\bullet$) and the upwind approximation (×). The exact solution is shown by a solid curve

is a second-order consistent approximation to $\mathcal{L}$. Note that the difference operator $\mathcal{L}_h$ has a 5–point stencil. Writing it in the form (10.30) with $\nu = 4$, we see that it is of positive type when

$$h|p| \le 2\varepsilon \quad \text{and} \quad h|q| \le 2\varepsilon, \qquad (10.50)$$

that is whenever $\mathrm{Pe}_h \le 1$, where $\mathrm{Pe}_h = h\max\{|p|, |q|\}/(2\varepsilon)$ is the *mesh* Peclet number. (So-called because it is based on specifying the length scale $L = h/2$ in (10.44)). Both inequalities (10.50) will be satisfied if h is sufficiently small, so that standard techniques can be used to establish that the numerical solution will converge at a quadratic rate as $h \to 0$.

Difficulties arise in practice when $\mathrm{Pe} \gg 1$ and it would be impractical to choose h so that $\mathrm{Pe}_h \le 1$. The resulting numerical solutions will invariably contain oscillations (often called "wiggles" in the literature after a famous paper of Gresho and Lee) as discussed in Example 6.14. They can also be seen in the right-hand plot in Fig. 10.17 ($\bullet$). A potential fix is pursued next.

Example 10.17 Show that the use of second-order central difference approximations of the advection term in the PDE $-\varepsilon(u_{xx} + u_{yy}) + u_x = 0$ is unsatisfactory if $h > 2\varepsilon$ and investigate ways in which the method may be improved.

A typical solution obtained from (10.49) with $p = 1$, $q = 0$ and $\mathrm{Pe}_h > 1$ is shown in Fig. 10.17 (right) with $\varepsilon = 0.02$ and $h = 1/16$. The reason for the oscillations is evident by taking the limit $\varepsilon \to 0$ in (10.49) in which case the method reduces to $U_{\ell+1,m} = U_{\ell-1,m}$. If M is even, this gives a numerical solution that is the same on alternate vertical grid lines so that

$$\sin(\pi hm) = U_{0,m} = U_{2,m} = U_{4,m} = \cdots = U_{M,m}$$

whereas

$$U_{1,m} = U_{3,m} = \cdots = U_{M+1,m} = 0.$$

One way of suppressing the wiggles is to replace central differencing of the advection terms by a one-sided "upwind" difference. The grounds for this strategy are that the solution of the reduced equation at any point P can be found by tracing the characteristic ($y = $ constant) through P backwards to the inflow boundary where the solution is known from the boundary condition.

Making this choice gives the difference operator

$$\mathcal{L}_h^- U_{\ell,m} := -\varepsilon h^{-2}(\delta_x^2 + \delta_y^2)U_{\ell,m} + h^{-1}\Delta_x^- U_{\ell,m} \tag{10.51}$$

which can be shown to be of positive type for all $h > 0$ (see Exercise 10.24). Note however that the improved stability comes at a price—the order of consistency of the approximation (10.51) is only linear. The solution of $\mathcal{L}_h^- U = 0$ with $\varepsilon = 0.02$ and $h = 1/16$ is shown in Fig. 10.17 (right) by crosses and, while it appears to be free of wiggles, the transition layer near $x = 1$ is smeared when compared to the exact solution.

The results of more extensive numerical testing are summarised in Fig. 10.18 for the central difference approximation scheme (left) and the upwind difference scheme (right). Data points corresponding to $Pe_h \leq 1$ are connected by solid lines and confirm that the central difference scheme ultimately converges at a second-order rate while the upwind scheme is only first-order accurate. A striking feature of the behaviour of the error is that it seems to increase roughly in proportion to $1/\varepsilon$ if h is kept fixed, in both cases. It can also be observed that the global error for the upwind scheme initially increases as h decreases (for fixed ε). Thus the upwind solution for large Peclet numbers is smooth but inaccurate. Eventually however, when h is small enough that $Pe_h < 1$, the upwind error decreases monotonically to zero.   ◊

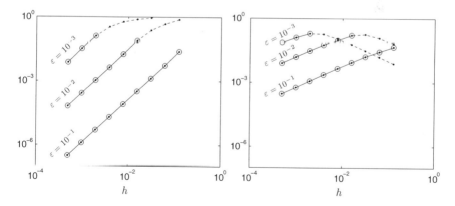

Fig. 10.18 Log-log plots of the global error $\|u - U\|$ versus h for solving the advection–diffusion problem in Example 10.17 using central differences (*left*), and upwind differences (*right*) for $\varepsilon = 10^{-1}$, 10^{-2} and 10^{-3}. The dashed lines connect data points where the mesh Peclet number $Pe_h > 1$

It is not appropriate to delve further into this fascinating and important topic here and we refer the interested reader to Stynes [22] who gives an authoritative survey of further techniques for tackling elliptic PDEs of this type. The books by Roos et al. [19] and Morton [15] also cover this material in greater depth as well as constructing approximation strategies for solving *unsteady* advection–diffusion equations.

Exercises

10.1 When the Poisson equation is solved over a rectangular, rather than a square domain, it may be appropriate to use a rectangular grid with spacings h_x and h_y in the x and y directions, respectively. Derive the following 5–point difference scheme,

$$2(\theta^{-1} + \theta)U_{\ell,m} - \theta^{-1}(U_{\ell+1,m} + U_{\ell-1,m}) - \theta(U_{\ell,m+1} + U_{\ell,m-1})$$
$$= h_x h_y f_{\ell,m},$$

for the grid point $(\ell h_x, m h_y)$, where $\theta = h_x / h_y$. Then show that the local truncation error, when appropriately scaled, is given by

$$-\tfrac{1}{12}\left(h_x^2 \partial_x^4 u + h_y^2 \partial_y^4 u\right)\big|_{(\ell h, m h)} + \mathcal{O}(h^4),$$

where $h = \max(h_x, h_y)$.

10.2 Suppose that, instead of a natural ordering of grid points, a red-black (or odd-even) ordering is used whereby the unknown grid values $U_{\ell,m}$ with $\ell + m$ even are ordered in a natural manner by columns to give a vector $\boldsymbol{u}_\mathrm{e}$ while those where $\ell + m$ are odd are similarly ordered to give a vector $\boldsymbol{u}_\mathrm{o}$. Using $M = 4$ show that the system may be written as $A\boldsymbol{u} = \boldsymbol{f}$, where

$$\boldsymbol{u} = \begin{bmatrix} \boldsymbol{u}_\mathrm{e} \\ \boldsymbol{u}_\mathrm{o} \end{bmatrix}, \qquad A = \begin{bmatrix} 4\,I_\mathrm{e} & B \\ B^\mathsf{T} & 4\,I_\mathrm{o} \end{bmatrix},$$

where I_e, I_o are identity matrices of dimension compatible with $\boldsymbol{u}_\mathrm{e}$ and $\boldsymbol{u}_\mathrm{o}$, respectively.

10.3 Consider the finite difference operator defined by

$$\mathcal{L}_h^\times U_{\ell,m} := \frac{1}{2h^2}[4U_{\ell,m} - U_{\ell+1,m+1} - U_{\ell-1,m+1} - U_{\ell-1,m-1} - U_{\ell+1,m-1}],$$

whose stencil is shown in Fig. 10.19. By first showing that

$$\mathcal{L}_h^\times U_{\ell,m} = \mathcal{L}_h^+ U_{\ell,m} - \tfrac{1}{2}h^{-2}\delta_x^2\delta_y^2 U_{\ell,m},$$

Fig. 10.19 The stencil of the
5–point finite difference
operator $\mathcal{L}_h^\times$ described in
Exercise 10.3

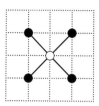

where $\mathcal{L}_h^+ := h^{-2}(\delta_x^2 + \delta_y^2)$ is the standard 5-point approximation of the Laplacian
(see (10.7)), or otherwise, verify that

$$\mathcal{L}_h^\times u_{\ell,m} = -\nabla^2 u\big|_{(\ell h, mh)} - \tfrac{1}{12}h^2\big(\partial_x^4 u + 6\partial_x^2\partial_y^2 u + \partial_y^4 u\big)\big|_{(\ell h, mh)} + \mathcal{O}(h^4)$$

so that $\mathcal{L}_h^\times u = f$ is consistent of second order with the Poisson equation (10.3).

10.4 Suppose that $\mathcal{L}_h^\times$ and $\mathcal{L}_h^+$ are as defined in Exercise 10.3. Show that there is a
value of λ for which

$$\lambda\mathcal{L}_h^\times u + (1-\lambda)\mathcal{L}_h^+ u = -\nabla^2 u - \tfrac{1}{12}h^2\,\nabla^4 u\big|_{(\ell h, mh)} + \mathcal{O}(h^4).$$

Use this result to construct a fourth-order 9–point difference approximation for
the Poisson equation. (Note: $\nabla^4 = (\nabla^2)^2 = \partial_x^4 + 2\partial_x^2\partial_y^2 + \partial_y^4$.)

10.5 Consider the solution of the Poisson equation (10.3) with a homogeneous
Dirichlet boundary condition on the unit square using the difference scheme $\mathcal{L}_h^\times U_{\ell,m} = f_{\ell,m}$ on a standard grid Ω_h where $\mathcal{L}_h^\times$ is defined in Exercise 10.3.

(a) Write the equations in matrix-vector form $A u = f$ using the standard columnwise ordering of grid points with $M = 4$. Identify the main features of the matrix
A and thereby generalise this to the case of an $M \times M$ grid.
(b) Suppose that an odd-even numbering of grid points is used as described in
Exercise 10.2. Show that the solution u_e on the "even" nodes can be calculated
independently of the unknowns u_o on the "odd" nodes and vice versa.

10.6 ☆ Suppose that $\mathcal{L}_h U \geq 0$ in Ω_h, where $\mathcal{L}_h$ is defined by (10.7) and Ω_h is a grid
imposed on the unit square. Deduce from Theorem 10.2 that either U is constant on
Ω_h or else it achieves its minimum value on $\partial\Omega_h$.
  If $\mathcal{B}_h$ is defined by (10.7) deduce that the operator $\mathcal{L}_h$ defined by (10.8) is inverse
monotone.

10.7 ☆ Suppose that the numerical solution U to a given BVP is known to converge at
a second-order rate and that solution values U_P^h and $U_P^{h/2}$ are computed at a particular
point P using grids of size h and $h/2$, respectively. Can you explain why the quantity
$(4U_P^h - U_P^{h/2})/3$ might provide a reliable estimate of the error in U_P^h?

10.8 Suppose that U denotes a finite difference approximation to Laplace's equation on the L-shaped region shown in Fig. 10.14 (right) with Dirichlet boundary conditions provided by the exact solution $u = r^{2/3} \sin(2\theta - \pi)/3$ (written in polar co-ordinates). The values of U_P^h for the point $P = (1/8, -1/8)$ are tabulated below.

h	1/8	1/16	1/32	1/64	1/128	1/256
U_P^h	0.31413	0.33203	0.33943	0.34206	0.34305	0.34343

Determine the EOC and give an estimate of the error in the computed value of U_P^h when $h = 1/256$.

10.9 ☆ Suppose that the Neumann boundary condition $-u_x = g(0, y)$ for the Poisson equation is approximated using a second-order central difference $-h^{-1}\Delta_x U_{0,m} = g(0, mh)$. Show that, if the fictitious grid value $U_{-1,m}$ is eliminated by making use of (10.6) with $\ell = 0$, then it reduces to the numerical boundary condition (10.22b).

10.10 By including remainder terms in the Taylor expansions (in the spirit of (6.5), say), show that the local truncation error in the boundary condition (10.22) may be expressed as

$$\mathcal{R}_h\big|_{0,m} = -\tfrac{1}{6}h^2 u_{xxx}(\xi, mh) - \tfrac{1}{24}h^2 u_{yyyy}(0, mh + \eta),$$

where $0 < \xi < h$ and $-h < \eta < h$.

10.11 Complete the details in the derivation of the Neumann condition (10.24) that is valid at the corner $(0, 1)$ of the unit square. Verify that the consistency error is second order with respect to h.

10.12 Suppose that the Poisson equation is solved using the standard 5–point difference scheme (10.6) over the rectangular domain shown in Fig. 10.6 (left) and that the solution is symmetric with respect to the y-axis. Give an explanation (based on the discussion immediately prior to Example 10.1) of why the time to compute the solution over the rectangle (without taking advantage of any symmetry) will be approximately four times longer than the time to compute the solution over the square part of the domain lying in $x \geq 0$.

How much longer would it take to solve the same problem over the full square domain shown in Fig. 10.6 (right) without taking advantage of any symmetry? Give reasons for your answer.

10.13 Consider the boundary value problem consisting of the Poisson equation in the unit square $0 < x, y < 1$ with source term $f(x, y) = xy$ and Dirichlet boundary condition $u(x, y) = x^2 + y^2$. Show that the solution u is symmetric with respect to the line $x = y$. That is, show that the problem is invariant to the change of variables $(x, y) \mapsto (y, x)$.

By exploiting the symmetry of the problem, show that the unknowns at grid points within the triangle $\{(x, y) : 0 < x < 1, 0 < y \leq x\}$ can be solved for independently of the remaining unknowns. What are the finite difference equations that hold at grid points $(\ell h, \ell h)$ lying on the diagonal of the square?

10.14 Consider the boundary value problem consisting of the Poisson equation $-\nabla^2 u = 1$ in the square $-1 < x, y < 1$ with a homogeneous Dirichlet boundary condition. Show that the solution u is symmetric with respect to the lines $x = 0$, $y = 0$, $y = x$ and $y = -x$.

Suppose that a grid is imposed on the domain by the grid lines $x = \ell h$ ($\ell = 0, \pm 1, \ldots, \pm M$) and $y = mh$ ($m = 0, \pm 1, \ldots, \pm M$) with $h = 1/M$. By exploiting these symmetries, show that the unknowns at grid points within the triangle $\{(x, y) : 0 < x < 1, 0 < y \le x\}$ can be solved for independently of the remaining unknowns. Write down the resulting finite difference matrix system in the case $h = 1/3$ and show that the equations may be scaled so that the coefficient matrix becomes symmetric.

10.15 Suppose that a grid is imposed on the domain $-1 < x, y < 1$ by the grid lines $x = -1 + \ell h$ ($\ell = 0, 1, \ldots, M$), and $y = -1 + mh$ ($m = 0, 1, \ldots, M$) with $h = 2/M$. Show how symmetry can be exploited so that a finite difference approximation to the solution of the boundary value problem described in the previous exercise can be computed by solving just three algebraic equations when $M = 5$. How many independent unknowns are there when M is an odd integer?

10.16* Suppose that the Neumann condition described in Sect. 10.2.1 is replaced by the Robin condition $-u_x + \sigma u = g(y)$ along $x = 0$.
    Construct the finite difference approximation analogous to (10.22).

10.17 Let Ω denote the interior of the triangle ABC for $A(0, 0)$, $B(0, 1)$ and $C(9/8, 0)$. Write down a finite difference replacement of the BVP

$$-\nabla^2 u + u = 0, \quad (x, y) \in \Omega$$
$$u = 0 \text{ on AB and BC}, \quad u = 4x(9 - 8x) \text{ on AC}$$

using a grid size $h = 1/4$, and assemble the resulting equations into a matrix-vector system.

10.18 Derive the finite difference approximation (10.29) of Laplace's equation in Example 10.7 that is valid at points $(x_\ell, 0)$ and show that, after appropriate scaling, it is consistent of order two with the Neumann boundary condition.

10.19 Show that the finite difference boundary operator (10.33b) may be written

$$\mathcal{B}_h U_{\ell,0} = -\Delta\theta^{-1} \Delta_\theta U_{\ell,0} - \tfrac{1}{2} r_\ell^2 \Delta\theta \mathcal{L}_h U_{\ell,0},$$

where $\mathcal{L}_h$ is defined by (10.33). Hence, verify that it is consistent of order two with the Neumann condition $-u_\theta(r, 0) = 0$.

10.20 Complete the proof of Theorem 10.13.

10.21 Show that the function

$$u(r, \theta) = \frac{1}{\pi}\left(\log(\frac{1}{r})\sin 2\theta + (\frac{\pi}{4} - \theta)\cos 2\theta - \frac{\pi}{4}\right)r^2$$

satisfies the boundary value problem in Example 10.15 with the nonzero Dirichlet condition $u(1, \theta) = \frac{1}{4}(1 - 4\theta/\pi) \cos 2\theta - \frac{1}{4}$ on the circular arc. Verify that $|u_{rr}| \to \infty$ as $r \to 0$ for all $0 < \theta < \frac{1}{2}\pi$.

10.22 ★ This exercise explores the construction of a finite difference approximation of the form (10.38) to the operator (10.36). To do this U is replaced by u in (10.38) and each of the eight terms $u(x_{\ell \pm 1}, y_{m \pm 1})$ is expanded in Taylor series about the point $P = (x_\ell, y_m)$. This results in an expression of the form

$$\mathcal{L}_h u|_P = C_0 u + h(C_1 u_x + C_2 u_y) + \tfrac{1}{2} h^2 C_3 u_{xx} + \cdots$$

where all the terms on the right are evaluated at P and where each of the coefficients C_k is a linear combination of $\{\alpha_j\}_{j=0}^8$ and is independent of h. Equating the terms in this expansion with those of $\mathcal{L}u$ up to (and including) all the terms in h^3 gives a total of ten linear equations, namely $C_k = 0$ for $k = 0, 1, 2, 6, 7, 8, 9$, and

$$h^2 C_3 = -a, \quad h^2 C_4 = -2b, \quad h^2 C_5 = -c,$$

in the nine constants $\alpha_0, \alpha_1, \ldots, \alpha_8$. Construct a 10×9 matrix C and a vector $\boldsymbol{a} \in \mathbb{R}^{10}$ such that these equations may be written in the form $C\boldsymbol{\alpha} = \boldsymbol{a}$.

Next, show that $\mathcal{L}v|_P = 0$ for all choices of $\{\alpha_j\}_{j=0}^8$ when $u = (x - x_{\ell-1})(x - x_\ell)(x - x_{\ell+1})$ and for a corresponding function involving y and y_m. Explain why this implies that the rank of C cannot exceed 8. By direct computation or otherwise, show that the rank of C is exactly 8 so that the system $C\boldsymbol{\alpha} = \boldsymbol{a}$ has a one parameter family of solutions. Deduce that this solution family gives the difference operator $\mathcal{L}_h U$ in (10.37) when the parameter is suitably chosen.

10.23 Consider the finite difference operator $\mathcal{L}_h$ defined by (10.41) when $h_x/h_y = \rho$. Show that it is possible to choose γ in such a way that $\mathcal{L}_h$ is of positive type when $\rho = \sqrt{a/c}$ and $b^2 < ac$.

10.24 ☆ Verify that the operator $\mathcal{L}_h^-$ defined by (10.51) is of positive type for all positive values of ε and h.

10.25 The upwind approximation of the PDE $-\varepsilon \nabla^2 u + p u_x = 0$ uses the backward difference approximation $u_x \approx h^{-1} \triangle_x^- u$ when $p > 0$ and the forward difference approximation $u_x \approx h^{-1} \triangle_x^+ u$ when $p < 0$. Show that the resulting difference operator can be written as

$$\mathcal{L}_h^\pm U_{\ell,m} := -\varepsilon h^{-2}(1 + |\mathrm{Pe}_h|)\delta_x^2 U_{\ell,m} - \varepsilon h^{-2}\delta_y^2 U_{\ell,m} + h^{-1} p \triangle_x U_{\ell,m},$$

where $\mathrm{Pe}_h = ph/(2\varepsilon)$. Generalize this result to construct an upwind difference approximation of the operator $\mathcal{L}$ in (10.43).

[Hint: Use Exercise 6.2 to establish the identity $\triangle_x^- = \triangle - \frac{1}{2}\delta_x^2$ with an analogous identity for $\triangle_x^+$.]

10.26 Express the advection–diffusion equation

$$-\varepsilon(u_{xx} + u_{yy}) - yu_x + xu_y = 0,$$

where $\varepsilon > 0$, in polar coordinates $x = r\cos\theta$, $y = r\sin\theta$. Construct a finite difference scheme that is consistent of second order with this PDE in a circle of unit radius centred on the origin using a polar grid with $r_\ell = \ell h$ ($h = 1/N$) and $\theta_m = m\Delta\theta$ ($\Delta\theta = 2\pi/M$). What, in particular, is the finite difference approximation at the origin?

Express the finite difference operator at a typical grid point with $\ell > 0$ in the form (10.30) and determine whether any restrictions on the grid parameters are needed in order that the discrete operator is of positive type.

Chapter 11
Finite Difference Methods for Parabolic PDEs

Abstract This self-contained chapter focuses on finite difference approximation of parabolic boundary value problems. Standard explicit and implicit time-stepping schemes are introduced and the quality of resulting numerical approximations is assessed using maximum principles as well as the classical Von Neumann stability framework. The development of method-of-lines software is discussed at the end of the chapter.

The focus of this chapter is on time dependent versions of the elliptic PDEs that feature in earlier chapters of the book. The associated parabolic PDE problems will take the specific form

$$u_t + \mathcal{L}u = f, \tag{11.1}$$

together with appropriate boundary and initial conditions, where $\mathcal{L}$ is an operator involving only spatial derivatives. For example, if there is only one space dimension then $\mathcal{L}$ is of the form (see Chap. 6)

$$\mathcal{L}u := -\varepsilon u_{xx} + au_x + bu, \tag{11.2}$$

where ε is a positive constant and a and b are functions of x and t.

The heat equation (that is (11.1) with $\mathcal{L}u := -u_{xx}$) will be employed as a prototype throughout the chapter to illustrate the main ideas relating to the construction and analysis of finite difference methods for parabolic equations. These ideas extend naturally to higher dimensions but the calculations are much more involved. To establish notation we suppose that (11.1) and (11.2) is to be solved on the domain

$$\Omega_\tau := \{(x, t) : 0 < x < 1, 0 < t \le \tau\} \tag{11.3}$$

with an initial condition $u(x, 0) = g(x)$ and Dirichlet boundary conditions

$$u(0, t) = g_0(t), \quad u(1, t) = g_1(t), \quad 0 < t \le \tau. \tag{11.4}$$

© Springer International Publishing Switzerland 2015
D.F. Griffiths et al., *Essential Partial Differential Equations*, Springer
Undergraduate Mathematics Series, DOI 10.1007/978-3-319-22569-2_11

In this setting, the value of u is specified on Γ_τ which comprises the boundary of Ω_τ excluding the top edge $t = \tau$ (shown by the lines marked with crosses in Fig. 11.1). Alternative boundary conditions will be considered later in the chapter. The value of τ is not usually specified in advance—computation of the solution may proceed, for example, until it is deemed to be sufficiently close to a steady state.

11.1 Time Stepping Algorithms

The closed domain $\overline{\Omega}_\tau$ (the shaded area in Fig. 11.1) is covered by a grid of lines parallel to the coordinate axes to give a set of grid points

$$\{(mh, nk) : m = 0, 1, 2, \ldots, M, n = 0, 1, 2, \ldots, N\}, \tag{11.5}$$

where $h = 1/M$ is the grid size in the x-direction and $k = \tau/N$ is the grid size in the t-direction. Our strategy for generating a numerical solution is to suppose that it has been computed at all time levels up to $t = nk$ ($n \geq 0$) and to devise an algorithm for determining the solution at the next time level $t = (n + 1)k$. The numerical solution at a typical grid point (x_m, t_n) will be denoted by U_m^n while the exact solution[1] at the same point is u_m^n. We shall also use U^n to denote the set of grid values at time level $t = nk$, the dot reminding us that the dependence on space has been suppressed. The simplest possible time stepping algorithm is studied in the next section.

11.1.1 An Explicit Method (FTCS)

The starting point for constructing a finite difference approximation to (11.1) and (11.2) is the Taylor expansion

$$u(x, t + k) = u(x, t) + k u_t(x, t) + \mathcal{O}(k^2). \tag{11.6}$$

Using the heat equation $u_t = u_{xx}$ as an illustration, we have

$$u(x, t + k) = u(x, t) + k u_{xx}(x, t) + \mathcal{O}(k^2).$$

If $x_m = mh$, $t_n = nk$ is a typical grid point in Fig. 11.1, then taking the centered difference approximation $u_{xx} = h^{-2} \delta_x^2 u + \mathcal{O}(h^2)$ gives

$$u_m^{n+1} = u_m^n + r \delta_x^2 u_m^n + \mathcal{O}(k^2) + \mathcal{O}(kh^2), \tag{11.7}$$

where $r = k/h^2$ is known as the *mesh ratio*. As usual, to develop a numerical method the remainder terms in (11.7) are discarded so the equation will no longer be satisfied

---

[1] A little care is needed to distinguish exponents from quantities (U, u, etc.) that are evaluated at time level n.

Fig. 11.1 Grid points $x = mh$, $t = nk$ divided into sets according to points (×) on Γ_τ where the solution is known from boundary or initial conditions, points (○) where the solution has previously been computed, and points (●) where it is about to be computed

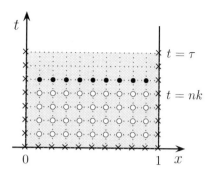

by the exact solution u of the PDE but instead by some grid function U. Thus the PDE $u_t = u_{xx}$ is replaced at (x_m, t_n) by the algebraic equation

$$U_m^{n+1} = U_m^n + r\delta_x^2 U_m^n. \tag{11.8a}$$

This is an example of a partial difference equation. It will be referred to as the FTCS scheme (an acronym for Forward Time, Centred Space). Recalling the definition of δ_x^2 from Table 6.1, the method may be written in the alternative form

$$U_m^{n+1} = rU_{m-1}^n + (1 - 2r)U_m^n + rU_{m+1}^n. \tag{11.8b}$$

The FTCS scheme provides a means of computing the approximate solution U^{n+1} at the advanced time level $t = (n + 1)k$ from known values U^n at the time level $t = nk$. For example, when $n = 0$ in (11.8b), the values of $U_1^1, U_2^1, \ldots, U_{M-1}^1$ may be calculated by setting $m = 1, 2, \ldots, M - 1$. The remaining values at the first time level are provided by the boundary conditions: $U_0^1 = g_0(t_1)$ and $U_M^1 = g_1(t_1)$. The computation then proceeds in an identical fashion with $n = 1$. Because the numerical solution at each grid point at the new time level is given directly in terms of known values the FTCS scheme is called an *explicit* method. It is associated with the *stencil* shown in Fig. 11.2. It is sometimes convenient to refer to the "anchor" and "target" points of a stencil. The anchor is the grid point at which the PDE is approximated and about which Taylor expansions are carried out—here the point (x_m, t_n) and the target (usually denoted by ○) is the grid point at which the solution is obtained—here the point (x_m, t_{n+1}).

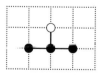

Fig. 11.2 The stencil for the FTCS method. The unknown value at the target point (x_m, t_{n+1}) (○) is computed from the three known grid values (●) at $t = t_n$

Example 11.1 Use the FTCS method to solve the heat equation on the domain Ω_τ with $\tau = 0.2$ with initial condition $u(x, 0) = \sin \pi x$ and end conditions $u(0, t) = u(1, t) = 0$. Compute the numerical solution with $h = 1/40$ and $k = rh^2$ with $r = 0.5$ ($k = 3.125 \times 10^{-4}$, $N = 640$ steps) and $r = 0.52$ ($k = 3.25 \times 10^{-4}$, $N = 616$ steps).

The exact solution is $u(x, t) = \exp(-\pi^2 t) \sin \pi x$ (see Fig. 8.1 with $n = 1$). The numerical solutions are computed at times $t = nk$ until the final time $\tau = 0.2$ and results are shown in Fig. 11.3 at every 80th time step. When $r = 0.5$ (left figure) the numerical and exact solutions are indistinguishable but, when the mesh ratio is increased to $r = 0.52$, oscillations appear which grow exponentially with time (when $t \approx 0.2$ the amplitude of the solution exceeds 3000, so the solution is not shown after $t_{480} = 0.156$). It will be shown later that this is a manifestation of an instability that occurs whenever $r > 0.5$.

A quantitative measure of the performance of the FTCS method is provided in Table 11.1 where the maximum error at $t \approx 0.2$ is shown for three grids ($h = 1/20, 1/40$ and $1/80$). In the case $r = 0.5$ the maximum global error reduces by a factor of 4 each time h is halved in accordance with an $\mathcal{O}(h^2)$ behaviour. In contrast, with $r = 0.52$ the global error for $h = 1/20$ is marginally larger than with $r = 0.5$ but, if h is reduced further then the error explodes!   ◇

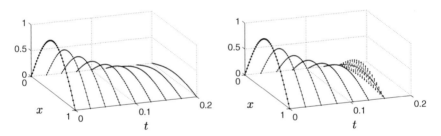

Fig. 11.3 FTCS solutions to the problem in Example 11.1 with $h = 1/40$, $r = 0.50$ (*left*) and $r = 0.52$ (*right*). The numerical solutions are shown by *dots* at every 80th time step. The *solid lines* show the corresponding exact solutions

Table 11.1 Results for the FTCS method in Example 11.1 with grids based on $M = 20, 40$ and 80 points and mesh ratios $r = 0.50, 0.52$

h	$r = 0.50$			$r = 0.52$	
	N	Max. error	Ratio	N	Max. error
1/20	160	0.0011	–	154	0.0012
1/40	640	0.00028	4.03	616	3360.3
1/80	2560	0.000070	4.01	2462	8.6×10^{64}

The maximum error is calculated after N time steps, when $t \approx 0.2$

In order to generalize the method to parabolic equations of the type (11.1), we suppose that the spatial operator $\mathcal{L}$ is approximated as described in Chap. 6 by a finite difference operator $\mathcal{L}_h$ that is consistent of order p. That is,

$$\mathcal{L}_h v = \mathcal{L}v + \mathcal{O}(h^p) \tag{11.9}$$

for any sufficiently smooth function v. By following the steps leading to (11.8) we find that

$$U_m^{n+1} = U_m^n - k\mathcal{L}_h U_m^n + kf_m^n. \tag{11.10}$$

On dividing both sides by k, it may be seen that the scheme could also be derived by making the finite difference replacements

$$\partial_t u_m^n \approx k^{-1}\Delta_t^+ U_m^n, \qquad \mathcal{L}u_m^n \approx \mathcal{L}_h U_m^n,$$

where Δ_t^+ denotes the forward difference in the t-direction. In view of this, the scheme (11.10) should be divided by k in order to be correctly scaled for checking its consistency. The resulting local truncation (or consistency) error is given by

$$\mathcal{R}_m^n = k^{-1}(u_m^{n+1} - u_m^n) + \mathcal{L}_h u_m^n - f_m^n \tag{11.11}$$

which leads, using (11.6) and (11.9), to the conclusion that $\mathcal{R}_m^n = \mathcal{O}(k) + \mathcal{O}(h^p)$. This tends to zero as $h, k \to 0$, establishing the consistency of the approximation with the PDE $u_t + \mathcal{L}u = f$.

In the specific case of the heat equation, the leading terms in the truncation error can be identified by expanding the Taylor series (as in Table 6.1) so that

$$u_m^{n+1} = u_m^n + ku_t\big|_m^n + \tfrac{1}{2}k^2 u_{tt}\big|_m^n + \mathcal{O}(k^3),$$
$$h^{-2}\delta_x^2 u_m^n = u_{xx}\big|_m^n - \tfrac{1}{12}h^2 u_{xxxx}\big|_m^n + \mathcal{O}(h^4).$$

Setting $u_t = u_{xx}$ then gives the estimate

$$\mathcal{R}_m^n = \tfrac{1}{2}ku_{tt}\big|_m^n - \tfrac{1}{12}h^2 u_{xxxx}\big|_m^n + \mathcal{O}(k^2) + \mathcal{O}(h^4) \tag{11.12}$$

for the local truncation error. This is first-order in k and second-order in h provided that u_{tt} and u_{xxxx} are bounded. Moreover, the error is "balanced" when $k \propto h^2$ (that is, when r is kept fixed) since the two leading terms both tend to zero at the same rate.

There is another, more fundamental, reason—which will surface in the next example—why k should tend to zero faster than h for an explicit method like FTCS.

Example 11.2 (Domain of dependence—the CFL condition) Show that the solution of the FTCS scheme (11.8) cannot converge, in general, to the solution of the heat equation when $h, k \to 0$ unless the ratio of the two parameters k/h also tends to 0.

The simplest situation concerns the pure initial value problem, where the heat equation is solved in the upper half plane $x \in \mathbb{R}, t > 0$ with given initial data $u(x, 0) = g(x)$.

Consider a point P having coordinates (X, T), independent of h and k, with $X > 0$, as shown in Fig. 11.4. In order to compute the numerical solution at P with the FTCS scheme (11.8) it is sufficient to first compute the solution at grid points marked with $\circ$ and these, in turn, depend on initial values along QR (marked with a ✗). The triangle PQR is the *domain of dependence* of P. (The numerical solution at the point P depends only on the numerical solution inside this triangle.) Similarly, the interval of initial values between Q and R is known as the *interval of dependence* of P. In contrast, it is known from the theory of characteristics (see Example 4.6) that the exact solution at the point P depends on the initial data along the entire x-axis. Thus, in general, for convergence of the numerical solution to the exact solution we require that the coordinates of Q and R should satisfy $x_Q \to -\infty$ and $x_R \to \infty$ as $h, k \to 0$. Setting $X = mh$ and $T = nk$, then

$$ x_Q = X - nh = X - \frac{h}{k}T, \qquad x_R = X + nh = X + \frac{h}{k}T, $$

so we deduce that k/h must tend to zero in the limit $h, k \to 0$.

To reinforce our assertion that this condition is necessary, let us suppose that there is a positive constant c such that $k/h \geq c$ as $h, k \to 0$. It follows that the interval QR remains of finite extent throughout the limit process. Thus, for an initial function $g(x)$ that vanishes in QR but is not identically zero, the numerical solution $U\big|_P$ at P is always zero while the exact solution $u\big|_P$ is non-zero, so $U\big|_P$ cannot converge to $u\big|_P$ as $h, k \to 0$.   ◇

This observation is expressed as a formal requirement in the next definition.

Definition 11.3 (*CFL condition*) Convergence of a finite difference approximation of a parabolic PDE cannot take place unless the domain of dependence of the numerical solution (at a generic point P) is identical to that of the exact solution in the limit $h, k \to 0$.

This requirement (which depends only on the shape of the stencil, not on the actual coefficients) provides a *necessary* condition for convergence of any explicit finite

Fig. 11.4 The domain of dependence (*shaded*) of a typical grid point P for a method with an FTCS-type stencil. The grid values marked ✗ are known from the initial condition and form the interval of dependence

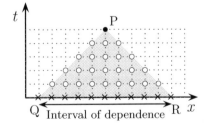

difference method for parabolic equations.[2] It is more widely applied to hyperbolic problems and it is in that context that we shall encounter it in the next chapter. The CFL condition identifies some specific finite difference methods (Exercise 11.6) or certain combinations of grid sizes as being nonconvergent. Note that there is no implication that satisfaction of the condition will lead to convergence—establishing this requires an assessment of the stability of the difference scheme (via inverse monotonicity, as done in earlier chapters).

Rather than considering the difficult issue of convergence to the solution on the whole domain Ω_τ, it is usual to establish convergence at a given time $t \in (0, \tau]$ as in following definition.

Definition 11.4 (*Convergence*) A finite difference solution U is said to converge to the exact solution u of a time dependent PDE at time t if

$$\lim_{\substack{k \to 0 \\ h \to 0}} \|u(\cdot, t) - U_\cdot^n\|_h = 0$$

when $nk = t$ is kept *fixed*[3]. Here $\| \cdot \|_h$ denotes any norm on functions defined on the spatial grid.

Having already established that the FTCS method (11.8) is *consistent* with the heat equation, convergence will follow if it can be shown that the method is *stable*. Our strategy will be the same as that used for Laplace's equation in the preceding chapter: a maximum principle implies inverse monotonicity which, with a suitable comparison function, will lead to ℓ_∞-stability. The most appropriate norm for establishing this chain of thought is the ℓ_∞ (maximum) norm (6.28), that is,

$$\|U_\cdot^n\|_{h,\infty} := \max_{0 \le m \le M} |U_m^n|. \tag{11.13}$$

Note that the index n is retained on the left hand side on (11.13) since the value of the norm is dependent on the specific time level t_n.

The theoretical results that follow lead to the establishment of stability and convergence with respect to the ℓ_∞ norm (11.13). They are presented in a manner that highlights the fact that the finite difference approximation satisfies properties analogous to those enjoyed by the exact solution of the heat equation (cf. Chap. 7).

Theorem 11.5 (Discrete maximum principle) *Suppose that the grid function U satisfies the inequalities*

$$-h^{-2}\delta_x^2 U_m^n + k^{-1}(U_m^{n+1} - U_m^n) \le 0 \tag{11.14}$$

for $(x_m, t_{n+1}) \in \Omega_\tau$ *and* $r = k/h^2 \le 1/2$ *then U is either constant or else attains its maximum value on* Γ_τ.

---

[2]It's importance was first recognised by Courant, Friedrichs and Lewy in 1928.

[3]Since k is determined in terms of h and r, it is unlikely that there is an integer n such that $nk = t$—it is sufficient that n be chosen so that $nk \to t$. Usually n is the largest integer such that $nk \le \tau$.

Proof First, we make the observation that if U is a constant function then (11.14) holds at every grid point with equality. Next, suppose that K_τ denotes the maximum value of U on Γ_τ, that is, define

$$K_\tau = \max_{(x_m, t_n) \in \Gamma_\tau} U_m^n.$$

The proof proceeds by induction on n. Since we know that $(x_m, t_0) \in \Gamma_\tau$, it follows that $U_m^0 \le K_\tau$ for $m = 0, 1, \ldots, M$. We suppose that U is bounded above by K_τ for all time levels up to, and including, t_j so that

$$U_m^j \le K_\tau \text{ for } m = 0, 1, \ldots, M.$$

Rearranging (11.14) with $n = j$ we find

$$U_m^{j+1} \le r U_{m-1}^j + (1 - 2r) U_m^j + r U_{m+1}^j, \quad m = 1, 2, \ldots, M - 1$$

and, because $U_{m-1}^j, U_m^j, U_{m+1}^j \le K_\tau$ and $r > 0$, $(1 - 2r) \ge 0$,

$$U_m^{j+1} \le \big(|r| + |1 - 2r| + |r|\big) K_\tau = K_\tau, \quad m = 1, 2, \ldots, M - 1.$$

Also, with $m = 0$ or $m = M$, U_m^{j+1} are boundary values so they are bounded above by K_τ. The induction hypothesis therefore holds at time level t_{j+1}. ☐

Corollary 11.6 *If the grid function U satisfies*

$$-h^{-2} \delta_x^2 U_m^n + k^{-1} (U_m^{n+1} - U_m^n) = 0$$

for $(x_m, t_{n+1}) \in \Omega_\tau$ and if $r \le 1/2$ then U attains its maximum and minimum values on Γ_τ. If the function U also satisfies a zero boundary condition ($U_0^n = U_M^n = 0$) then

$$\|U_\cdot^n\|_{h,\infty} \le \|U_\cdot^0\|_{h,\infty} \quad n = 1, 2, \ldots. \tag{11.15}$$

Proof See Exercise 11.7. ☐

The inequality (11.15) establishes stability with respect to initial conditions. Next, suppose that we define an operator $\mathcal{L}_h$ in terms of the left hand side of (11.14), that is,

$$\mathcal{L}_h U_m^n := -h^{-2} \delta_x^2 U_m^n + k^{-1} (U_m^{n+1} - U_m^n). \tag{11.16}$$

Is the FTCS operator $\mathcal{L}_h$ of *positive type*? Rearranging (11.16) to give

$$\mathcal{L}_h U\big|_P = \alpha_0 U\big|_P - \sum_{j=1}^{3} \alpha_j U\big|_{Q_j} \tag{11.17}$$

(as in Definition 10.8) with the target point $P(x_m, t_{n+1})$ and neighbours (x_m, t_n), $(x_{m\pm 1}, t_n)$, the coefficients are $\alpha_0 = k^{-1}$, $\alpha_1 = \alpha_3 = h^{-2}$ and $\alpha_2 = k^{-1}(1 - 2r)$. Checking the inequality conditions (10.31) confirms that $\mathcal{L}_h$ is a positive type operator whenever $r \leq 1/2$.

Corollary 11.7 (Inverse monotonicity) *The operator $\mathcal{L}_h$ defined by*

$$\mathcal{L}_h U_m^n = \begin{cases} -h^{-2}\delta_x^2 U_m^n + k^{-1}(U_m^{n+1} - U_m^n) & \text{for } (x_m, t_{n+1}) \in \Omega_\tau, \\ U_m^n & \text{for } (x_m, t_n) \in \Gamma_\tau \end{cases}$$

is inverse monotone whenever $r \leq 1/2$.

Proof Inverse monotonicity (that is, $\mathcal{L}_h U \geq 0$ implies that $U \geq 0$) is covered in Exercise 11.8, while stability follows since $\Phi_m^n := 1 + t_n$ satisfies $\mathcal{L}_h \Phi \geq 1$ on both Ω_τ and Γ_τ and therefore provides a suitable comparison function (see Lemma 6.8).[4]   □

Corollary 11.8 *Suppose that the partial derivatives u_{tt} and u_{xxxx} of the solution to the heat equation together with Dirichlet boundary conditions and a specified initial condition, are bounded on the domain Ω_τ. If $r \leq 1/2$ then the FTCS approximation (11.8) is convergent at the rate $\mathcal{O}(k) + \mathcal{O}(h^2)$.*

Proof This follows from Theorem 6.7 and the results established above.   □

Before discussing other (possibly superior) finite difference approximation strategies, the complications that arise when a linear reaction term is added to the basic heat equation model will be studied.

Example 11.9 Investigate the convergence of the FTCS method

$$U_m^{n+1} = rU_{m-1}^n + (1 - 2r + \gamma k)U_m^n + rU_{m+1}^n \qquad (11.18)$$

for solving the PDE $u_t = u_{xx} + \gamma u$ (where γ is a constant) for $0 < x < 1, t > 0$, given an initial condition and homogeneous Dirichlet boundary conditions.

The consistency of (11.18) is assessed in Exercise 11.2 so the focus of attention switches to the discrete maximum principle. The first prerequisite for extending the proof of Theorem 11.5 is that the coefficients on the right hand side of (11.18) should be non-negative. This requires that $1 - 2r + \gamma k \geq 0$ which is automatically satisfied if $\gamma k \geq 2r$, that is, if γ is big enough so that $\gamma h^2 \geq 2$. (Note that this cannot hold as $h \to 0$.) Otherwise, if $\gamma h^2 < 2$, the inequality can be rearranged to give the time

---

[4]Note that the stability constant $C := \max_{m,n} \Phi_m^n = 1 + \tau$ grows with τ, which precludes the use of this type of analysis to study behaviour as $t \to \infty$. With homogeneous Dirichlet boundary conditions it is sufficient to take $\Phi_m^n = 1$ and the inequality (11.15) follows.

step restriction

$$k \le \frac{h^2}{2 - \gamma h^2},\tag{11.19}$$

which is a small perturbation of the corresponding limit $r \le 1/2$ for the heat equation.

When $\gamma \le 0$ the argument proceeds as in Theorem 11.5 but, when $\gamma > 0$, the PDE has the potential for growing solutions so neither it, nor the finite difference approximation satisfies a maximum principle. In this case the question of stability is tackled directly. Taking the absolute value of both sides of (11.18), using the triangle inequality and assuming that $1 - 2r + \gamma k \ge 0$ gives

$$\begin{aligned}
|U_m^{n+1}| &\le |r||U_{m-1}^n| + |1 - 2r + \gamma k||U_m^n| + |r||U_{m+1}^n| \\
&\le (|r| + |1 - 2r + \gamma k| + |r|) \, \|U_\cdot^n\|_{h,\infty} \\
&= (1 + \gamma k)\|U_\cdot^n\|_{h,\infty}.
\end{aligned}$$

The inequality holds for the value of m for which $U_m^{n+1} = \|U_\cdot^{n+1}\|_{h,\infty}$ (and trivially at the endpoints, since $U_0^{n+1} = U_M^{n+1} = 0$). Thus,

$$\|U_\cdot^{n+1}\|_{h,\infty} \le (1 + \gamma k)\|U_\cdot^n\|_{h,\infty}$$

and so $\|U_\cdot^n\|_{h,\infty} \le (1 + \gamma k)^n\|U_\cdot^0\|_{h,\infty}$. Using the inequality $1 + z \le e^z$ (valid for all real z), with $z = \gamma k$ then gives

$$\|U_\cdot^n\|_{h,\infty} \le e^{\gamma t_n}\|U_\cdot^n\|_{h,\infty} \le e^{\gamma \tau}\|U_\cdot^0\|_{h,\infty}$$

since $t_n \le \tau$. We therefore have stability in the maximum norm in the sense that

$$\|U_\cdot^n\|_{h,\infty} \le C\|U_\cdot^0\|_{h,\infty}\tag{11.20}$$

with stability constant $C = \exp(\gamma \tau)$ that is independent of both h and k provided that (11.19) holds.   ◊

The proof leading up to (11.20) (with $\gamma = 0$) could also have been used to establish (11.15) directly—indeed, it is probably the most common approach adopted in the literature.

In summary, the FTCS method has in its favour its inherent simplicity and the fact that it is second-order convergent in space. The big issue is the restriction $r \le 1/2$ on the time step—if there are M grid points in space then in excess of $2\tau M^3$ grid points are required in order to integrate the heat equation over the interval $(0, \tau]$. The method requires five arithmetic operations per grid point so the cost of the method is also proportional to the number of grid points. In order to increase the accuracy of the results by one decimal place, one would expect to increase M by a factor $\sqrt{10}$ and the cost would increase by a factor of $10^{3/2} > 30$; two decimal places would increase the cost by a factor of 1000! The alternative difference scheme introduced in the next section is much more stable—it will not be necessary to fix the ratio k/h^2 when the spatial grid is refined.

11.1.2 An Implicit Method (BTCS)

In order to construct a second method for advancing the heat equation solution from t_n to t_{n+1} we use

$$u(x, t - k) = u(x, t) - ku_t(x, t) + \mathcal{O}(k^2) \tag{11.21}$$

instead of the Taylor expansion (11.6). Applying the PDE leads to

$$u(x, t) - ku_{xx}(x, t) = u(x, t - k) + \mathcal{O}(k^2)$$

and, when (x, t) is the grid point (x_m, t_{n+1}) and u_{xx} is approximated by $h^{-2}\delta_x^2 u$, we arrive at the finite difference method

$$U_m^{n+1} - r\delta_x^2 U_m^{n+1} = U_m^n, \tag{11.22a}$$

which can also be written as

$$-rU_{m-1}^{n+1} + (1 + 2r)U_m^{n+1} - rU_{m+1}^{n+1} = U_m^n. \tag{11.22b}$$

The resulting difference approximation involves three consecutive unknown values of U. The method is *implicit* since it requires the solution of a set of linear algebraic equations at each time level in order to determine the set of values U^{n+1}. The approximation method (11.22) will be referred to as the BTCS scheme (an acronym for Backward Time, Centred Space) because it can also be derived by making the finite difference replacements

$$\partial_t u_m^{n+1} \approx k^{-1}\triangle_t^- U_m^{n+1}, \qquad u_{xx}\big|_m^{n+1} \approx h^{-2}\delta_x^2 U_m^{n+1}$$

where $\triangle_t^-$ denotes the backward difference in the t-direction—see Table 6.1. The associated finite difference stencil is shown in Fig. 11.5.

Fig. 11.5 The stencil for the BTCS method. The target and anchor points (denoted by o) are coincident at the point (x_m, t_{n+1})

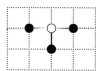

The local truncation error is given by

$$\mathcal{R}_m^n = k^{-1}(u_m^{n+1} - u_m^n) - h^{-2}\delta_x^2 u_m^{n+1} \tag{11.23}$$

which, by using Taylor expansions centred on $x = x_m$, $t = t_{n+1}$, leads to

$$\mathcal{R}_m^n = -\tfrac{1}{2}k u_{tt}\big|_m^{n+1} - \tfrac{1}{12}h^2 u_{xxxx}\big|_m^{n+1} + \mathcal{O}(k^2) + \mathcal{O}(h^4). \tag{11.24}$$

This implies that the BTCS scheme is consistent of order $\mathcal{O}(k) + \mathcal{O}(h^2)$. (The same order as the FTCS scheme.)

In contrast to the FTCS method (where unknown grid values are calculated one at a time), the unknown values at $t = t_{n+1}$ are all computed simultaneously in the BTCS approach (as shown in Fig. 11.1). To interpret this using linear algebra, suppose that the heat equation is supplemented by end conditions $u(0, t) = g_0(t)$ and $u(1, t) = g_1(t)$ for $t \geq 0$, where g_0 and g_1 are given continuous functions, together with an initial condition $u(x, 0) = g(x)$ for $0 < x < 1$. Then, defining

$$\boldsymbol{u}^n = [U_1^n, U_2^n, \ldots, U_{M-1}^n]^\mathsf{T}$$

to be a vector containing grid values at time level $t = nk$, with a similar definition for $\boldsymbol{u}^{n+1}$, and with the $(M-1) \times (M-1)$ tridiagonal matrix A and vector $\boldsymbol{f}^{n+1}$ so that

$$A = \begin{bmatrix} 1+2r & -r & 0 & \cdots & 0 \\ -r & 1+2r & -r & & \\ 0 & -r & \ddots & \ddots & \\ & & \ddots & & -r \\ 0 & & & -r & 1+2r \end{bmatrix}, \quad \boldsymbol{f}^{n+1} = \begin{bmatrix} r\,g_0(t_{n+1}) \\ 0 \\ \vdots \\ 0 \\ r\,g_1(t_{n+1}) \end{bmatrix},$$

the BTCS scheme (11.22) can be expressed as the matrix-vector system

$$A\boldsymbol{u}^{n+1} = \boldsymbol{u}^n + \boldsymbol{f}^{n+1}, \quad n = 0, 1, 2, \ldots. \tag{11.25}$$

In (11.25) the vector $\boldsymbol{f}^{n+1}$ contains the contributions from the boundary data and the vector $\boldsymbol{u}^0$ is known from the initial condition $U_m^0 = g(mh)$. The matrix A is positive definite (cf. Lemma 6.1)—and therefore nonsingular—so the systems $A\boldsymbol{u}^1 = \boldsymbol{u}^0 + \boldsymbol{f}^1$, $A\boldsymbol{u}^2 = \boldsymbol{u}^1 + \boldsymbol{f}^2$, ... have unique solutions that may be determined sequentially to provide the numerical solution at any time t_n.

It may appear that a considerable amount of work has to be carried out for each value of n but this is not the case in this simple one-dimensional setting. The system of linear equations at each time step involves the same matrix A which, being positive definite, has a Cholesky factorization $A = R^\mathsf{T} R$, where R is upper triangular and *bidiagonal* (only the entries $R_{i,i}$ and $R_{i,i+1}$ are nonzero). The factorization is only

Table 11.2 Maximum errors after N time steps (at $t = 0.2$) for the BTCS solution in Example 11.15 for mesh ratios $r = 0.5$ and $r = 5$

h	$r = 0.5$				$r = 5$		
	Max. error				Max. error		
	N	FTCS	BTCS	Ratio	N	BTCS	Ratio
1/20	160	0.0011	0.0023	–	16	0.017	–
1/40	640	0.00028	0.00056	4.00	64	0.0043	3.94
1/80	2560	0.000070	0.00014	4.00	256	0.0011	3.98

performed once, then on every step, the two systems

$$R^{\mathsf{T}} v = u^n + f^{n+1}, \qquad R u^n = v$$

are solved in which v is an intermediate vector. In this case, the work per time step is roughly twice that of the FTCS method.[5] This means that the two schemes, FTCS and BTCS, generate numerical solutions having the same local truncation error at a comparable cost. Is one better than the other?

Example 11.10 Use the BTCS method to solve the heat equation on the domain Ω_τ with $\tau = 0.2$ with initial condition $u(x, 0) = \sin \pi x$ and with end conditions $u(0, t) = u(1, t) = 0$. Compute the numerical solution with $h = 1/40$ and $k = rh^2$ with $r = 0.5$ ($k = 3.125 \times 10^{-4}$, $N = 640$ steps) and $r = 5$ ($k = 3.125 \times 10^{-3}$, $N = 64$ steps) and compare the errors at the final time with those obtained using the FTCS method (cf. Example 11.1).

When $r = 0.5$ the BTCS errors reported in Table 11.2 are approximately twice as large as the corresponding FTCS errors (a consequence of the respective local truncation errors being in the ratio $(r + 1/6)/(r - 1/6)$—see Exercise 11.1 and Exercise 11.9). This means that the BTCS scheme is not cost effective (compared to FTCS with the same grid ratio) since the computation is roughly twice the cost. The advantage of BTCS is seen when larger values of r are used. The numerical solutions remain perfectly smooth and show no indication of instability. Unfortunately, the global error is proportional to $r + 1/6$ so, as shown in Table 11.2, it grows by a factor of 8 with this increase in r.   ◇

The stability of the BTCS approach is confirmed by the following theoretical analysis.

---

[5]The cost of the initial Cholesky factorization is approximately $5M$ arithmetic operations and the cost per step is subsequently approximately $6M$ operations.

Theorem 11.11 (Discrete maximum principle) *Suppose that the grid function U satisfies the inequalities*

$$-h^{-2}\delta_x^2 U_m^{n+1} + k^{-1}(U_m^{n+1} - U_m^n) \leq 0 \tag{11.26}$$

for $(x_m, t_{n+1}) \in \Omega_\tau$ then U is either constant or else attains its maximum value on Γ_τ.

Proof The proof is by induction on n with the hypothesis that $U_m^n \leq K_\tau$, where K_τ is defined in Theorem 11.5 and where the hypothesis is seen to be true at $n = 0$. Rewriting the inequality (11.26) with $n = j$ as

$$(1 + 2r) U_m^{j+1} \leq U_m^j + r (U_{m+1}^{j+1} + U_{m-1}^{j+1})$$

and, supposing that $U_m^j \leq K_\tau$, we find

$$(1 + 2r) U_m^{j+1} \leq K_\tau + 2r \max_{0 \leq i \leq M} U_i^{j+1}, \quad m = 1, 2, \ldots, M - 1.$$

The right hand side is independent of m so this inequality holds when the left hand side is maximized with respect to m and therefore

$$(1 + 2r) \max_{0 < m < M} U_m^{j+1} \leq K_\tau + 2r \max_{0 \leq i \leq M} U_i^{j+1}.$$

If the maximum occurs at an interior point, $m = m^*$ say, we deduce that

$$(1 + 2r) U_{m^*}^{j+1} \leq K_\tau + 2r U_{m^*}^{j+1},$$

from which it follows that
$$U_m^{j+1} \leq U_{m^*}^{j+1} \leq K_\tau.$$

Otherwise the maximum at the time level $(j + 1)$ occurs on the boundary $m = 0$ or $m = M$. In either case the induction hypothesis holds with $n = j + 1$. □

Corollary 11.12 *If the grid function U satisfies*

$$-h^{-2}\delta_x^2 U_m^{n+1} + k^{-1}(U_m^{n+1} - U_m^n) = 0$$

for $(x_m, t_{n+1}) \in \Omega_\tau$ then U attains its maximum and minimum values on Γ_τ. If the function U also satisfies a zero boundary condition ($U_0^n = U_M^n = 0$) then

$$\|U_\cdot^n\|_{h,\infty} \leq \|U_\cdot^0\|_{h,\infty}, \quad n = 1, 2, \ldots. \tag{11.27}$$

Proof See Exercise 11.11. □

Corollary 11.13 (Inverse monotonicity) *The operator $\mathcal{L}_h$ defined by*

$$\mathcal{L}_h U_m^n = \begin{cases} -h^{-2}\delta_x^2 U_m^{n+1} + k^{-1}(U_m^{n+1} - U_m^n) & \text{for } (x_m, t_{n+1}) \in \Omega_\tau, \\ U_m^n & \text{for } (x_m, t_n) \in \Gamma_\tau \end{cases}$$

is inverse monotone and stable for all $r > 0$.

Proof Follows from the proof of Corollary 11.7. □

Corollary 11.14 (Convergence) *Suppose that the partial derivatives u_{tt} and u_{xxxx} of the solution to the heat equation together with Dirichlet boundary conditions and a specified initial condition, are bounded on the domain Ω_τ. The BTCS approximation (11.8) is convergent at the rate $\mathcal{O}(k) + \mathcal{O}(h^2)$.*

Proof This follows from Theorem 6.7 and the results established above. □

To summarise, the BTCS scheme overcomes the main deficiency of the FTCS scheme—its conditional stability. In the next section we target a second deficiency—the first-order accuracy with respect to the time step k.

11.1.3 The θ-Method

The next idea is to look at linear combinations of the FTCS and BTCS methods that were analysed in the preceding sections. This idea will be developed for the general PDE $u_t + \mathcal{L}u = f$, where $\mathcal{L}$ is given by (11.2), assuming (to simplify the notation) that the coefficients of $\mathcal{L}$ are independent of t.

When $\mathcal{L}$ is replaced by a finite difference approximation $\mathcal{L}u = \mathcal{L}_h u + \mathcal{O}(h^p)$, the FTCS method applied to $u_t + \mathcal{L}u = f$ leads to (11.10)

$$U_m^{n+1} = U_m^n - k\mathcal{L}_h U_m^n + kf_m^n, \tag{11.28}$$

whose local truncation error (given by (11.11) and relabelled $\mathcal{R}^+$) is given by

$$\mathcal{R}^+\big|_m^n = \tfrac{1}{2}k u_{tt}\big|_m^n + \tfrac{1}{6}k^2 u_{ttt}\big|_m^n + \mathcal{O}(k^3) + \mathcal{O}(h^p).$$

The BTCS method applied to the same PDE results in

$$U_m^{n+1} = U_m^n - k\mathcal{L}_h U_m^{n+1} + kf_m^{n+1} \tag{11.29}$$

whose local truncation error is

$$\begin{aligned} \mathcal{R}^-\big|_m^n &= k^{-1}(u_m^{n+1} - u_m^n) + \mathcal{L}_h u_m^{n+1} - f_m^{n+1} \\ &= -\tfrac{1}{2}k u_{tt}\big|_m^{n+1} + \tfrac{1}{6}k^2 u_{ttt}\big|_m^{n+1} + \mathcal{O}(k^3) + \mathcal{O}(h^p), \end{aligned}$$

where, following the derivation of (11.23), the Taylor series expansions have been centred on $x = x_m$ and $t = t_{n+1}$. We now form a family of methods by taking a convex combination[6] of the two methods. That is, a multiple $(1 - \theta)$ of (11.28) is added to a multiple θ of (11.29) to give

$$U_m^{n+1} = U_m^n - k\left[\theta(\mathcal{L}_h U_m^{n+1} - f_m^{n+1}) + (1 - \theta)(\mathcal{L}_h U_m^n - f_m^n)\right], \qquad (11.30)$$

which we will refer to as the θ–method. The local truncation error is, by virtue of its construction, given by $\mathcal{R}^\theta = \theta\mathcal{R}^- + (1 - \theta)\mathcal{R}^+$, although a little care is required in combining $\mathcal{R}^-$ and $\mathcal{R}^+$ because they are evaluated at different times. If we let $t_{n+1/2} = t_n + \frac{1}{2}k$, then the Taylor expansions

$$\phi_m^n = \phi_m^{n+1/2} - \tfrac{1}{2}k\phi_t\big|_m^{n+1/2} + \mathcal{O}(k^2)$$
$$\phi_m^{n+1} = \phi_m^{n+1/2} + \tfrac{1}{2}k\phi_t\big|_m^{n+1/2} + \mathcal{O}(k^2)$$

applied to a function $\phi(x, t)$ leads to

$$u_{tt}\big|_m^{n+1} - u_{tt}\big|_m^n = k u_{ttt}\big|_m^{n+1/2} + \mathcal{O}(k^2),$$

when $\phi = u_{tt}$, and to

$$u_{ttt}\big|_m^{n+1} + u_{ttt}\big|_m^n = 2u_{ttt}\big|_m^{n+1/2} + \mathcal{O}(k^2),$$

when $\phi = u_{ttt}$. Combining these results gives the asymptotic estimate

$$\mathcal{R}^\theta\big|_m^{n+1/2} = (\tfrac{1}{2} - \theta)k u_{tt}\big|_m^{n+1/2} - \tfrac{1}{12}k^2 u_{ttt}\big|_m^{n+1/2} + \mathcal{O}(k^3) + \mathcal{O}(h^p), \qquad (11.31)$$

which is, in general, $\mathcal{O}(k) + \mathcal{O}(h^p)$, the same as its constituent parts. The choices $\theta = 0$ and $\theta = 1$ reduce to the FTCS and BTCS methods, respectively. However, choosing $\theta = 1/2$ increases the temporal accuracy to second order and the resulting scheme is called the *Crank–Nicolson* method.[7] It is given by

$$(1 + \tfrac{1}{2}k\mathcal{L}_h)\, U_m^{n+1} = (1 - \tfrac{1}{2}k\mathcal{L}_h)\, U_m^n + \tfrac{1}{2}k(f_m^{n+1} + f_m^n), \qquad (11.32)$$

and has the local truncation error

$$\mathcal{R}^{\frac{1}{2}}\big|_m^{n+1/2} = -\tfrac{1}{12}k^2 u_{ttt}\big|_m^{n+1/2} + \mathcal{O}(k^3) + \mathcal{O}(h^p). \qquad (11.33)$$

---

[6]A convex combination is a linear combination in which the coefficients are non-negative and sum to one.

[7]Phyllis Nicolson (1917–1968) was another pioneer of numerical PDEs who made a name for herself while living in Manchester. She led a very interesting life.

This scheme can also be derived by evaluating the PDE $u_t + \mathcal{L}u = f$ midway between time levels at $x = x_m$, $t = t_n + \frac{1}{2}k$ and employing the finite difference replacements

$$\partial_t u_m^{n+1/2} = k^{-1}\delta_t u_m^{n+1/2} + \mathcal{O}(k^2), \quad \mathcal{L}u_m^{n+1/2} = \frac{1}{2}\mathcal{L}_h(u_m^{n+1} + u_m^n) + \mathcal{O}(k^2) + \mathcal{O}(h^p),$$

together with $f_m^{n+1/2} = \frac{1}{2}(f_m^{n+1} + f_m^n) + \mathcal{O}(k^2)$, each of which is second-order accurate in time. Note that when the coefficients in $\mathcal{L}$ depend on t, the second of these should be written as

$$\mathcal{L}u_m^{n+1/2} = \frac{1}{2}\big[(\mathcal{L}_h u)_m^{n+1} + (\mathcal{L}_h u)_m^n\big] + \mathcal{O}(k^2).$$

For the heat equation $\mathcal{L}u = -u_{xx}$, $\mathcal{L}_h u = -h^{-2}\delta_x^2 u$, $f = 0$, the Crank–Nicolson method takes the simple form

$$\left(1 - \tfrac{1}{2}r\delta_x^2\right) U_m^{n+1} = \left(1 + \tfrac{1}{2}r\delta_x^2\right) U_m^n, \tag{11.34a}$$

which can be also be written as

$$-\tfrac{1}{2}r U_{m-1}^{n+1} + (1+r) U_m^{n+1} - \tfrac{1}{2}r U_{m+1}^{n+1} = \tfrac{1}{2}r U_{m-1}^n + (1-r) U_m^n + \tfrac{1}{2}r U_{m+1}^n. \tag{11.34b}$$

Since it involves three values of U at the next time level the Crank–Nicolson method is *implicit*. The associated difference stencil is shown in Fig. 11.6.

Note that the coefficients in the expression (11.34b) are the same as those for the BTCS method in (11.22b) with r replaced by $r/2$ and $-r/2$ on the left and right, respectively. Thus, if we denote the matrix in (11.25) by $A(r) := A$, then the equation (11.34b) can be expressed as the matrix-vector system

$$A(r/2)\,\boldsymbol{u}^{n+1} = A(-r/2)\,\boldsymbol{u}^n + \boldsymbol{f}^{n+1/2}, \tag{11.35}$$

where, if $f \neq 0$, the source term is given by $\boldsymbol{f}^{n+1/2} = \frac{1}{2}(\boldsymbol{f}^{n+1} + \boldsymbol{f}^n)$. The matrix $A(r/2)$ is positive definite (because $A(r)$ is) so the system (11.35) has a unique solution at each time level.

To evaluate the computational expense of (11.35), we note that the cost of evaluating the right-hand side of (11.35) is about the same as that of one step of FTCS, while the cost of solving the linear system is the same as that a single BTCS step. Since a BTCS step is about twice the cost of a FTCS step, the computational expense

Fig. 11.6 The stencil for the Crank–Nicolson method. The ∘ symbol indicates the target point (x_m, t_{n+1})

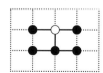

is effectively three times that of FTCS. The question is: does Crank–Nicolson lead to a better approximation?

Example 11.15 Use the Crank–Nicolson method to solve the heat equation on the domain Ω_τ with $\tau = 0.2$ with initial condition $u(x, 0) = \sin \pi x$ and with end conditions $u(0, t) = u(1, t) = 0$. Compute the numerical solution with $r = 5$ ($k = 5h^2$) and $k = h$ with grid sizes $h = 1/20, 1/40$ and $1/80$ and compare the errors at the final time with those obtained using the BTCS method (cf. Example 11.10).

The Crank–Nicolson results are presented in Table 11.3, together with the corresponding BTCS results (the errors for $r = 5$ are taken from Table 11.2). When $r = 5$ the global errors for Crank–Nicolson are significantly smaller than those for BTCS. A possible explanation for the error not reducing by a factor of 4 when h is halved is that it is dominated by temporal errors on the coarsest grid but, as h is reduced, it is dominated by spatial error (the temporal error is proportional to $k^2 = 25h^4$, making it negligibly small). It is when $k = h$ (so that the time and space contributions to the local truncation error are balanced) that the Crank–Nicolson method really comes into its own. Relatively few time steps are required to achieve high accuracy and a second-order rate of convergence is achieved without any trace of instability. The BTCS method is only first-order accurate in this comparison.   ◊

Is this convergence behaviour typical, or is it specific to this example? The answer to this question can be found by investigating the theoretical properties of the Crank–Nicolson approximation. This will turn out to be more subtle than for either of the first-order schemes.

Theorem 11.16 (Discrete maximum principle) *Suppose that the grid function U satisfies the inequalities*

$$- \tfrac{1}{2}h^{-2}\delta_x^2(U_m^{n+1} + U_m^n) + k^{-1}(U_m^{n+1} - U_m^n) \le 0 \qquad (11.36)$$

for $(x_m, t_{n+1}) \in \Omega_\tau$ and $r = k/h^2 \le 1$ then U is either constant or else attains its maximum value on Γ_τ.

Table 11.3 Maximum errors after N time steps (when $t = 0.2$) for the Crank–Nicolson (C–N) solution in Example 11.15 when $r = 5$ and $r = 1/h$

$1/h$	$r = 5$				$k = h$				
	Max. error				Max. error			Max. error	
	N	BTCS	C–N	Ratio	N	BTCS	Ratio	C–N	Ratio
20	16	0.017	0.00022	–	4	0.063	–	0.0051	–
40	64	0.0043	0.00012	1.82	8	0.033	1.92	0.0013	4.05
80	256	0.0011	0.000034	3.52	16	0.017	1.96	0.00031	4.01

Proof This result can again be established by induction on n with the hypothesis that $U_m^n \leq K_\tau$, where K_τ is defined in Theorem 11.5 and where the hypothesis is seen to be true at $n = 0$. We suppose that an intermediate grid function $U^{j+1/2}$ is computed using the right hand side of (11.34b) (with $n = j$), that is,

$$U_m^{j+1/2} = \tfrac{1}{2} r U_{m-1}^j + (1 - r) U_m^j + \tfrac{1}{2} r U_{m+1}^j. \tag{11.37}$$

This corresponds to one step of FTCS with a time step $k/2$ so, using the argument from the proof of Theorem 11.5, we have that

$$\max_{0 < m < M} U_m^{j+1/2} \leq K_\tau$$

provided $r \leq 1$. The inequality (11.36) then becomes

$$- \tfrac{1}{2} r U_{m-1}^{j+1} + (1 + r) U_m^{j+1} - \tfrac{1}{2} r U_{m+1}^{j+1} \leq U_m^{j+1/2} \tag{11.38}$$

which corresponds to the inequality (11.26) for the BTCS method with a time step $k/2$. The desired result then follows by applying the same argument used in the proof of Theorem 11.11. □

Corollary 11.17 *If the grid function U satisfies*

$$- \tfrac{1}{2} h^{-2} \delta_x^2 (U_m^{n+1} + U_m^n) + k^{-1} (U_m^{n+1} - U_m^n) = 0$$

for $(x_m, t_{n+1}) \in \Omega_\tau$ and $r \leq 1$ then U attains its maximum and minimum values on Γ_τ. Furthermore, if U is subject to homogeneous Dirichlet BCs, then

$$\|U_.^n\|_{h,\infty} \leq \|U_.^0\|_{h,\infty}, \quad n = 1, 2, \ldots. \tag{11.39}$$

Proof See Exercise 11.15. □

Corollary 11.18 (Inverse monotonicity) *The operator $\mathcal{L}_h$ defined by*

$$\mathcal{L}_h U_m^n = \begin{cases} - \tfrac{1}{2} h^{-2} \delta_x^2 (U_m^{n+1} + U_m^n) + k^{-1} (U_m^{n+1} - U_m^n) & \text{for } (x_m, t_{n+1}) \in \Omega_\tau, \\ U_m^n & \text{for } (x_m, t_n) \in \Gamma_\tau \end{cases}$$

is inverse monotone and stable for $0 < r \leq 1$.

Proof Follows from the proof of Corollary 11.7. □

The restriction $r \leq 1$ raises a concern about the stability of the scheme, although the need to enforce it is not evident from the numerical results in Example 11.15. In fact, the restriction $r \leq 1$ turns out to be necessary and sufficient for the Crank–Nicolson scheme to be inverse monotone (see Exercise 11.16). To explore this issue further, we recall that the generic definition of stability in Definition 6.6 for the equation $\mathscr{L}_h U_m^n = \mathscr{F}_h|_m^n$, with $\mathscr{L}_h$ given in Corollary 11.18 and source term

$$\mathscr{F}_h|_m^n = \begin{cases} U_m^0, & n = 0 \\ 0 & n > 0 \end{cases},$$

requires that

$$\|U_\cdot^n\|_{h,\infty} \leq C\|U_\cdot^0\|_{h,\infty}, \quad n = 1, 2, \ldots \tag{11.40}$$

with a positive constant C (the stability constant) that is independent of h and k. Such a bound can in fact be established for the Crank–Nicolson scheme for the heat equation (with $C \leq 23$, but we would conjecture a value of $C \leq 3$ as being more realistic), although the proof is beyond the scope of this book. This result implies that the restriction $r \leq 1$ needed for (11.39) is *sufficient* but not necessary for stability with respect to the maximum norm.

The above discussion highlights a weakness in seeking a theory based on positive type operators. A simple way of circumventing the issue is to analyse approximation methods using a different norm—which is what we will do in the next section. Before that, an example is constructed that reveals a defect in the Crank–Nicolson method: a propensity for generating oscillatory solutions when the mesh ratio is large and the initial data is not smooth.

Example 11.19   Use the Crank–Nicolson method to solve the heat equation on the domain Ω_τ with initial and Dirichlet boundary data taken from the exact solution $u(x, t) = \mathrm{erf}((x - 1/2)/\sqrt{4t})$. Compute the numerical solution with $h = k = 1/40$ (that is, for $r = 1$) and show that the numerical solution is highly oscillatory.

The initial condition has a discontinuity at $x = 1/2$ and is shown in Fig. 11.7 along with the numerical solution (dots connected by dashed lines) and exact solution (solid curves) for the first three time steps. Note that the discontinuity leaves an artefact in the numerical solution that oscillates in time around the exact solution!   ◊

The unwanted oscillations can be removed by making a simple modification to the Crank–Nicolson method at the first time step. This correction is based on the observation made in the proof of Theorem 11.16 that the Crank–Nicolson update is equivalent to the consecutive applications of the FTCS and BTCS methods with time steps $k/2$. The fix is simply to replace the first occurrence of the half-step FTCS method (at $t = 0$) by a half-step of the BTCS method. Since the same coefficient matrix is involved at all time steps only one Cholesky factorization is required so the cost of the modification is negligible. The results are shown on the right of Fig. 11.7. The improvement is dramatic: the numerical and exact solutions actually become indistinguishable.[8]

---

[8]Our fix is a refinement of a suggestion due to Rannacher [17] that the time integration be initiated with two steps of BTCS before reverting to the Crank–Nicolson method.

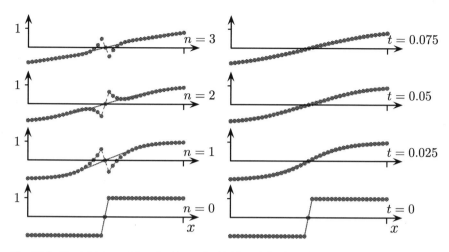

Fig. 11.7 The initial condition for Example 11.19 and the first three time steps of the Crank–Nicolson solution shown by *dots* connected by *dashed lines*. The exact solution is shown by the *solid curve*. On the *right hand side* the first half time step of FTCS in the Crank–Nicolson method is replaced by a half step of BTCS

11.2 Von Neumann Stability

This section of the book is devoted to "classical" numerical analysis. The aim is to construct a relatively simple test for distinguishing between stable and unstable finite difference methods.[9] This will be done by measuring the stability of a grid function in the ℓ_2 (or "root mean square") norm (as opposed to the ℓ_∞ or maximum norm)

$$\|U_\cdot^n\|_{h,2} := \left(h \sum_{m=0}^{M}{}'' |U_m^n|^2 \right)^{1/2}, \tag{11.41}$$

where the double prime on the summation sign indicates that the first and last terms should be halved, so

$$\|U_\cdot^n\|_{h,2}^2 = h \left(\tfrac{1}{2}|U_0^n|^2 + |U_1^n|^2 + \cdots + |U_{M-1}^n|^2 + \tfrac{1}{2}|U_M^n|^2 \right).$$

---

[9]The test is due to John von Neumann (1903–1957). Apart from being a brilliant mathematician, his name is synonymous with the early development of computers—motivated by the need to solve PDEs.

Note that the presence of the factor h ensures that

$$\|u\|_{h,2} \rightarrow \left(\int_0^1 |u(x)|^2 \, dx \right)^{1/2}$$

for any continuous function u as $h \rightarrow 0$.

Stability in this norm—known as ℓ_2 stability—requires a constant C, independent of both h and k, such that

$$\|U_{\cdot}^n\|_{h,2} \leq C\|U_{\cdot}^0\|_{h,2}, \quad n = 1, 2, \ldots. \tag{11.42}$$

A consequence of Exercise 11.17 is the bound $\|U_{\cdot}^n\|_{h,2} \leq \|U_{\cdot}^n\|_{h,\infty}$. Combining this with (11.40) leads to stability in the least squares norm for bounded initial data

$$\|U_{\cdot}^n\|_{h,2} \leq \|U_{\cdot}^n\|_{h,\infty} \leq C\|U_{\cdot}^0\|_{h,\infty}. \tag{11.43}$$

Note that (11.43) implies that ℓ_2 stability is a weaker requirement than maximum norm stability.

The von Neumann approach is to check to see if the finite difference scheme is stable in this weaker norm. The essence of the test is to look for a function $\xi(\kappa)$ of some real variable κ (the wavenumber) so that the finite difference equations have a solution of the form[10]

$$U_m^n = \xi^n e^{i\kappa m h}, \quad i = \sqrt{-1}. \tag{11.44}$$

The ansatz functions are called *discrete Fourier modes*. Since this expression is unchanged if an integer multiple of $2\pi M$ is added to κ, it is sufficient to consider values of κ in the interval $[-\pi M, \pi M]$, that is, $\kappa h \in [-\pi, \pi]$. We observe that $\kappa = 0$ corresponds to a grid function that is constant in space, while $\kappa h = \pm\pi$ gives $U_m^n = \xi^n(-1)^m$ which represents a wave of wavelength $2h$, the shortest that can be represented on the spatial grid. Next, if we substitute (11.44) into (11.41) and assume a periodic boundary condition then we find that

$$\|U_{\cdot}^n\|_{h,2} = |\xi(\kappa)|^n \tag{11.45}$$

so a solution of the form (11.44) will grow or decay in time depending on whether $|\xi| > 1$ or $|\xi| < 1$. For this reason, ξ is known as the *amplification factor* of the scheme being examined.

The von Neumann test is valid only when the expression for ξ does not depend on either m or n when (11.44) is substituted into the target difference scheme. In practice this means that the test is only applicable to finite difference approximation of problems with constant coefficients. In problems where the exact solution decays in time it is usually appropriate to require that $|\xi(\kappa)| \leq 1$ for all $\kappa h \in [-\pi, \pi]$ in

---

[10]The quantity ξ^n is the nth power of ξ and not a superscript. It could be written as $(\xi)^n$ were it not so ugly.

which case the stability inequality (11.42) holds with stability constant $C = 1$ (since $\|U^0\|_{h,2} = 1$). For problems where the exact solution may grow in time, we require that a constant c be found, independent of h and k, such that

$$|\xi| \le 1 + ck, \tag{11.46}$$

for all $\kappa h \in [-\pi, \pi]$. It follows, using the inequality $1 + z \le e^z$ (valid for any positive real number z), that

$$|\xi|^n \le (1 + ck)^n \le e^{cnk} \le e^{c\tau},$$

since $nk = t_n \le \tau$. In this case (11.42) holds with stability constant $C = e^{c\tau}$.

Definition 11.20 (*von Neumann stability*) A finite difference approximation of a parabolic PDE with a solution of the form $U_m^n = \xi^n e^{i\kappa mh}$ is said to be *von Neumann* (or ℓ_2) *stable*, if a non-negative constant c can be found, independent of both h and k, such that $|\xi| \le 1 + ck$ for all $\kappa h \in [-\pi, \pi]$.

The following relationships prove useful when testing the stability of difference schemes

$$\begin{aligned}
U_m^{n+1} &= \xi^{n+1} e^{i\kappa mh} &&= \xi U_m^n, \\
U_{m\pm 1}^n &= \xi^n e^{i\kappa(m\pm 1)h} &&= e^{\pm i\kappa h} U_m^n, \\
U_{m\pm 1}^{n+1} &= \xi^{n+1} e^{i\kappa(m\pm 1)h} &&= \xi e^{\pm i\kappa h} U_m^n.
\end{aligned} \tag{11.47a}$$

In particular, we note the following results

$$\delta_x^2 U_m^n = [e^{-i\kappa h} - 2 + e^{i\kappa h}] U_m^n = -4 \sin^2(\tfrac{1}{2}\kappa h)\, U_m^n, \tag{11.47b}$$

$$\Delta_x U_m^n = [e^{i\kappa h} - e^{-i\kappa h}]/2 = i \sin(\kappa h)\, U_m^n. \tag{11.47c}$$

The following examples will show the usefulness of the approach.

Example 11.21 Investigate the ℓ_2 stability of the FTCS approximation of the heat equation, that is

$$U_m^{n+1} = U_m^n + r\delta_x^2 U_m^n. \tag{11.48}$$

Substituting $U_m^n = \xi^n e^{i\kappa mh}$ into (11.48) and using (11.47b) we find that

$$\xi = 1 - 4r \sin^2(\tfrac{1}{2}\kappa h). \tag{11.49}$$

Since solutions of the heat equation decay in time we simply take $c = 0$ in (11.46), so the von Neumann condition for stability is that

$$|\xi| \le 1, \quad \text{for all real } \kappa h \in [-\pi, \pi].$$

Thus, since ξ in (11.49) is real, the condition for stability is that

$$-1 \leq 1 - 4r \sin^2(\tfrac{1}{2}\kappa h) \leq 1.$$

The right inequality is satisfied for all $r > 0$ while the left inequality requires that

$$r \sin^2(\tfrac{1}{2}\kappa h) \leq \tfrac{1}{2},$$

for all values of $\kappa h \in [-\pi, \pi]$. The left hand side is maximized when $\kappa h = \pm\pi$ and so it will be satisfied for all κ provided $0 < r \leq \tfrac{1}{2}$. This is, coincidentally, the same condition that was derived earlier for stability in the maximum norm. The above theory seems to be consistent with the computational results presented in Example 11.1. Instability is associated with the highest possible wavenumber $\kappa h = \pi$, which explains the sawtooth appearance of the FTCS solution shown in Fig. 11.3 when $r = 0.52 > 0.5$. There is, however, a hidden subtlety in the final column of Table 11.1. For $r = 0.52$ the amplification factor $\xi = 1 - 4r \sin^2(\tfrac{1}{2}\pi h)$ actually satisfies $|\xi| < 1$ for all the tabulated values of h! This suggests that the numerical results should be stable for this value of r.[11]   ◇

Example 11.22   Show that Crank–Nicolson approximation of the heat equation (11.34a) is unconditionally stable in an ℓ_2 sense.

Using (11.47b) it is readily shown that the amplification factor is given by

$$\xi = \frac{1 - 2r \sin^2(\tfrac{1}{2}\kappa h)}{1 + 2r \sin^2(\tfrac{1}{2}\kappa h)} \tag{11.50}$$

which satisfies $-1 \leq \xi \leq 1$ for all $\kappa h \in [-\pi, \pi]$ and for all $r > 0$ and is therefore unconditionally stable. This contrasts with Corollary 11.17 where stability with respect to the maximum norm could only be established under the condition $r \leq 1$.   ◇

Example 11.23   Investigate the ℓ_2 stability of the FTCS approximation of the heat equation with a reaction term $u_t = u_{xx} + \gamma u$, that is

$$U_m^{n+1} = U_m^n + r\delta_x^2 U_m^n + \gamma k U_m^n. \tag{11.51}$$

---

[11] The explanation for this apparent paradox is rounding error. The computations would be stable if done in exact arithmetic. If we add the term $2\sin(\kappa x_m) \times 10^{-16}$ with $\kappa = \pi/h$ to the initial condition to simulate round-off effects, then the amplification factor (11.49) predicts growth by a factor $|1 - 4r| = 1.08$ (when $r = 0.52$) at each time step, which is consistent with the tabulated results.

Substituting $U_m^n = \xi^n e^{i\kappa m h}$ into (11.51) and using (11.47b) we find that

$$\xi = 1 - 4r \sin^2(\tfrac{1}{2}\kappa h) + \gamma k. \tag{11.52}$$

In the case $\gamma > 0$ there is the possibility (dependent on initial data and boundary conditions) that the exact solutions may grow in time, in which case the appropriate von Neumann stability condition is

$$|\xi| \leq 1 + \gamma k, \qquad \kappa h \in [-\pi, \pi].$$

Since ξ in (11.52) is real, this leads to the conditions

$$-1 - \gamma k \leq 1 - 4r \sin^2(\tfrac{1}{2}\kappa h) + \gamma k \leq 1 + \gamma k.$$

The right inequality is satisfied for all $r > 0$ while the left inequality requires that $2r \sin^2(\tfrac{1}{2}\kappa h) \leq 1 + \gamma k$ for all values of $\kappa h \in [-\pi, \pi]$. The worst case occurs when $\kappa h = \pm\pi$, so the condition that must be satisfied for von Neumann stability is $0 < r \leq \tfrac{1}{2}(1 + \gamma k)$. This is the same condition that was established in Example 11.9 for stability in the maximum norm.   ◊

A final comment on the previous example is that although restrictions on the maximum time step are critically important in numerical computations, it is the behaviour as $h, k \to 0$ that is decisive when considering stability in the context of convergence theory. This requires only that $r < 1/2$, regardless of the magnitude of γ or whether it is positive or negative. Thus, as far as convergence is concerned, the addition of the term γu to the PDE has no effect on the relationship between h and k. This is true more generally: stability in the limit $h, k \to 0$ is dictated by the approximations to the highest derivatives that occur in each independent variable, lower derivative (or undifferentiated) terms serve as small perturbations.

The amplification factor was real in all the examples considered thus far. The final example illustrates the possibility of having a complex-valued ξ.

Example 11.24   Investigate the ℓ_2 stability of the FTCS approximation of the advection–diffusion equation $u_t = \varepsilon u_{xx} + u_x$, that is

$$U_m^{n+1} = U_m^n + r\delta_x^2 U_m^n + \rho \triangle_x U_m^n \tag{11.53}$$

where $r = \varepsilon k/h^2$, $\varepsilon > 0$ and $\rho = k/h$.

Substituting $U_m^n = \xi^n e^{i\kappa m h}$ into (11.53) and using (11.47b)–(11.47c) we find that

$$\xi = 1 - 4r \sin^2(\tfrac{1}{2}\kappa h) + i \rho \sin(\kappa h) \tag{11.54}$$

which is clearly complex. The appropriate von Neumann condition for stability is $|\xi| \leq 1$ $(\kappa h \in [-\pi, \pi])$, which is more conveniently rewritten as

$$|\xi|^2 - 1 \leq 0. \tag{11.55}$$

To make progress, we set $s = \sin^2(\frac{1}{2}\kappa h)$ and use the half-angle formula $\sin(\kappa h) = 2\sin(\frac{1}{2}\kappa h)\cos(\frac{1}{2}\kappa h)$, which leads to

$$\begin{aligned}|\xi|^2 - 1 &= (1 - 4rs)^2 - 1 + \rho^2 \sin^2(\kappa h) \\ &= -8rs(1 - 2rs) + 4\rho^2 s(1 - s) \\ &= -4s\left[2r(1 - 2rs) - \rho^2(1 - s)\right],\end{aligned}$$

which is required to be non-positive for all $s \in [0, 1]$. Next, since the expression in square brackets is linear in s, it will be non-negative for $s \in [0, 1]$ if, and only if, it is non-negative at the endpoints $s = 0, 1$. This immediately leads to the following condition on the parameters

$$\tfrac{1}{2}\rho^2 \leq r \leq \tfrac{1}{2}, \tag{11.56}$$

which must be satisfied if the difference scheme (11.53) is to be ℓ_2 stable.

It seems sensible to insist[12] that $\rho = k/h \leq 1$. Setting $r = \varepsilon k/h^2 > 0$ and introducing the mesh Peclet number $\text{Pe}_h = h/(2\varepsilon)$ (cf. Example 10.16), we find that the condition (11.56) translates to a time step restriction $k \leq h^2/2\varepsilon$ when $\text{Pe}_h \leq 1$ (with a h-independent restriction $k \leq 2\varepsilon$ otherwise). The importance of the condition $\text{Pe}_h \leq 1$ in establishing a maximum principle for the FTCS scheme is explored in Exercise 11.21. The advection term dominates diffusion when $\text{Pe}_h > 1$ so it may be advisable in these situations to base finite difference schemes on the underlying hyperbolic, rather than the parabolic, PDE as exemplified by Leith's scheme in Exercise 12.11.   ◊

We note once again that, in the limit $h, k \to 0$, it is only the highest derivative terms in each variable that dictate stability. The restrictions on the time step have to be respected in calculations with finite h and k however—otherwise the FTCS scheme is bound to show instability as time evolves.

The justification for the von Neumann approach in more general situations is a complex and subtle issue that will not be pursued here. We refer the interested reader to the books of Strikwerda [21, Chap. 2] or LeVeque [14, Sect. 9.6].

---

[12]Recall the CFL condition: if the FTCS scheme is to have any chance of converging then $\rho \to 0$ when $h, k \to 0$.

11.3 Advanced Topics and Extensions

Some important extensions are discussed in the following sections. While we touch on some important practical issues, the coverage of topics will inevitably be somewhat limited.

11.3.1 Neumann and Robin Boundary Conditions

The content of this section complements the discussion of boundary conditions for elliptic PDEs in Sect. 10.2.1. To illustrate the main idea, we consider the simplest FTCS approximation (11.8b) of the heat equation on the interval $0 < x < 1$ subject to a Neumann boundary $-u_x(0, t) = g_0(t)$ condition at $x = 0$ (the negative sign indicates that it is the outward normal derivative that is prescribed). Our aim is to construct a numerical boundary condition that is second-order accurate. We begin with the centred difference

$$-u_x(0, nk) = -h^{-1}\Delta_x u_0^n + \mathcal{O}(h^2)$$

at $t = nk$ so the boundary condition reads

$$-\tfrac{1}{2}h^{-1}(-U_{-1}^n + U_1^n) = g_0(nk). \tag{11.57}$$

Note that this involves the value of U at a grid point $(-h, nk)$. This is shown by an asterisk in Fig. 11.8 and is referred to as a fictitious grid point because it lies outside the domain. However, by applying the FTCS method (11.8b) at $x = 0$ we have

$$U_0^{n+1} = rU_{-1}^n + (1 - 2r)U_0^n + rU_1^n$$

and the contribution of the fictitious grid point can be eliminated by solving this equation for U_{-1}^n and substituting the result into (11.57). This results in the approximation

$$-\frac{1}{2hr}(U_0^{n+1} - (1 - 2r)U_0^n - 2rU_1^n) = g_0(nk). \tag{11.58}$$

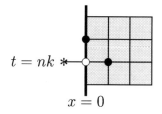

Fig. 11.8 The grid in the vicinity of the boundary point $(0, nk)$ where the stencil of the FTCS method involves a fictitious grid point (indicated by $*$) at $(-h, nk)$

$t = nk$

$x = 0$

which leads to to the explicit update formula

$$U_0^{n+1} = (1 - 2r)U_0^n + 2rU_1^n - 2hrg_0(nk). \tag{11.59}$$

The truncation error in the boundary condition should be based on (11.58), since it is correctly scaled. Using Taylor series in the usual way gives

$$
\begin{aligned}
\mathcal{R}_0^n &= \frac{1}{2rh}\left(u_0^{n+1} - (1 - 2r)u_0^n - 2ru_1^n - 2hrg_0(nk)\right) \\
&= \frac{h}{2k}\left([u + ku_t + \tfrac{1}{2}k^2u_{tt} + \cdots] - \right. \\
&\qquad \left. (1 - 2r)u - 2r[u + hu_x + \tfrac{1}{2}h^2u_{xx} + \tfrac{1}{6}h^3u_{xxx} + \cdots] - 2hrg_0(nk)\right) \\
&= [-u_x - g_0(nk)] + \tfrac{1}{2}h[u_t - u_{xx}] - \tfrac{1}{6}h^2u_{xxx} + \tfrac{1}{4}hku_{tt} + \cdots,
\end{aligned}
$$

where all terms on the right hand side are evaluated at $(0, nk)$. This establishes the second-order consistency, $\mathcal{R}_0^n = \mathcal{O}(h^2)$, of the approximation to the numerical boundary condition. Moreover, the theoretical results developed for the FTCS method with Dirichlet boundary conditions in Sect. 11.1.1 will continue to hold for the new boundary condition (with the same restriction, $r \leq 1/2$).

11.3.2 Multiple Space Dimensions

How easy is it to extend the time stepping strategies introduced earlier in the chapter to PDEs with more than one space dimension? To answer this question, we now turn our focus onto the two-dimensional heat equation $u_t = u_{xx} + u_{yy}$ defined on a general domain Ω in the x–y plane (as shown in Fig. 10.1).

Numerical methods for solving the parabolic PDE $u_t = \mathcal{L}u$ may be obtained by combining a time stepping method (as in Sect. 11.1) with a finite difference approximation of $\mathcal{L}u := -(u_{xx} + u_{yy})$ on a subdivision Ω_h (as described in Chap. 10). In the following we assess the stability of some of the resulting schemes.[13]

We also introduce some simple modifications that are designed to help cope with the massive increase in grid points from M to M^2 at every time level.[14] The simplest approximation scheme is discussed first.

---

[13] The definition of the ℓ_2 norm (11.41) has to be modified to read

$$\|U^n_\cdot\|_{h,2} := \left(h^2 \sum_{\ell=0}^{M}{}'' \sum_{m=0}^{M}{}'' |U_{\ell,m}^n|^2\right)^{1/2}, \tag{11.60}$$

in two space dimensions.

[14] The modifications are therefore especially relevant when solving a PDE problem defined in three space dimensions.

Example 11.25 Investigate the stability of the FTCS approximation of the two-dimensional heat equation on the unit square with a grid of size $h \times h$, with $h = 1/M$.

Combining (11.10) with the standard 5-point approximation of the Laplacian (see Sect. 10.1) gives the FTCS method

$$U_{\ell,m}^{n+1} = U_{\ell,m}^n + r(\delta_x^2 + \delta_y^2)U_{\ell,m}^n, \tag{11.61a}$$

where $r = k/h^2$. This can also be written as

$$U_{\ell,m}^{n+1} = r(U_{\ell+1,m}^n + U_{\ell-1,m}^n + U_{\ell,m+1}^n + U_{\ell,m-1}^n + U_{\ell,m}^n) + (1 - 4r)U_{\ell,m}^n. \tag{11.61b}$$

The coefficients on the right hand side are positive, and the corresponding operator will thus be of positive type whenever $0 < r \leq 1/4$. This implies that the largest allowable time step is half the corresponding value in one dimension. To assess the ℓ_2 stability, we need as appropriate ansatz in two dimensions,

$$U_{\ell,m}^n = \xi^n e^{i(\kappa_x \ell + \kappa_y m)h}, \quad i = \sqrt{-1} \tag{11.62}$$

which, when substituted into (11.61), gives the amplification factor

$$\xi = 1 - 4r(\sin^2 \tfrac{1}{2}\kappa_x h + \sin^2 \tfrac{1}{2}\kappa_y h).$$

A straightforward calculation reveals that $-1 \leq \xi \leq 1$ if, and only if, $0 < r \leq 1/4$, which is the same as the condition for stability in the maximum norm. ◇

It is relatively easy to modify the basic FTCS scheme (11.61b) so as to recover the one-dimensional time step restriction. The ploy is simply to add the term $r^2 \delta_x^2 \delta_y^2 U_{\ell,m}^n$ to the right hand side of (11.61a). Since this represents a finite difference approximation of the term $k^2 u_{xxyy}$, the local truncation error remains $\mathcal{O}(k) + \mathcal{O}(h^2)$. However, the right hand side can then be factorized into two components

$$U_{\ell,m}^{n+1} = (1 + r\delta_x^2)(1 + r\delta_y^2)U_{\ell,m}^n, \tag{11.63a}$$

naturally leading to a two-stage solution process

$$V_{\ell,m}^n = (1 + r\delta_y^2)U_{\ell,m}^n, \quad U_{\ell,m}^{n+1} = (1 + r\delta_x^2)V_{\ell,m}^n. \tag{11.63b}$$

The refined FTCS method (11.63) is a *locally one-dimensional* scheme. The first stage involves computing V^n at all grid points except those on horizontal boundaries (where $\delta_y^2 U_{\ell,m}^n$ is not defined). The second stage computes U^{n+1} at all the internal grid points. While the overall cost per grid point is marginally greater than the unmodified scheme, the advantage of the two-stage method is that it is stable in both the ℓ_2 and maximum norms for $0 < r \leq 1/2$. The details are left to Exercise 11.28 and Exercise 11.29.

Repeating the construction in Sect. 11.1.2 leads to the BTCS approximation in two dimensions

$$U_{\ell,m}^{n+1} - r(\delta_x^2 + \delta_y^2)U_{\ell,m}^{n+1} = U_{\ell,m}^n, \tag{11.64a}$$

which can also be written as

$$(1 + 4r)U_{\ell,m}^{n+1} - r(U_{\ell+1,m}^{n+1} + U_{\ell-1,m}^{n+1} + U_{\ell,m+1}^{n+1} + U_{\ell,m-1}^{n+1}) = U_{\ell,m}^n. \tag{11.64b}$$

When (11.64b) is used to approximate the heat equation on the unit square with a Dirichlet boundary condition, the equations can be expressed in the form of a matrix-vector system

$$A\boldsymbol{u}^{n+1} = \boldsymbol{u}^n + r\boldsymbol{f}^{n+1}, \quad n = 0, 1, 2, \ldots, \tag{11.65}$$

where $\boldsymbol{f}^{n+1}$ contains nonzero boundary values of $U_{\cdot}^{n+1}$, and the matrix A has the same block tridiagonal structure as the matrix (10.10) that arises when computing a finite difference solution of Laplace's equation on a square. To generate a numerical solution the Cholesky factorization $A = R^\mathsf{T} R$ should first be computed (with computational work proportional to M^4) then, at each subsequent time step, the solution $\boldsymbol{u}^{n+1}$ can be determined via

$$R^\mathsf{T}\boldsymbol{v} = \boldsymbol{u}^n + r\boldsymbol{f}^{n+1}, \qquad R\boldsymbol{u}^{n+1} = \boldsymbol{v}, \tag{11.66}$$

where $\boldsymbol{v}$ is an intermediate vector. If this is done then, as discussed in Sect. 10.1.1, the cost of each time step (11.65) will be proportional to M^3.

To generate a locally one-dimensional version of the BTCS scheme, the term $r^2\delta_x^2\delta_y^2 U_{\ell,m}^{n+1}$ must be added to the left hand side of (11.64). As for the FTCS scheme, the modified scheme can then be factorised so that

$$(1 - r\delta_x^2)(1 - r\delta_y^2)\, U_{\ell,m}^{n+1} = U_{\ell,m}^n, \tag{11.67a}$$

without increasing the order of the local truncation error. The factorisation leads to the following two-stage solution process

$$(1 - r\delta_y^2)V_{\ell,m}^n = U_{\ell,m}^n, \quad (1 - r\delta_x^2)U_{\ell,m}^{n+1} = V_{\ell,m}^n, \tag{11.67b}$$

where at each stage the solution vector can be constructed by performing a sequence of one-dimensional linear solves (to generate a single row or a column of the grid solution values) with the tridiagonal matrix A that appears in (11.25). The overall cost of computing U^{n+1} via (11.67) will thus be proportional to M^2 rather than M^3 for (11.64)—a substantial gain in efficiency!

The Crank–Nicolson scheme may also be readily extended to multiple space dimensions. The locally one-dimensional implementation of Crank–Nicolson is also referred to as the Alternating Direction Implicit method.[15]

Example 11.26 Investigate the stability of the locally one-dimensional Crank–Nicolson approximation of the two-dimensional heat equation on the unit square with a grid of size $h \times h$, with $h = 1/M$.

The Crank-Nicolson approximation is given by (11.32), where $\mathcal{L}_h$ is the standard 5-point approximation of the Laplacian. Written out explicitly, it is

$$\left(1 - \tfrac{1}{2}r(\delta_x^2 + \delta_y^2)\right)U_{\ell,m}^{n+1} = \left(1 + \tfrac{1}{2}r(\delta_x^2 + \delta_y^2)\right)U_{\ell,m}^n, \qquad (11.68)$$

with local truncation error $\mathcal{O}(k^2) + \mathcal{O}(h^2)$. The locally one-dimensional variant is obtained by adding $\tfrac{1}{4}r^2\delta_x^2\delta_y^2 U_{\ell,m}^{n+1}$ to the left hand side and $\tfrac{1}{4}r^2\delta_x^2\delta_y^2 U_{\ell,m}^n$ to the right hand side, giving

$$(1 - \tfrac{1}{2}r\delta_x^2)(1 - \tfrac{1}{2}r\delta_y^2)U_{\ell,m}^{n+1} = (1 + \tfrac{1}{2}r\delta_x^2)(1 + \tfrac{1}{2}r\delta_y^2)U_{\ell,m}^n. \qquad (11.69)$$

The contribution that the additional terms make to the local truncation error is

$$\frac{1}{4k}r^2\delta_x^2\delta_y^2(u_{\ell,m}^{n+1} - u_{\ell,m}^n) = \frac{1}{4}k^2 u_{xxyyt} + \text{higher-order terms}$$

which does not alter the second-order consistency of the original method.

The amplification factor of the modified scheme is given by

$$\xi = \left(\frac{1 - 2r\sin^2(\tfrac{1}{2}\kappa_x h)}{1 + 2r\sin^2(\tfrac{1}{2}\kappa_x h)}\right)\left(\frac{1 - 2r\sin^2(\tfrac{1}{2}\kappa_y h)}{1 + 2r\sin^2(\tfrac{1}{2}\kappa_y h)}\right)$$

which, being a product of amplification factors of one-dimensional Crank–Nicolson operators (see (11.50), satisfies $-1 \leq \xi \leq 1$ for all κ_x, κ_y and $r > 0$. This means that the scheme (11.69) is unconditionally ℓ_2 stable. $\Diamond$

The modified scheme method is usually implemented as a two-stage process

$$\left.\begin{array}{l} (1 - \tfrac{1}{2}r\delta_y^2)V_{\ell,m}^{n+1} = (1 + \tfrac{1}{2}r\delta_x^2)U_{\ell,m}^n \\ (1 - \tfrac{1}{2}r\delta_x^2)U_{\ell,m}^{n+1} = (1 + \tfrac{1}{2}r\delta_y^2)V_{\ell,m}^{n+1} \end{array}\right\} \qquad (11.70)$$

again involving an intermediate grid function V. This pair of equations defines the "classic" ADI method. The calculation of V^{n+1} from the first of these equations proceeds columnwise but requires boundary conditions on the horizontal edges of

---

[15]The ADI method was devised for the oil industry by Peaceman and Rachford in 1955. It is still used in anger, over 50 years later!

the domain. These are obtained by subtracting the second equation from the first to give

$$V_{\ell,m}^{n+1} = \tfrac{1}{2}\big(U_{\ell,m}^{n+1} + U_{\ell,m}^{n} - \tfrac{1}{2}r\delta_x^2(U_{\ell,m}^{n+1} - U_{\ell,m}^{n})\big). \tag{11.71}$$

When $m = 0, M$ and $\ell = 1, 2, \ldots, M - 1$ these equation express the boundary values of V on horizontal edges in terms of the known boundary values of U^n and U^{n+1}. The calculation of U^{n+1} then proceeds row by row. The cost of a single ADI step is about twice that of the BTCS method. The ADI method is significantly more efficient however—the second-order accuracy in k allows much larger time steps to be taken to achieve the same solution accuracy.

11.3.3 The Method of Lines

The *method of lines* is a popular strategy for solving time-dependent PDEs that exploits techniques (and, particularly, software) for solving initial value problems for ordinary differential equations. We give an example to convey the flavour of the method.

Example 11.27 (Example 8.7 revisited) Solve the heat equation $u_t = u_{xx}$ in the semi-infinite strip $\{(x, t) : 0 < x < 1, t > 0\}$ with end conditions $u_x(0, t) = 0$, $u(1, t) = \sin 2\pi t$ and with a homogeneous initial condition $u(x, 0) = 0$ using the method of lines.

The basic philosophy of the method of lines is that a finite difference grid is only defined for the spatial variables. In this one-dimensional setting, we have grid points $x_m = mh$ for $m = 0, 1, \ldots, M$ (with $h = 1/M$) and we note that the unknown grid point values U_m will be functions of time: $U_m(t)$. To give a specific example, if the spatial approximation is the standard centred finite difference δ_x^2 (as in (11.8) or (11.22)) then we generate the ordinary differential equation

$$\frac{\mathrm{d}}{\mathrm{d}t}U_m = \frac{1}{h^2}\big(U_{m-1}(t) - 2U_m(t) + U_{m+1}(t)\big), \tag{11.72}$$

at every interior grid point $m = 1, 2, \ldots, M - 1$. Imposing the Dirichlet boundary condition at $x = 1$ gives $U_M(t) = \sin 2\pi t$, which feeds into the ODE (11.72) at the last interior grid point. Next, if we take a centred difference approximation of the Neumann boundary condition at $x = 0$, then the fictitious grid point value $U_{-1}(t)$ (that arises when putting $m = 0$ in (11.72), cf. Sect. 11.3.1) is replaced by the value $U_1(t)$, leading to the following boundary equation

$$\frac{\mathrm{d}}{\mathrm{d}t}U_0 = \frac{2}{h^2}\big(U_1(t) - U_0(t)\big). \tag{11.73}$$

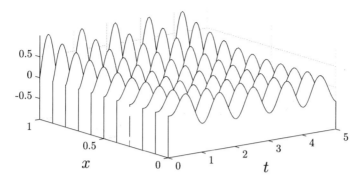

Fig. 11.9 A method of lines solution to Example 11.27 for $0 \leq t \leq 5$ using a second-order finite difference approximation in space with $h = 1/10$

To summarise, the method of lines solution is the system of coupled ODEs (11.72) and (11.73) for the M unknown functions $U_m(t)$, supplemented by the initial conditions $U_m(0) = 0, m = 0, 1, \ldots, M - 1$. ◊

Systems of ODEs can readily be solved up to a given error tolerance using professional (or open-source) software packages. A sample solution[16] in the case of $M = 10$ is illustrated in Fig. 11.9. Note that for each grid point there is a continuous solution curve $(x_m, t, U_m(t))$. The book by Ascher [1] is a good starting point for anyone looking for more information. One drawback of the method is that the decoupling of the spatial and temporal approximations makes it difficult to effectively balance the component discretization errors. It is very easy to generate time-accurate solutions which are completely dominated by spatial error!

Exercises

11.1 ☆ By differentiating the PDE, show that the leading terms in the local truncation error (11.12) of consistency of the FTCS approximation of the heat equation may be combined, using the PDE to give $\frac{1}{2}h^2(r - \frac{1}{6})u_{xxt}\big|_m^n$. What is the order of consistency of the FTCS method when $r = 1/6$?

11.2 Construct FTCS approximations of the PDEs (a) $u_t = u_{xx} + f(x, t)$ and (b) $u_t = u_{xx} - u$. Examine the local truncation errors and show that, in neither case, is there a value of $r = k/h^2$ that leads to a higher order of consistency than $\mathcal{O}(k) + \mathcal{O}(h^2)$.

11.3 ☆ Suppose that the FTCS method (11.8b) is used to solve the heat equation with initial condition $u(x, 0) = g(x), 0 < x < 1$ and end conditions $u(0, t) = g_0(t), u(1, t) = g_1(t)$. Show that the solution $\boldsymbol{u}^{n+1}$ at the $(n + 1)$st time level may be expressed as

---

[16]Computed by solving the ODE system using the MATLAB function `ode15s`.

$$u^{n+1} = A(-r)u^n + f^n,$$

where $A(r)$ is the matrix appearing in the BTCS scheme (11.25).

11.4 Suppose that the point P in Fig. 11.4 has the coordinates $X = mh$, $T = nk$ and let Q and R denote the points $(x_{m-n}, 0)$ and $(x_{m+n}, 0)$. If the grid function U is determined by the FTCS scheme (11.8b) and if $U_j^0 = (-1)^j$ ($j = m - n, m - n + 1, \ldots, m + n$), show that

$$U_j^1 = -(4r - 1)(-1)^j, \quad j = m - n + 1, m - n + 2, \ldots, m + n - 1$$

and that $U_m^n = (4r - 1)^n (-1)^m$.

Next, suppose that the grid sizes are reduced so that h becomes $h/2$, k becomes $k/4$ (leaving r unchanged) and $X = 2mh$, $T = 4nk$ so that P remains fixed. Deduce that this method is unstable as $h, k \to 0$ with $r \geq 1/2 + \varepsilon$, where $\varepsilon > 0$ is any number independent of both h and k.

11.5 Show that the finite difference scheme

$$U_m^{n+1} = \tfrac{1}{3} r U_{m-2}^n + (1 - r)U_m^n + \tfrac{2}{3} r U_{m+1}^n$$

is a consistent approximation of the heat equation $u_t = u_{xx}$. What is the stencil of the method? Sketch the domain of dependence of the method and specify the interval of dependence of a typical grid point P having coordinates (mh, nk).

11.6 Show that the finite difference scheme

$$U_m^{n+1} = U_m^n + r(U_{m-2}^n - 2U_{m-1}^n + U_m^n)$$

is a consistent approximation of the heat equation $u_t = u_{xx}$. Sketch the stencil of the method and use the domain of dependence argument to show that its solution cannot converge to that of the differential equation as $h \to 0$ and $k \to 0$.

11.7 Prove Corollary 11.6 by replacing U by $-U$ in Theorem 11.5.

11.8 Prove that the operator $\mathscr{L}_h$ defined in Corollary 11.7 is inverse monotone when $r \leq 1/2$.

11.9 By taking Taylor series expansions about the point $x = mh$, $t = (n + 1)k$, or otherwise, show that the local truncation error of the BTCS method (11.22) is given by (11.24). Show also that the leading terms may be combined to give $-\tfrac{1}{2} h^2 (r + \tfrac{1}{6}) u_{txx}|_m^{n+1}$.

11.10 Use Lemma 6.1 to show that the matrix A in (11.25) is positive definite.

11.11 Complete the details of the proof of Corollary 11.12. Hence prove that the BTCS method converges as $h, k \to 0$ without restriction on the mesh ratio r.

11.12 Determine the leading term in the local truncation error of the scheme

$$U_m^{n+1} = U_m^n + r\delta_x^2 U_m^n - \tfrac{1}{2}k\gamma(U_m^{n+1} + U_m^n)$$

for solving the parabolic equation $u_t = u_{xx} - \gamma u$.

11.13 Write down the finite difference scheme defining the θ-method for solving the heat equation with initial condition $u(x, 0) = g(x), 0 < x < 1$ and end conditions $u(0, t) = g_0(t), u(1, t) = g_1(t)$. Show that the solution $\boldsymbol{u}^{n+1}$ at the $(n + 1)$st time level may be expressed as

$$A(\theta r)\,\boldsymbol{u}^{n+1} = A\big((\theta - 1)r\big)\boldsymbol{u}^n + \boldsymbol{f}^n,$$

where $A(r)$ is the matrix A appearing in the BTCS scheme (11.25). Can you express $\boldsymbol{f}^n$ in terms of the boundary functions $g_0(t)$ and $g_1(t)$?

11.14 Use the formula (B.11) for the eigenvalues of a tridiagonal matrix to determine the largest and smallest eigenvalues of the matrix $A(\theta r)$ of the preceding exercise. Use this to prove that $A(\theta r)$ is positive definite for $\theta \geq 0$.

11.15 Complete the details of the proof of Corollary 11.17.

11.16 ☆ Suppose that the Crank–Nicolson scheme (11.34b) is used with a grid with $h = 1/2$, boundary conditions $U_0^n = U_2^n = 0$ (for all $n \geq 0$) and initial condition $U_1^0 = 1$. Determine U_1^1 and deduce that the difference scheme cannot be inverse monotone if $r > 1$. This establishes $r \leq 1$ as being a necessary condition for inverse monotonicity. (Corollary 11.18 shows that it is also a sufficient condition.)

11.17 Suppose that the norms $\| \cdot \|_{h,\infty}$ and $\| \cdot \|_{h,2}$ are defined by (11.13) and (11.41), respectively. Prove that

$$\tfrac{1}{2}\sqrt{h}\|U.\|_{h,\infty} \leq \|U.\|_{h,2} \leq \|U.\|_{h,\infty}$$

for any grid function U. Investigate the relative sizes of the terms in these inequalities when,

(a) $U_m = 1$ for each $m = 0, 1, ..., M$,
(b) $U_0 = 1$ and $U_m = 0$ for $0 < m \leq M$.

11.18 ☆ Show that the amplification factor of the BTCS scheme (11.22a) is

$$\xi - \frac{1}{1 + 4r\sin^2(\tfrac{1}{2}\kappa h)}$$

and hence show that the BTCS scheme is unconditionally ℓ_2 stable, that is, stable for all values of $r > 0$.

11.19 Show that the amplification factor for the θ-method applied to the one-dimensional heat equation is given by

$$\xi = \frac{1 - 4r(1 - \theta)\sin^2(\tfrac{1}{2}\kappa h)}{1 + 4r\theta \sin^2(\tfrac{1}{2}\kappa h)}.$$

Hence show that the θ–scheme is ℓ_2 stable if $(1 - 2\theta)r \le \tfrac{1}{2}$ (so there is no restriction on r if $\theta \ge 1/2$).

11.20 Examine the ℓ_2 stability of the explicit finite difference scheme

$$U_m^{n+1} = \left[1 + r\delta_x^2\right]U_m^n - \tfrac{1}{2}k(U_{m-1}^n + U_{m+1}^n)$$

for solving the PDE $u_t = u_{xx} - u$. What conditions on h and k ensure that the method is stable in the sense of von Neumann?

Is there any advantage in using the above method in preference to the more standard FTCS method (11.18) with $\gamma = -1$?

11.21 Show that the FTCS approximation (11.53) of the advection–diffusion equation may be written in the form

$$U_m^{n+1} = \alpha_{-1}U_{m-1}^n + \alpha_0 U_m^n + \alpha_1 U_{m+1}^n$$

for coefficients α_1, α_2 and α_3 that are non-negative and sum to 1 when $\tfrac{1}{2}\rho \le r \le \tfrac{1}{2}$. Express these as restrictions on k in terms of h and compare with the corresponding conditions given in (11.56) for ℓ_2-stability.

Prove that a discrete maximum principle in the sense of Theorem 11.5 holds when U satisfies

$$U_m^{n+1} \le \alpha_{-1}U_{m-1}^n + \alpha_0 U_m^n + \alpha_1 U_{m+1}^n$$

for $(x_m, t_{n+1}) \in \Omega_\tau$.

11.22 The advection–diffusion equation $u_t = u_{xx} - 2u_x$ is approximated by the "semi-implicit" finite difference scheme

$$U_m^{n+1} - U_m^n = r\delta_x^2 U_m^{n+1} - c\left(U_{m+1}^n - U_{m-1}^n\right),$$

where $r = k/h^2$ and $c = k/h$. Show that the von Neumann amplification factor is given by

$$\xi = \frac{1 - 2ic \sin \kappa h}{1 + 4r \sin^2 \tfrac{1}{2}\kappa h}.$$

Hence prove that $|\xi| \le 1$ for all real κ if $k \le \tfrac{1}{2}$.

11.23 Show that the amplification factor of the finite difference scheme in Exercise 11.5 is

$$\xi = 1 - r + \tfrac{1}{3}r(2e^{i\kappa h} + e^{-2i\kappa h}).$$

Determine the largest mesh ratio r for which it is ℓ_2-stable.

11.24 Necessary conditions for ℓ_2-stability may be obtained by sampling the amplification factor ξ at certain wavenumbers ($\kappa h = 0, \pi$, for example) or by examining the MacLaurin expansion of either ξ or $|\xi|^2$ in powers of h. Investigate these possibilities for the amplification factors (a) the FTCS method (11.49), (b) the advection–diffusion scheme (11.54) and (c) the semi-implicit scheme in Exercise 11.22.

11.25 Describe how one might modify the update formula (11.59) in order to accommodate the Robin end condition $-u_x(0, t) + \sigma u(0, t) = g_0(t)$, where σ is a positive constant. Check that the resulting formula has a local truncation error of $\mathcal{O}(h^2)$.

11.26 * Determine a second-order accurate numerical boundary condition for the BTCS approximation of the heat equation subject to the Neumann condition $-u_x(0, t) = g_0(t)$ at $x = 0$.
    [Hint: start from the condition (11.57) with $n + 1$ replacing n.]

11.27 Determine a second-order accurate numerical boundary condition for the Crank–Nicolson approximation of the heat equation subject to the Neumann condition $-u_x(0, t) = g_0(t)$ at $x = 0$.

11.28 Show that the locally one-dimensional scheme (11.63a) for the two-dimensional heat equation has amplification factor

$$\xi = (1 - 4r \sin^2 \tfrac{1}{2}\kappa_x h)(1 - 4r \sin^2 \tfrac{1}{2}\kappa_y h).$$

Deduce that the scheme is ℓ_2 stable whenever $0 < r \le 1/2$.

11.29 * Show that the locally one-dimensional scheme (11.63a) for the two-dimensional heat equation in the unit square with a homogeneous Dirichlet boundary condition is stable in the maximum norm, that is $\|U^n\|_{h,\infty} \le \|U^0\|_{h,\infty}$, if $r \le 1/2$.

11.30 Suppose that $V_{\ell,m}^{n+1}$ is replaced by $U_{\ell,m}^{n+1/2}$ (the approximation to u at $x = mh$, $y = \ell h$, $t = (n + 1/2)k$). Show that the individual equations in the ADI method (11.70) are both consistent with the two-dimensional heat equation.

11.31 The heat equation in polar coordinates with circular symmetry is given by $u_t = \frac{1}{r}(ru_r)_r$ and may be approximated by the explicit finite difference scheme (see (10.33)

$$U_m^{n+1} = U_m^n + \frac{k}{h^2} r_m \delta_r (r_m \delta_r U_m^n),$$

where $r_m = mh > 0$. Show that

(a) this scheme is identical to that obtained by writing the PDE as $u_t = u_{rr} + \frac{1}{r} u_r$ and using a FTCS style approximation involving second-order centred differences for the terms u_{rr} and u_r.

(b) the scheme is of positive type when a time step restriction $k \leq h^2/2$ is imposed.

Use l'Hôpital's rule to show that the PDE reduces to $u_t = 2u_{rr}$ in the limit $r \to 0$. Write down a standard FTCS approximation of the limiting PDE and determine the largest mesh ratio k/h^2 for which it is of positive type. [Hint: look at Exercise 8.7 first.]

11.32 In polar coordinates the heat equation in two dimensions is given by $u_t + \mathcal{L}u = 0$ with $\mathcal{L}u = -u_{rr} - \frac{1}{r} u_r - \frac{1}{r^2} u_{\theta\theta}$.

Suppose that the PDE is approximated by the explicit scheme (11.10) so that $-\mathcal{L}_h$ is the expression given in (10.33a). Show that the resulting scheme will be of positive type whenever[17]

$$k \leq \frac{h^2 \Delta\theta^2}{2(1 + \Delta\theta^2)}.$$

11.33 Consider the PDE $u_t = \nabla^2 u$ in the triangular domain having vertices at $(0, 0)$, $(0, 1)$ and $(9/8, 0)$ together with the same Dirichlet conditions as in Exercise 10.17. An explicit finite difference approximation of the PDE is provided by the FTCS scheme

$$U_{\ell,m}^{n+1} = U_{\ell,m}^{n} - k\mathcal{L}_h U_{\ell,m}^{n}$$

(cf. (11.10)), where $\mathcal{L}_h$ is an approximation to $-\nabla^2$. Write down the set of finite difference equations when $h = 1/4$ using the spatial approximation described in Exercise 10.17. What restriction needs to be imposed on the time step k in order for these equations to be of positive type? How does this compare to the corresponding limit $k \leq h^2/4$ for a regular grid?

11.34 Suppose that the finite difference equations at the 3 internal grid points adjacent to the hypotenuse in the previous example are replaced by the BTCS equations

$$U_{\ell,m}^{n+1} + k\mathcal{L}_h U_{\ell,m}^{n+1} = U_{\ell,m}^{n}.$$

Show that the approximation may be calculated at each time level without the need to solve linear equations (so they remain explicit) and that the equations are of positive type provided that $k \leq h^2/4$.

---

[17]This bound is so small as to make the scheme impractical.

Chapter 12
Finite Difference Methods for Hyperbolic PDEs

Abstract This self-contained chapter focuses on finite difference approximation of hyperbolic boundary value problems. A number of explicit and implicit time-stepping schemes are introduced and their stability, dissipation and dispersion is analysed. State-of-the-art schemes for hyperbolic PDEs that involve flux limiters are discussed at the end of the chapter.

The construction and analysis of difference approximations for hyperbolic equations will mirror the discussion of parabolic equations in the previous chapter. Properties such as local truncation error, stability and convergence of schemes can be defined in exactly the same way. One aspect that is different is the Courant–Friedrichs–Lewy (CFL) condition, which has a hyperbolic variant. A point that should be emphasised right at the start is that explicit methods for hyperbolic equations are relatively appealing. Thus the emphasis of the chapter will be on explicit schemes. We will show that practical methods are perfectly stable for time steps that scale with the spatial grid size, that is $k \sim h$.

Parabolic PDEs contain diffusive terms so that the initial data becomes smoother over time (see Example 4.7) and perturbations—such as local truncation errors or rounding errors—are damped out as time evolves. This contrasts with hyperbolic PDEs such as $pu_x + qu_y = 0$ which has a constant solution along characteristics, so any perturbation of the solution will persist indefinitely. When solving nonlinear PDEs even smooth initial data can evolve to form a shock. This inherent lack of smoothness is what makes the numerical solution of hyperbolic PDE problems so challenging.

Despite its apparent simplicity, we shall focus almost exclusively on schemes for solving the advection equation (pde.1),

$$u_t + au_x = 0, \qquad (12.1)$$

in which a (the wave speed) is constant. It should also be emphasised at the outset that (12.1) is not an end in itself, but merely the first step on the way to solving hyperbolic systems of PDEs that are nonlinear or else have variable coefficients.

© Springer International Publishing Switzerland 2015
D.F. Griffiths et al., *Essential Partial Differential Equations*, Springer Undergraduate Mathematics Series, DOI 10.1007/978-3-319-22569-2_12

To avoid the issue of dealing with boundary conditions, we will suppose that (12.1) is posed on the half-space $\{x \in \mathbb{R}, t \geq 0\}$ with an initial condition $u(x, 0) = g(x)$. To enable computation, we will additionally suppose that the initial condition is periodic of period L so that $g(x + L) = g(x)$ for all x. This implies that the solution u will also have a periodic profile, which in turn means that it will be sufficient to solve the initial-value problem over $0 \leq x \leq L$ with grid size $h = L/M$. More realistic boundary conditions will be explored later, in Sect. 12.4.2. We will discover that there are no particular obstacles to constructing stable methods (at least for linear problems). The real difficulty is in constructing approximation methods that keep the numerical solution constant on characteristics, especially in the presence of discontinuities in the initial condition or its derivatives.

The assessment of numerical methods in earlier chapters was done by giving tables of results that showed how the global error behaves as the grid size(s) tend to zero. Such an approach is not appropriate for hyperbolic problems, since even a small discrepancy in wave speed can lead to a 100 % error at later times. We shall therefore use graphical evidence to compare alternative approximation schemes. The performance of methods depends on many factors and rather than display a range of test problems using different data, we shall borrow an idea introduced by Leonard (1991) and use an initial condition made up of pulses of increasing levels of continuity (discontinuous, continuous, continuously differentiable) with peaks of different widths. The variety in the regularity of the initial data will turn out to be useful in ascertaining the strengths and (particularly the) weaknesses of numerical methods.

Example 12.1 Describe the solution of the one-way wave equation (12.1) on the interval $0 \leq x \leq 3$ with $a = 1$, with periodic boundary conditions and the initial condition

$$g(x) = \begin{cases} 1, & \frac{1}{4} < x < \frac{3}{4} \quad \text{(square wave)} \\ 1 - |4x - 6|, & \frac{5}{4} < x < \frac{7}{4} \quad \text{(triangular wave)} \\ \cos^2 \pi(2x - 5) & \frac{9}{4} < x < \frac{11}{4} \quad \text{(cosine squared wave)} \\ 0, & \text{otherwise.} \end{cases} \tag{12.2}$$

The characteristics for $0 \leq x \leq 3$ are shown in Fig. 12.1 over a single period, along with the exact solution at times $t = 0, 1, 2, 3$. A characteristic (and the information it conveys) that exits the domain at $x = 3$ immediately reappears (along with its information) at $x = 0$.   ◊

The exact solution of the one-way wave equation (12.1) is constant along the characteristics $x - at = $ constant. Thus the solution u_m^{n+1} at the point (x_m, t_{n+1}) may be expressed in terms of the solution at the earlier time $t = t_n$ via

$$u_m^{n+1} = u(x_m - ch, t_n), \tag{12.3}$$

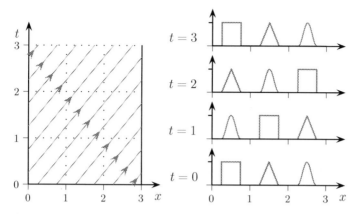

Fig. 12.1 Characteristics for the test problem in Example 12.1 (*left*) and the exact solution at times $t = 0, 1, 2, 3$ (*right*)

where $c = ak/h$ is known as the *Courant number*.[1] Note that the solution translates to the right a distance ch at each time step k. Indeed, if the grid size is constructed so that the Courant number is an integer, then the solution satisfies the exact relation

$$u_m^{n+1} = u((m - c)h, t_n) = u_{m-c}^n, \tag{12.4}$$

which, together with the initial condition and periodic boundary condition, characterises a simple algorithm for computing the exact solution at every grid point! This shows that explicit finite difference schemes that *follow the characteristics* can be very effective. This avenue is explored in detail in the first section.

12.1 Explicit Methods

We start by supposing that (12.1) is approximated by an explicit finite difference method of the form

$$U_m^{n+1} = \sum_{j=-\mu}^{\nu} \alpha_j U_{m+j}^n, \tag{12.5}$$

where μ and ν are nonnegative integers. We shall refer to this as a (μ, ν)-scheme. A typical stencil is illustrated in Fig. 12.2. The stencil uses μ grid points upwind of

---

[1]Richard Courant (1888–1972) founded a world-leading institute for applied mathematics in New York University. He also pioneered the idea of using finite element approximation methods to solve elliptic PDEs defined on irregularly shaped domains.

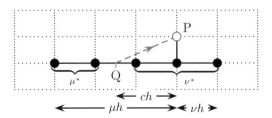

Fig. 12.2 A (μ, ν)–stencil with $\mu = 3$ and $\nu = 1$. The *dashed line* PQ shows the characteristic through P when $a > 0$. Integers μ^* and ν^* refer to the number of points situated upwind and downwind of the intersection point Q

P and ν points downwind of P. The right hand side of (12.5) approximates the right hand side of (12.3) by a weighted average of values $\{U^n_{m+j}\}^\nu_{j=-\mu}$ from the previous time level.

An explicit difference scheme will not converge to the exact solution of a parabolic PDE in the limit $h, k \to 0$ unless the CFL condition given in Definition 11.3 is satisfied. The hyperbolic PDE analogue is just as important. Moreover since it depends on the geometry of the stencil, not on the actual coefficients, the condition applies to general hyperbolic problems. The details are presented below.

Consider a grid point P with coordinates (X, T) with $T > 0$ fixed as $h, k \to 0$ as shown in Fig. 12.3. Setting $X = Jh$ and $T = Nk$ and applying (12.5) with $n = N - 1$, the difference solution U^N_J at P will clearly depend on the $(\mu + \nu + 1)$ grid point values U^{N-1}_{J+j}, $j = -\mu, \ldots, \nu$, which, themselves depend on the values U^{N-2}_{J+j}, $j = -2\mu, \ldots, 2\nu$. Continuing the recursion back to $t = 0$, it can be seen that the solution value U^N_J is solely determined by the initial data in the interval $[x_A, x_B]$ between points A and B in Fig. 12.3, where

$$x_A = X - \mu\frac{h}{k}, \qquad x_B = X + \nu\frac{h}{k}. \tag{12.6}$$

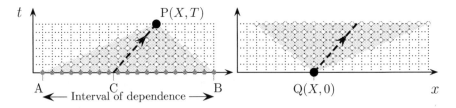

Fig. 12.3 The domain of dependence (*left*) of a grid point $P(X, T)$ and domain of influence (*right*) of $Q(X, 0)$ for an explicit scheme with a $(2, 1)$-stencil. The grid values along AB form the interval of dependence of $P(X, T)$. The broken lines show the characteristics through P and Q

The triangle PAB shown in Fig. 12.3 defines the *domain of dependence* of the numerical solution at P for the scheme (12.5). The interval $x_A \leq x \leq x_B$ is the associated *interval of dependence*. In contrast, the domain of dependence of the exact PDE solution is simply the characteristic through P, backwards in time, that intersects the x-axis at C, say.

Next, imagine that the grid is successively refined keeping $c = ak/h$ fixed. In this scenario the points A and B do not change as the grid is refined and the numerical solution at the point P is only ever dependent on initial data in the fixed interval $[x_A, x_B]$. Critically, if C does not lie in the interval $[x_A, x_B]$, then the numerical and exact solutions at P depend on different data for all grids in the sequence—so convergence is out of the question. Such a situation is formalised in the following definition.

Definition 12.2 (*CFL condition*) Convergence of a finite difference approximation of a hyperbolic PDE cannot take place if the characteristic through a generic grid point P does not intersect the x–axis within the interval of dependence of P.

Example 12.3 Show that a (μ, ν)-method finite difference solution of the advection equation will not converge when $h, k \to 0$ with $c = ak/h$ kept fixed, unless $-\nu \leq c \leq \mu$.

The CFL condition simply requires that $x_A \leq x_C \leq x_B$ where the intersection point C has coordinate $x_C = X - c\frac{h}{k}$. Using (12.6) gives the condition

$$X - \mu\frac{h}{k} \leq X - c\frac{h}{k} \leq X + \nu\frac{h}{k}$$

which can be rearranged to give the condition $-\nu \leq c \leq \mu$.   ◊

Note that the triangle PAB and the outline of the difference stencil are similar triangles in the case of a constant coefficient PDE, so the CFL condition can be often checked by inspection of the stencil (as shown in Fig. 12.2).

It is also instructive to focus on the behaviour of the (μ, ν)-method looking forward in time. The initial condition at a grid point $Q(X, 0)$ (with $X = Jh$) will be used in the computation of the values U^1_{J+j}, $j = -\nu, \ldots, \mu$, (note the relative positions of μ and ν in this sequence) which, themselves, will be used to compute U^2_{J+j}, $j = -2\nu, \ldots, 2\mu$. Thus the initial value at Q will affect the numerical solution in the shaded region shown in Fig. 12.3 (right), known as the *domain of influence* of Q. The corresponding domain of influence of the exact solution is the characteristic line $x - at = X$ through Q. This knowledge is of particular interest when Q is a distinguished point in the initial data—for example, when Q is the location of a discontinuity (either in the solution or in one of its derivatives). The domain of influence shows that this will ultimately affect the numerical solution over the entire domain, while the effect for the exact solution is confined to a single characteristic.

12.1.1 Order Conditions

We would like to choose the coefficients in (12.5) so as to maximise the order of consistency of the associated (μ, ν)-method. To make progress in this direction, the local truncation error of (12.5) is defined by

$$\mathcal{R}_m^n := \frac{1}{k}\left(u_m^{n+1} - \sum_{j=-\mu}^{\nu} \alpha_j u_{m+j}^n\right) \tag{12.7}$$

(the factor $1/k$ ensures it is correctly scaled). Expanding u_m^{n+1} and u_{m+j}^n about (x_m, t_n) using standard Taylor series gives

$$
\begin{aligned}
\mathcal{R}_m^n = k^{-1}\left(u + ku_t + \tfrac{1}{2}k^2 u_{tt} + \cdots\right)\big|_m^n \\
- \sum_{j=-\mu}^{\nu} \alpha_j k^{-1}\left(u + jhu_x + \tfrac{1}{2}(jh)^2 u_{xx} + \cdots\right)\big|_m^n.
\end{aligned} \tag{12.8}
$$

Differentiating the PDE $u_t = -au_x$ with respect to t we find

$$u_{tt} = -au_{xt} = -a\partial_x(u_t) = a^2 u_{xx}. \tag{12.9}$$

Repeating this process gives $\partial_t^\ell u = (-a\partial_x)^\ell u$ and substituting into (12.8) gives

$$\mathcal{R}_m^n = k^{-1}(C_0 u + hC_1 u_x + h^2 C_2 u_{xx} + \cdots)\big|_m^n, \tag{12.10}$$

where

$$C_\ell = \frac{1}{\ell!}\left((-c)^\ell - \sum_{j=-\mu}^{\nu} j^\ell \alpha_j\right), \qquad \ell = 0, 1, \ldots. \tag{12.11}$$

Thus, assuming that the refinement strategy is to keep c fixed (so that h/k is a constant) the scheme (12.5) will be consistent of order p if

$$C_0 = C_1 = \cdots = C_p = 0. \tag{12.12}$$

The constraints (12.12) generate a system of $(p + 1)$ linear algebraic equations in $(\mu + \nu + 1)$ unknowns (the α's). Let $q := \mu + \nu$ (this will be the maximum possible order). When $p = q$ the number of equations matches the number of unknowns. Writing the system in matrix-vector form gives $V\boldsymbol{\alpha} = \boldsymbol{c}$, where $\boldsymbol{\alpha} = [\alpha_{-\mu}, \alpha_{-\mu+1}, \ldots, \alpha_\nu]^{\mathsf{T}}$ and

$$V = \begin{bmatrix} 1 & 1 & \cdots & 1 \\ -\mu & -\mu+1 & \cdots & \nu \\ \vdots & & & \vdots \\ (-\mu)^q & (-\mu+1)^q & \cdots & \nu^q \end{bmatrix}, \qquad c = \begin{bmatrix} 1 \\ -c \\ \vdots \\ (-c)^q \end{bmatrix}. \tag{12.13}$$

The matrix V is called a Vandermonde matrix and is nonsingular (since no two columns are the same; see Trefethen and Bau [26, p. 78]). This implies that there is a *unique* difference scheme of maximum order.[2]

The construction of the method of order q can be accomplished without the need to solve linear equations. This will be explored in the next example.

Example 12.4 Show how polynomial interpolation may be used to construct the (μ, ν)-method of maximal order $q = \mu + \nu$.

Since V is independent of c we deduce that the coefficients of the method are polynomials of degree q. Suppose that $\Phi(x)$ denotes the polynomial of degree q that interpolates the $q+1$ solution values at the current time level $\{(x_{m+j}, U^n_{m+j})\}^\nu_{j=-\mu}$, so that

$$\Phi(x_m + jh) = U^n_{m+j}, \qquad j = -\mu, \ldots, \nu. \tag{12.14}$$

We now consider a numerical method that mimics the behaviour (12.3) of the exact solution, that is,

$$U^{n+1}_m = \Phi(x_m - ch). \tag{12.15}$$

Using the Lagrange form of the interpolating polynomial (see, for instance, Süli and Mayers [23, Definition 6.1]), we can write

$$\Phi(x) = \sum_{j=-\mu}^{\nu} L_j(x) U^n_{m+j}, \tag{12.16a}$$

where

$$L_j(x) = \prod_{\substack{\ell=-\mu \\ \ell \neq j}}^{\nu} \frac{x_{m+\ell} - x}{x_{m+\ell} - x_{m \mid j}}. \tag{12.16b}$$

The key property is that $L_j(x_{m+j}) = 1$ while $L_j(x_{m+i}) = 0$ when $i \neq j$. Combining (12.15) with (12.16a) and then directly comparing coefficients with (12.5) we find that

$$\alpha_j = L_j(x_m - ch) = \prod_{\substack{\ell=-\mu \\ \ell \neq j}}^{\nu} \frac{x_{m+\ell} - x_m + ch}{x_{m+\ell} - x_{m+j}} = \prod_{\substack{\ell=-\mu \\ \ell \neq j}}^{\nu} \frac{\ell + c}{\ell - j}, \tag{12.17}$$

so that the coefficients are polynomials in c (of degree q) and $\alpha_j := \alpha_j(c)$.

---

[2]It also follows from the nonsingularity of V that the equation (12.12) have full rank for $p \leq q$ so that there are $(\mu + \nu - p)$ families of methods of order p.

The local truncation error of our new method can also be readily calculated. This will require the introduction of a (different) polynomial $\phi(x)$ of degree q that interpolates the *exact* PDE solution at the points $\{x_{m+j}\}_{j=-\mu}^{\nu}$, that is

$$\phi(x) = \sum_{j=-\mu}^{\nu} L_j(x) u_{m+j}^n.$$

Starting from (12.15) and using the exact solution characterisation (12.3), the local truncation error of the new method is given by

$$\mathcal{R}_m^n = \frac{1}{k}\left(u_m^{n+1} - \phi(x_m - ch)\right) = \frac{1}{k}\left(u(x_m - ch, t_n) - \phi(x_m - ch, t_n)\right),$$

The quantity in brackets on the right is just the error in polynomial interpolation at the point $x_m - ch$. An analytical expression for this error can be found in many numerical analysis textbooks, for example in Süli and Mayers [23, Theorem 6.2]). If the polynomial interpolation error term is bounded (by defining M_{q+1} to be the maximum value of $|\partial_x^{q+1} u(x - ch, t_n)|$ over the interval $x_{m-\mu} < x - ch < x_{m+\nu}$) then we end up with the following bound

$$|\mathcal{R}_m^n| \le \frac{M_{q+1}}{(q+1)!} \left(\prod_{j=-\mu}^{\nu} (j + c)\right) \frac{1}{k} h^{q+1} t. \tag{12.18}$$

The ratio h/k is constant, so (12.18) implies that the order of the new method is q. Since there is a unique method of optimal order, the construction (12.17) must coincide with the method described earlier where the coefficients were found by solving the system $V\alpha = c$. $\Diamond$

12.1.2 Stability Conditions

The existence of (μ, ν)-methods of order $\mu + \nu$ raises the prospect of finding high-order explicit schemes of *positive type*. The next example shows that searching for such a scheme is futile.

Example 12.5 (Stability barrier) Suppose that (12.5) defines an (μ, ν)-operator of positive type (that is, $\alpha_j \ge 0$ for $j = -\mu, \ldots, \nu$, see Definition 10.8). Show that the consistency of the associated explicit difference scheme cannot be higher than first order.

The aim is to show that the assumption of a positive type scheme of order two leads to a contradiction. As discussed already, the conditions $C_0 = C_1 = C_2 = 0$ in (12.11) represent three linearly independent equations in $(\mu + \nu + 1)$ unknowns. Since the coefficients are nonnegative by assumption, we start by defining two real vectors

$$\boldsymbol{x} = (\alpha_j^{1/2})_{j=-\mu}^{\nu}, \quad \boldsymbol{y} = (j\alpha_j^{1/2})_{j=-\mu}^{\nu}.$$

Next, writing out the equation $C_1 = 0$ from (12.11), taking the absolute value of both sides and then applying the Cauchy–Schwarz inequality (B.3) gives

$$|c| = \Big| \sum_{j=-\mu}^{\nu} j\alpha_j \Big| = \Big| \boldsymbol{x}^{\mathsf{T}}\boldsymbol{y} \Big| \leq \|\boldsymbol{x}\| \, \|\boldsymbol{y}\|$$

$$= \underbrace{\Big(\sum_{j=-\mu}^{\nu} \alpha_j \Big)^{1/2}}_{1} \underbrace{\Big(\sum_{j=-\mu}^{\nu} j^2 \alpha_j \Big)^{1/2}}_{c^2} = |c|,$$

where the terms on the right simplify using the equations $C_0 = 0$ and $C_2 = 0$. This is a contradiction: if $\boldsymbol{x}^{\mathsf{T}}\boldsymbol{y} = \|\boldsymbol{x}\| \, \|\boldsymbol{y}\|$ then the vectors $\boldsymbol{x}$ and $\boldsymbol{y}$ have to be parallel (that is a multiple of one another) which is not possible because of their construction. ◇

A first-order positive type scheme will be discussed in the next section. The stability barrier makes it impossible to use elementary arguments to establish the stability of explicit second-order (or higher-order) schemes in the ℓ_∞ (or maximum) norm.

To make any progress one has to assess the stability in a weaker norm, as was done in Sect. 11.2. A theoretical result that is especially useful in this regard is the following.

Theorem 12.6 (von Neumann stability) *A necessary condition for a finite difference scheme (12.5) to have order p and be ℓ_2 stable, is that the stencil must have at least[3] $\lceil p/2 \rceil$ grid points on either side of the point where the characteristic through the target point (x_m, t_{n+1}) intersects the previous time level.*

Proof A proof can be found in Jeltsch and Smit [9] and builds on earlier work of Iserles and Strang.[4] (The proof relies on the concept of order stars so it is beyond the scope of this textbook.) □

An alternative interpretation of the von Neumann stability condition is that if μ^* and ν^* are the number of upwind and downwind points relative to the intersection point Q (see Fig. 12.2), then the maximum order of the scheme will satisfy

---

[3]This is known as the *ceiling* function. It implies that $p/2$ should be rounded up when p is an odd integer.

[4]The historical background to this theorem is discussed in the seminal paper by Iserles and Nørsett [8].

$$p \leq \begin{cases} 2 \min(\mu^*, \nu^*), & \text{when } p \text{ is even} \\ 2 \min(\mu^*, \nu^*) - 1, & \text{when } p \text{ is odd.} \end{cases}$$

We are now ready to turn our attention to specific methods. One way to construct explicit schemes is to choose integers μ, ν, and then to determine the coefficients in the (μ, ν)-stencil by applying the order relations (12.12). We will not take this path. Instead we will follow the methodology used to approximate parabolic equations in Sect. 11.1. This approach reveals more of the structure of the resulting methods and generalises more readily to PDEs with variable coefficients, reaction terms or source functions.

12.1.3 First-Order Schemes

We start with the lowest-order case. The simplest approximation of (12.1) derives from the Taylor expansion

$$u(x, t + k) = u(x, t) + k u_t(x, t) + \mathcal{O}(k^2). \tag{12.19}$$

Using the differential equation, we have $u_t = -a u_x$ so

$$u(x, t + k) = u(x, t) - a k u_x(x, t) + \mathcal{O}(k^2).$$

Approximating the spatial derivative with the backward difference formula $u_x = h^{-1} \triangle_x^- u + \mathcal{O}(h)$ then leads to

$$u_m^{n+1} = u_m^n - (ak/h)\triangle_x^- u_m^n + \mathcal{O}(h) + \mathcal{O}(k^2).$$

The discrete version of this is the FTBS scheme (Forward Time, Backward Space),

$$U_m^{n+1} = U_m^n - c \triangle_x^- U_m^n = c U_{m-1}^n + (1 - c) U_m^n, \tag{12.20}$$

where, as discussed earlier, $c = ak/h$ is the Courant number. The associated stencil is shown in Fig. 12.4 (left). The figure also shows why the FTBS scheme is referred to as the first-order *upwind scheme* in the case $c > 0$. The restriction $0 < c \leq 1$ is a consequence of the CFL condition discussed earlier.

The local truncation error of the FTBS scheme is given by

$$\begin{aligned} \mathcal{R}_m^n &= \frac{1}{k} [u(x_m, t_n + k) - u(x_m, t_n)] + \frac{a}{h} \triangle_x^- u(x_m, t_n) \\ &= \left(u_t + a u_x + \tfrac{1}{2} k u_{tt} - \tfrac{1}{2} a h u_{xx} \right) \big|_m^n + \mathcal{O}(h^2) + \mathcal{O}(k^2), \end{aligned} \tag{12.21}$$

so it is $\mathcal{O}(h) + \mathcal{O}(k)$, first order in space and time.

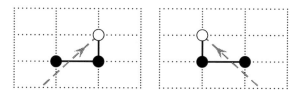

Fig. 12.4 Stencils for the FTBS method (12.20) (*left*) and the FTFS method (12.25) (*right*). The *dashed lines* show the characteristic through the target point (x_m, t_{n+1}) (o) when $0 < c < 1$ (*left*) and $-1 < c < 0$ (*right*)

Returning to the issue of stability. The coefficients on the right of (12.20) are both positive for $0 < c \leq 1$. Thus the scheme is ℓ_∞ stable under this condition. It is important to appreciate that (because of the CFL condition) the FTBS method is unstable whenever $c < 0$, that is, when the characteristics move from right to left. Another interesting feature is that (12.20) reduces to $U_m^{n+1} = U_{m-1}^n$ when $c = 1$. This means that the FTBS scheme exactly mimics the relation (12.4) satisfied by the exact solution when the characteristics $x - at = $ constant pass diagonally through the grid. This feature is called the *unit CFL property*.

Before looking at the ℓ_2 stability of the FTBS scheme, we recall that the advection equation $u_t + au_x = 0$ has separation of variables solutions of the form $u(x, t) = A(t)\,e^{i\kappa x}$, where κ is the *wavenumber* (that is, the number of waves per 2π units of distance) with coefficients $A(t) = e^{-ia\kappa t}$ that are periodic in time. This leads to fundamental solutions[5] of the form $u(x, t) = e^{-i\omega t}e^{i\kappa x}$, where $\omega = a\kappa$ is the frequency (the number of waves per 2π units of time), and the relation between ω and κ is known as the *dispersion relation*. Note that $|u(x, t)| = |e^{-i\omega t}e^{i\kappa x}| = 1$, so a solution of the advection equation must have constant modulus. This property will not, in general, be satisfied by finite difference approximations of the exact solution.

To test for ℓ_2 stability using the von Neumann approach, we simply substitute $U_m^n = \xi^n e^{i\kappa mh}$ into (12.20) (cf. Sect. 11.2). The resulting amplification factor is

$$\xi = 1 - c + ce^{-i\kappa h}. \tag{12.22}$$

Note that ξ depends on κh and c. Writing $\xi = (1 - c + c\cos \kappa h) - i\sin \kappa h$, taking the square of its modulus and simplifying the trigonometric functions using half-angle formulae, leads to

$$|\xi|^2 - 1 = -4c(1 - c)\sin^2 \tfrac{1}{2}\kappa h. \tag{12.23}$$

Hence, $|\xi| \leq 1$ for all wavenumbers κ if, and only if, $0 < c \leq 1$—the same condition as for maximum norm stability. Stability is not the whole story however. Since $|\xi| < 1$ for almost all wavenumbers, components of the numerical solution are highly likely to be damped out, as the following example will illustrate.

---

[5]The advection equation is a linear PDE with constant coefficients, so the real and imaginary parts provide linearly independent (real) solutions.

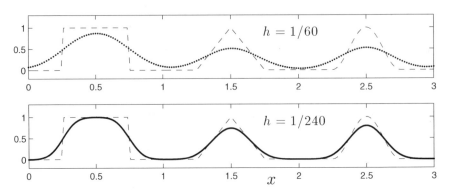

Fig. 12.5 The FTBS solution to the problem in Example 12.1 (*dots*) at $t = 3$ and the exact solution (*dashed line*) for Courant number $c = 0.45$

Example 12.7 Use the FTBS method to solve the advection equation on the interval $0 \leq x \leq 3$ with $a = 1$, together with the initial condition (12.2) and periodic boundary conditions. Compute the numerical solution for one complete period ($\tau = 3$) with a Courant number[6] $c = 0.45$ using two spatial grids—a coarse grid ($h = 1/60$, 400 time steps) and a fine grid ($h = 1/240$, 1600 time steps).

When (12.20) is used at the left boundary ($m = 0$) to determine U_0^{n+1} it involves the value U_{-1}^n corresponding to a grid point lying outside the domain. However, periodicity implies that $U_{m+M}^n = U_m^n$ for all m and so the required value is equal to U_{M-1}^n. Also, periodicity implies that $U_M^{n+1} = U_0^{n+1}$ so that formula (12.20) is not required at $m = M$.

The numerical results shown in Fig. 12.5 illustrate the demanding nature of this test. The three pulses are heavily damped (smeared) to such an extent that the evolved triangular and cosine-squared waves are almost indistinguishable. Computing an accurate solution to this model problem will be a challenge!   ◊

Insight into this disappointing behaviour may be obtained by revisiting the expression (12.21) for the local truncation error of the FTBS scheme. Differentiating the PDE as done in (12.9) and substituting the result into (12.21) gives the expression

$$R_m^n = \left(u_t + au_x - \tfrac{1}{2}ah(1 - c)u_{xx} \right)\Big|_m^n + \mathcal{O}(h^2) + \mathcal{O}(k^2). \tag{12.24}$$

Thus, the FTBS method is not only a first-order approximation of the advection equation $u_t + au_x = 0$, it is also a *second-order* approximation of the advection–diffusion equation

$$u_t + au_x = \tfrac{1}{2}ah(1 - c)u_{xx},$$

in which the diffusion coefficient ($\tfrac{1}{2}ah(1 - c)$) is positive for $0 < c < 1$ and tends to zero with h. The FTBS method, being a stable, consistent approximation of

---

[6]This is carefully chosen so as not to flatter schemes with the unit CFL property.

this modified PDE will have solutions that behave like those of advection–diffusion equations (see Example 4.7) and as a result, will tend to become smoother as time evolves! This approach of linking the solution of a numerical method to that of a different PDE to the one it was designed to solve is known as the method of *modified equations*. Its application to initial value problems for ODEs is described in some detail by Griffiths and Higham [7, Chap. 13].

To obtain a stable scheme for the advection equation when $a < 0$ the backward space difference is replaced in (12.20) by a forward difference to give the FTFS method,

$$U_m^{n+1} = U_m^n - c\triangle_x^+ U_m^n = (1 + c)U_m^n - cU_{m+1}^n \tag{12.25}$$

whose stencil is shown in Fig. 12.4 (right). The order of local accuracy is again $\mathcal{O}(h) + \mathcal{O}(k)$ but in this case we have stability in both the ℓ_∞ and the ℓ_2 sense whenever $-1 \le c < 0$ (see Exercise 12.2).

The FTBS results in Fig. 12.5 illustrate why first-order methods are generally regarded as being too inefficient for practical use (even for problems with smooth solutions), motivating the construction of higher-order schemes. One obvious way of increasing the local accuracy in space is to use a second-order central difference operator $\triangle_x$. Doing this leads to the FTCS method, given by

$$U_m^{n+1} = (1 - c\triangle_x)U_m^n = \tfrac{1}{2}c\,U_{m-1}^n + U_m^n - \tfrac{1}{2}c\,U_{m+1}^n. \tag{12.26}$$

Unfortunately, while the local truncation error is $\mathcal{O}(h^2) + \mathcal{O}(k)$, the FTCS scheme is unconditionally unstable (see Exercise 12.3). More successful ways of increasing the order of accuracy are described in the next two sections.

12.1.4 Second-Order Schemes

The first scheme that we discuss is, perhaps, the most celebrated of all finite difference methods for hyperbolic PDEs. To improve the spatial accuracy we need to include the k^2 term in the Taylor expansion used in the previous section, that is

$$u(x, t + k) = u(x, t) + ku_t(x, t) + \tfrac{1}{2}k^2 u_{tt}(x, t) + \mathcal{O}(k^3). \tag{12.27}$$

The time derivatives can again be replaced by space derivatives by observing that $u_t = -au_x$ and $u_{tt} = a^2 u_{xx}$ ((12.1) and (12.9)), to give

$$u(x, t + k) = u(x, t) - aku_x(x, t) + \tfrac{1}{2}a^2 k^2 u_{xx}(x, t) + \mathcal{O}(k^3).$$

Fig. 12.6 Stencils for the second-order Lax–Wendroff method (*left*) and for the leapfrog method (*right*)

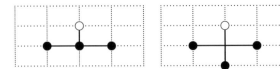

Taking second-order approximations in space

$$u_x = h^{-1}\Delta_x u + \mathcal{O}(h^2), \qquad u_{xx} = h^{-2}\delta_x^2 u + \mathcal{O}(h^2),$$

and neglecting the remainder terms, then leads to the *Lax–Wendroff* method[7]

$$U_m^{n+1} = \left[1 - c\Delta_x + \tfrac{1}{2}c^2\delta_x^2\right] U_m^n, \tag{12.28a}$$

which, when written out explicitly, takes the form

$$U_m^{n+1} = \tfrac{1}{2}c(1+c)U_{m-1}^n + (1-c^2)U_m^n + \tfrac{1}{2}c(c-1)U_{m+1}^n. \tag{12.28b}$$

The associated finite difference stencil is shown in Fig. 12.6 (left). The local truncation error is given by

$$\mathcal{R}_m^n := \frac{1}{k}\left(u_m^{n+1} - u_m^n + c\Delta_x u_m^n - \tfrac{1}{2}c^2\delta_x^2 u_m^n\right), \tag{12.29}$$

which, using the definitions in Table 6.1, can be written as

$$\mathcal{R}_m^n = \left(u_t + au_x + \tfrac{1}{6}k^2 u_{ttt} + \tfrac{1}{6}ah^2 u_{xxx}\right)\Big|_m^n + \mathcal{O}(kh^2) + \mathcal{O}(k^3).$$

Differentiating the PDE a second time, and rearranging gives

$$\mathcal{R}_m^n = \tfrac{1}{6}ah^2(1-c^2)u_{xxx}\Big|_m^n + \mathcal{O}(kh^2) + \mathcal{O}(k^3), \tag{12.30}$$

which shows that the method is consistent of second order.

There is no choice of c (other than the isolated cases $c = 0, \pm 1$) for which the coefficients in the scheme are all nonnegative, so we cannot prove stability in the maximum norm by elementary means (a strict maximum principle does not hold). By applying the von Neumann stability test the scheme can, however, be shown to be ℓ_2 stable for values of c satisfying $-1 \le c \le 1$ (see Exercise 12.7). This is significant because it implies that the Lax–Wendroff scheme is capable of accommodating both right- and left-moving waves ($a > 0$ and $a < 0$).

---

[7]Peter Lax (1926–), a renowned mathematician and a pioneer of the numerical analysis of PDEs, is closely associated with the Courant Institute in New York.

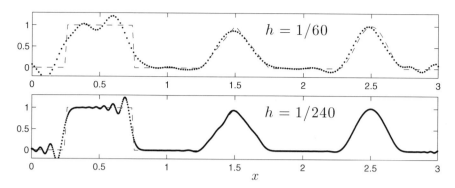

Fig. 12.7 The Lax–Wendroff solution to the problem in Example 12.8 at $t = 3$ and the exact solution (*dashed line*) for Courant number $c = 0.45$

Example 12.8 Use the Lax–Wendroff method to solve the advection equation on the interval $0 \leq x \leq 3$ with $a = 1$, together with the initial condition (12.2) and periodic boundary conditions. Compute the numerical solution with $c = 0.45$ for one complete period ($\tau = 3$) using two spatial grids—a coarse grid ($h = 1/60$, 400 time steps) and a fine grid ($h = 1/240$, 1600 time steps).

The numerical results are presented in Fig. 12.7. The most striking feature is that the method does not cope well with the discontinuous nature of the square wave on either grid. An "oscillating tail" accompanies each break in continuity. This behaviour is typical of solutions computed using the Lax–Wendroff method. The rounded "corners" of the triangular wave are an indication that the second-order method is also adding some extra diffusion. Close inspection of the solution for $h = 1/60$ reveals that the numerical wave (dots) travels slightly slower than the exact solution. Compared to the results for the first-order FTCS scheme in Fig. 12.5, a substantial improvement in the resolution of the solution is obtained using the Lax–Wendroff method. ◇

An alternative second-order method is known as *leapfrog*.[8] It is generated by approximating the derivatives in the advection equation using simple centered differences

$$u_t|_m^n = k^{-1} \Delta_t u_m^n + \mathcal{O}(k^2) = \tfrac{1}{2} k^{-1}(u_m^{n+1} - u_m^{n-1}) + \mathcal{O}(k^2)$$
$$u_x|_m^n = h^{-1} \Delta_x u_m^n + \mathcal{O}(h^2) = \tfrac{1}{2} h^{-1}(u_{m+1}^n - u_{m-1}^n) + \mathcal{O}(h^2),$$

leading to the *three-level* difference scheme

$$U_m^{n+1} = U_m^{n-1} - c(U_{m+1}^n - U_{m-1}^n). \tag{12.31}$$

---

[8]The method is named after the children's playground game.

It is called the leapfrog method because U_m^{n+1} is obtained by adjusting the value of U_m^{n-1} without reference to the 'centre' value U_m^n, as shown in Fig. 12.6. The method has, by construction, a local truncation error that is $\mathcal{O}(h^2)$ (for fixed c), and it can be shown to be ℓ_2 stable for $-1 \le c \le 1$ (see Exercise 12.12). Another significant feature of leapfrog is that the (von Neumann) amplification factor has the property

$$|\xi(\kappa h)| = 1, \quad -1 \le c \le 1$$

for all wave numbers κ—so it displays no damping (like the PDE). This makes it a *nondissipative* method, of which we shall have more to say later (in Sect. 12.4.1). Although this may seem to be an attractive property, we shall see from the numerical experiments that some level of damping (especially of high frequency components of the solution) is often beneficial.

Since the method requires values at two previous levels t_n and t_{n-1} in order to compute the solution at the next time level t_{n+1}, two levels of initial data are needed to start the time stepping process. The initial condition provides the data at $t_0 = 0$ while the second-order Lax–Wendroff scheme may be used to provide the data at t_1.

Example 12.9   Use the leapfrog method to solve the advection equation on the interval $0 \le x \le 3$ with $a = 1$, together with the initial condition (12.2) and periodic boundary conditions. Compute the numerical solution with $c = 0.45$ for one complete period ($\tau = 3$) using two spatial grids—a coarse grid ($h = 1/60$, 400 time steps) and a fine grid ($h = 1/240$, 1600 time steps), and compare the results with those obtained using Lax–Wendroff in Example 12.8.

The numerical results are presented in Fig. 12.8. The leapfrog solution profile (shown dotted) shows a considerable level of noise. This is mostly generated by the discontinuities in the square wave. If the computation was repeated with the square wave

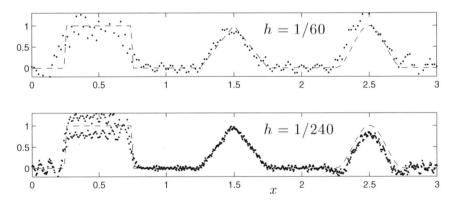

Fig. 12.8 The leapfrog method solution to the problem in Example 12.9 at $t = 3$ and the exact solution (*dashed line*) for Courant number $c = 0.45$

removed from the initial condition, then the accuracy would be comparable to that
of the Lax–Wendroff method (shown in Fig. 12.7).   ◊

The origin of the high level of numerical noise may be clarified by labelling the
grid points as being "odd" or "even" depending on whether $m + n$ is odd or even.
The leapfrog update involves only even or only odd grid points. This means that the
solution on even grid points can be calculated independently of the solution on the
odd grid points (similar to the situation outlined in Exercise 10.4). As a result, if the
initial condition involves a discontinuity, then consecutive points might have very
different values and this difference will persist for all time.

12.1.5 A Third-Order Scheme

An ideal scheme for solving the advection equation would retain the good features of the Lax–Wendroff method but would be less diffusive and less dispersive.
To construct such a scheme we follow a procedure—discussed in Sect. 6.4.2 and
Example 10.6—which aims to raise the order of the Lax–Wendroff scheme by including a finite difference approximation of its local truncation error.

The local truncation error for the Lax–Wendroff scheme was derived earlier,

$$\mathcal{R}_m^n = \tfrac{1}{6}ah^2(1 - c^2)u_{xxx}\big|_m^n + \mathcal{O}(kh^2) + \mathcal{O}(k^3). \tag{12.32}$$

The leading term in this expansion involves the third derivative of the solution in
space, so a finite difference approximation of this term must involve at least 4 consecutive grid points. These cannot be placed symmetrically about $x = x_m$ and so
a decision has to made about whether to bias the scheme in an upwind or downwind direction. Intuition (and the first-order scheme results in Sect. 12.1.3) suggests
a biasing of the stencil in the upwind direction. Thus, when $a > 0$, $\mathcal{R}_m^n$ in (12.32) is
approximated by $-\tfrac{1}{6}ah^{-1}(1 - c^2)\triangle_x^- \delta_x^2 U_m^n$. Otherwise, when $a < 0$, the backward
difference $\triangle_x^-$ should be replaced by the forward difference $\triangle_x^+$. Multiplying by k (to
account for the normalization) and subtracting the approximation from (12.28) gives
a variant of the QUICK scheme (Quadratic Upstream Interpolation for Convective
Kinematics) originally devised by B.P. Leonard [13]:

$$U_m^{n+1} = [1 - c\triangle_x + \tfrac{1}{2}c^2\delta_x^2 + \tfrac{1}{6}c(1 - c^2)\triangle_x^- \delta_x^2] U_m^n \tag{12.33a}$$

(in the case $a > 0$) which shall refer to as the *third-order upwind scheme*. This is
a well-defined (μ, ν)-method of the form (12.5) with $\mu = 2$ and $\nu = 1$ having the
following coefficients

$$\alpha_{-2} = -\tfrac{1}{6}c(1 - c^2), \qquad \alpha_{-1} = \tfrac{1}{2}c(1 + c)(2 - c),$$
$$\alpha_0 = \tfrac{1}{2}(1 - c^2)(2 - c), \qquad \alpha_1 = \tfrac{1}{6}c(c - 1)(2 - c). \tag{12.33b}$$

Fig. 12.9 Stencils for the third-order upwind scheme (12.33) for $a > 0$ (*left*) and for $a < 0$ (*right*)

The coefficients are cubic polynomials: this is a necessary condition for the method to be of order 3 (see Example 12.4). The two possible alternative stencils are illustrated in Fig. 12.9. A consequence of the inbuilt bias is that the scheme (12.33) is only ℓ_2-stable for positive Courant numbers—specifically only when $0 \leq c \leq 1$. This and other aspects of the third-order scheme are explored in Exercise 12.13.

Example 12.10 Solve the advection equation using the third-order upwind method on the interval $0 \leq x \leq 3$ with $a = 1$, together with the initial condition (12.2) and periodic boundary conditions. Compute the numerical solution with $c = 0.45$ for one complete period ($\tau = 3$) using two spatial grids—a coarse grid ($h = 1/30$, 200 time steps) and a fine grid ($h = 1/120$, 800 time steps).

The numerical results are presented in Fig. 12.10. These computations appear to be exceptionally accurate, not least because the computational grids have half as many grid points as those used for the second-order methods earlier in the chapter.   ◇

12.1.6 Quasi-implicit Schemes

Implicit methods are needed when solving parabolic PDEs—they can follow the physics by avoiding the stability restrictions of explicit methods. The physics of

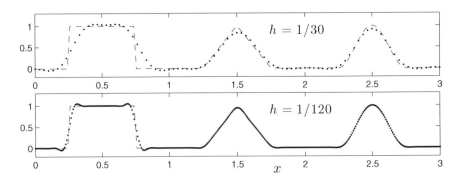

Fig. 12.10 The third-order scheme solution to the problem in Example 12.10 at $t = 3$ and the exact solution (*dashed line*) for Courant number $c = 0.45$

Fig. 12.11 Stencils for the BTBS scheme (12.34) (*left*) and the box scheme (12.38) (*right*). No target points are shown since they cannot be uniquely defined

hyperbolic PDEs is different—so the case for using implicit methods is less clear-cut. Two implicit schemes are described below. We will refer to them as "quasi-implicit" since they only involve two unknowns at the new time level and (except when periodic boundary conditions apply) may be applied without having to solve systems of algebraic equations.

The simplest quasi-implicit method is obtained when the forward difference in the FTBS scheme (12.20) is replaced by a backward difference, so that

$$U_m^{n+1} = U_m^n - c\Delta_x^- U_m^{n+1}, \tag{12.34a}$$

or, alternatively

$$-c\,U_{m-1}^{n+1} + (1+c)U_m^{n+1} = U_m^n. \tag{12.34b}$$

We will refer to (12.34) as the BTBS method. The associated finite difference stencil is shown in Fig. 12.11 (left). When the scheme is used to solve (12.1) with periodic boundary conditions, the solution vector

$$\boldsymbol{u}^{n+1} = [U_1^{n+1}, U_2^{n+1}, \ldots, U_M^{n+1}]^\mathsf{T}$$

at time t_{n+1} is obtained by solving the system $C\boldsymbol{u}^{n+1} = \boldsymbol{u}^n$ where

$$C = \begin{bmatrix} (1+c) & 0 & \cdots & -c \\ -c & (1+c) & & \\ & & \ddots & \ddots \\ 0 & & & -c\ (1+c) \end{bmatrix}. \tag{12.35}$$

The matrix C is a *circulant* matrix. Each row is obtained by translating the preceding row one entry to the right in a cyclic fashion.[9] Also, all the entries of C^{-1} are nonzero, so each component of $\boldsymbol{u}^{n+1}$ depends on every component of $\boldsymbol{u}^n$. This is perfect—it implies that there is no CFL time step restriction. (This is not the case when the scheme is used to solve problems with more realistic boundary conditions—see Exercise 12.23.) To apply the von Neumann stability test, we substitute $U_m^n = \xi^n \exp(i\kappa mh)$ into (12.34b). The resulting amplification factor is the reciprocal of

---

[9]The action of the inverse of a circulant matrix may be efficiently computed using a Fast Fourier Transform (FFT), see Strang [20].

the FTBS scheme amplification factor (12.22) with $-c$ instead of $+c$, that is,

$$\xi = \frac{1}{1 + c - ce^{-i\kappa h}}. \tag{12.36}$$

Thus, since the FTBS scheme is ℓ_2 stable for $-1 \leq -c < 0$, the BTBS scheme will be ℓ_2 stable if, and only if $c > 0$ or $c \leq -1$ (see Exercise 12.23). The set of stable Courant numbers is thus composed of two disjoint sets.

The order of accuracy of the FTBS scheme can be increased (while preserving a compact stencil) by approximating the leading term in the local truncation error and feeding it back into the scheme. To this end, we note that (see Exercise 12.14) the local truncation error of the scheme (12.34) is given by

$$\mathcal{R}_m^n = \frac{1}{k}\left((1+c)u_m^{n+1} - cu_{m-1}^{n+1} - u_m^n\right) = \tfrac{1}{2}h(1+c)\,u_{xt}|_m^{n+1} + \mathcal{O}(h^2) + \mathcal{O}(k^2) \tag{12.37}$$

The leading term can be approximated by $\tfrac{1}{2}k^{-1}(1+c)\triangle_x^-\triangle_t^- U_m^{n+1}$ and, when multiplied by k (to compensate for the scaling factor in $\mathcal{R}_m^n$) and subtracted from the left hand side of (12.34), we obtain the second-order accurate *box* scheme

$$(1-c)U_{m-1}^{n+1} + (1+c)U_m^{n+1} = (1+c)U_{m-1}^n + (1-c)U_m^n, \tag{12.38}$$

whose stencil is shown in Fig. 12.11 (right).

The associated von Neumann amplification factor is given by

$$\xi = \frac{1 - c + (1+c)e^{-i\kappa h}}{1 + c + (1-c)e^{-i\kappa h}}, \tag{12.39}$$

from which it can be shown (see Exercise 12.19) that the box scheme is nondissipative: $|\xi| = 1$ for all wavenumbers κ. It is also unconditionally ℓ_2 stable.

Example 12.11 Solve the advection equation using the box scheme on the interval $0 \leq x \leq 3$ with $a = 1$, together with the initial condition (12.2) and periodic boundary conditions. Compute the numerical solution with $c = 0.45$ for one complete period ($\tau = 3$) using two spatial grids—a coarse grid ($h = 1/60$, 400 time steps) and a fine grid ($h = 1/240$, 1600 time steps)—and compare the performance with that of the second-order explicit schemes.

The numerical results are presented in Fig. 12.12. The overall resolution of the profiles can be seen to be similar to that of the Lax–Wendroff method in Fig. 12.7, but contain more noise due to the lack of any damping (but not as much as the leapfrog scheme). The behaviour at discontinuities is strikingly similar to the Lax–Wendroff method except that spurious oscillations are being generated downstream of the discontinuity, rather than upstream.   ◇

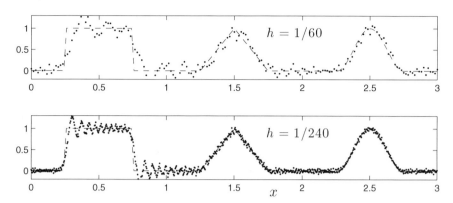

Fig. 12.12 The box scheme solution to the problem in Example 12.11 at $t = 3$ and the exact solution (*dashed line*) for Courant number $c = 0.45$

12.2 The Two-Way Wave Equation

The second prototypical hyperbolic PDE is the wave equation (pde.5),

$$u_{tt} = a^2 u_{xx} \tag{12.40}$$

with constant wave speed a. The simplest direct approximation is to use a second-order centred difference for each of the derivative terms,

$$u_{tt}|_m^n = k^{-2}\delta_t^2 u_m^n + \mathcal{O}(k^2), \quad u_{xx}|_m^n = h^{-2}\delta_h^2 u_m^n + \mathcal{O}(k^2)$$

leading to,

$$k^{-2}\delta_t^2 U_m^n = a^2 h^{-2}\delta_h^2 U_m^n \tag{12.41a}$$

or, when written out explicitly,

$$U_m^{n+1} = 2(1 - c^2)U_m^n + c^2(U_{m+1}^n + U_{m-1}^n) - U_m^{n-1}, \tag{12.41b}$$

where $c = ak/h$.

The stencil of this scheme is the same as the 5-point approximation of the Laplacian (Fig. 10.2, right), the main difference is that the new time value in (12.41b) can be explicitly updated from the three known values on the previous time levels, while maintaining stability. The method (12.41) requires two levels of starting values: this mirrors the requirement that the PDE has two initial conditions in order to be well posed. Suitable initial conditions, taken from Example 4.5, are given by

$$u(x, 0) = g_0(x), \quad u_t(x, 0) = g_1(x). \tag{12.42}$$

The first of these provides one of the initial conditions for the difference scheme: $U_m^0 = g_0(x_m)$. For the second initial condition we use a process similar to that used when approximating a Neumann boundary condition in Sect. 11.3.1. The first step is to replace the t–derivative in (12.42) by the central time difference $\triangle_t$ which gives the second-order approximation $k^{-1}\triangle_t U_m^0 = g_1(x_m)$, that is,

$$U_m^1 - U_m^{-1} = 2kg_1(x_m).$$

This involves a fictitious grid point (x_m, t_{-1}) but, when combined with (12.41b) at $n = 0$, leads to the modified update formula

$$U_m^1 = kg_1(x_m) + (1 - c^2)g_0(x_m) + \tfrac{1}{2}c^2\big(g_0(x_{m+1}) + g_0(x_{m-1})\big) \qquad (12.43)$$

which we use to compute the numerical solution at $t = t_1$.

The CFL condition for (12.41) imposes the condition $|c| \leq 1$. To assess the ℓ_2 stability of the method, we simply substitute $U_m^n = \xi^n \mathrm{e}^{\mathrm{i}\kappa mh}$ into (12.41). With some algebra, the amplification factor can be shown to satisfy the quadratic equation

$$\xi^2 - 2(1 - 2c^2 s^2)\xi + 1 = 0, \qquad (12.44)$$

where $s = \sin(\tfrac{1}{2}\kappa h)$. It may be also be shown (see Exercise 12.20) that the roots of the quadratic in (12.44) form a complex conjugate pair with product equal to 1 for all wave numbers. Since the roots are real if $c^2 > 1$, this means that the 5-point scheme is ℓ_2 stable only when $-1 \leq c \leq 1$.

We shall return to the two-way wave equation in Example 12.18 where it is treated as a system of first-order PDEs.

12.3 Convergence Theory

The von Neumann stability test is a hidden gem. All the requisite information for studying finite difference approximations of constant coefficient PDE problems with periodic boundary conditions is contained in the amplification factor ξ. Indeed, when solving constant coefficient hyperbolic PDEs, the factor ξ is not only used to establish convergence of methods but also to provide insight into the qualitative nature of the resulting numerical solutions.

We want to study the propagation of a single Fourier mode of the form

$$u(x, t) = \mathrm{e}^{\mathrm{i}(\kappa x - \omega t)}, \quad \kappa \in \mathbb{R}, \qquad (12.45)$$

where κ and ω represent the wavenumber and frequency, respectively. The Fourier mode describes a (complex valued) travelling wave of unit amplitude and wavelength $2\pi/\kappa$. The function u will be a solution of the one-way wave equation $u_t + au_x = 0$, if the initial condition satisfies $u(x, 0) = \exp(\mathrm{i}\kappa x)$ *and* if there is a linear *dispersion relation* relating the frequency to the wavenumber, that is, whenever

$$\omega = a\kappa. \tag{12.46}$$

In this case the velocity $a = \omega/\kappa$ is known as the *phase speed*. More generally, a typical constant coefficient (hyperbolic) PDE will have solutions (12.45) whenever a nonlinear dispersion relation is satisfied. (For the two-way wave equation $u_{tt} = a^2 u_{xx}$ we require that $\omega^2 = a^2 \kappa^2$ so the speed of propagation $a = \omega(\kappa)/\kappa$ is different for different wavenumbers.)

The fact that the speed a is the same for waves of all frequencies is what makes the one-way wave equation so special. When a typical Fourier mode is evaluated at a grid point (mh, nk) we have

$$u_m^n = e^{i(\kappa mh - \omega nk)} = [e^{-i\omega k}]^n e^{i\kappa mh}$$

and, using the dispersion relation leads to

$$u_m^n = [e^{-ic\theta}]^n e^{im\theta}, \tag{12.47}$$

where $\theta = \kappa h$ and $c = ak/h$. The von Neumann test checks the stability of a finite difference solution with exactly the same spatial behaviour as u_m^n in (12.47): that is, a grid solution that takes the specific form

$$U_m^n = \xi(\theta)^n e^{im\theta}, \tag{12.48}$$

where we have written $\xi = \xi(\theta)$ here to emphasise its dependence on the scaled wavenumber θ. The relationship between (12.47) and (12.48) means that the order of consistency of a scheme can be identified by simply comparing the MacLaurin expansion of $e^{-ic\theta}$ with that of the von Neumann amplification factor $\xi(\theta)$. The process is illustrated in the following example.

Example 12.12 Suppose that the Lax–Wendroff method is used to solve the one-way wave equation on a periodic domain, together with the initial condition $u(x, 0) = \exp(i\kappa x)$. Express the local truncation error in terms of the von Neumann amplification factor, and hence deduce the order of consistency of the finite difference approximation scheme.

The local truncation error of the Lax–Wendroff method is (from (12.29))

$$\mathcal{R}_m^n = \frac{1}{k}\left(u_m^{n+1} - u_m^n + c\triangle_x u_m^n - \tfrac{1}{2}c^2\delta_x^2 u_m^n\right). \tag{12.49}$$

Substituting the exact solution (12.47) into this expression gives the alternative representation

$$\mathcal{R}_m^n = \frac{1}{k}\left(e^{-ic\theta} - 1 + ic\sin\theta + 2c^2\sin^2\tfrac{1}{2}\theta\right)u_m^n.$$

However, from Exercise 12.7, the amplification factor of the Lax–Wendroff scheme is given by $\xi(\theta) = 1 - ic \sin \theta - 2c^2 \sin^2 \frac{1}{2}\theta$, so that

$$R_m^n = \frac{1}{k}\left(e^{-ic\theta} - \xi(\theta)\right)u_m^n. \tag{12.50}$$

The order of consistency of the scheme is concerned with the behaviour of R_m^n in the limit $h \to 0$, that is, when $\theta = \kappa h \to 0$. Thus, comparing the Maclaurin expansions

$$\xi(\theta) = 1 - ic\theta - \tfrac{1}{2}c^2\theta^2 + \tfrac{1}{6}ic\theta^3 + \mathcal{O}(\theta^4)$$
$$e^{-ic\theta} = 1 - ic\theta - \tfrac{1}{2}c^2\theta^2 + \tfrac{1}{6}ic^3\theta^3 + \mathcal{O}(\theta^4),$$

together with $c = ak/h$ and $\theta = \kappa h$, we find

$$e^{-ic\theta} - \xi(\theta) = -\tfrac{1}{6}ic(1 - c^2)\theta^3 + \cdots = \mathcal{O}(kh^2) + \mathcal{O}(k^3).$$

We deduce that the order of consistency is $R_h = \mathcal{O}(h^2)$ when c is kept fixed.   ◇

It can be shown that that the relationship (12.50) holds for all explicit methods of (μ, ν)-type (see Exercise 12.21). In this simplified periodic setting the global error $E = u - U$ can also be expressed as a function of the amplification factor ξ. As a result, one can directly establish the convergence of the approximation scheme. Such a proof is constructed in the following lemma.

Lemma 12.13 (consistency + stability = convergence) *Suppose that a finite difference method with order of consistency $p > 0$ is used to solve the one-way wave equation on a periodic domain, together with initial condition $u(x, 0) = \exp(i\kappa x)$. If the method is ℓ_2 stable (that is, if $|\xi| \leq 1$), and if c is kept fixed, then the numerical solution will converge to the exact solution in the limit $h \to 0$.*

Proof Using (12.47)–(12.48) the global error is

$$E_m^n = \left(\zeta^n - \xi^n\right)e^{im\theta},$$

where $\zeta(\theta) = \exp(-ic\theta)$. Using the algebraic identity

$$\zeta^n - \xi^n = (\zeta - \xi)(\zeta^{n-1} + \zeta^{n-2}\xi + \cdots + \zeta\xi^{n-2} + \xi^{n-1})$$

with $|\zeta| = 1$ (from its definition) and $|\xi| \leq 1$ (from ℓ_2 stability) we deduce that

$$|E_m^n| \leq n\,|\zeta - \xi|\,|e^{im\theta}| = n\,|\zeta - \xi|.$$

It follows from (12.50) (see Exercise 12.21) that $|\zeta - \xi| = k|R_m^n|$, hence

$$|E_m^n| \leq kn\,|R_m^n| = t_n\,|R_m^n| = \mathcal{O}(h^p), \tag{12.51}$$

since the method is consistent of order p. This shows that the global error tends to zero with h at exactly the same rate as the consistency error.   □

More generally, discrete solutions of the one-way wave equation on a periodic domain of length L will be linear combinations of the fundamental grid solution (12.48) (see Appendix E). Such solutions can be written as

$$U_m^n = \sum_{|j|\leq\lfloor M/2\rfloor}^{*} A_j \xi(\theta_j)^n e^{im\theta_j}, \tag{12.52}$$

with $\theta_j = 2\pi jh/L$, where the coefficients $\{A_j\}$ depend on the initial condition. In this case, the global error in the approximation of the PDE can also be written as a sum of contributions, so that

$$|E_m^n| \leq \sum_{|j|\leq\lfloor M/2\rfloor}^{*} A_j \left|\zeta(\theta_j)^n - \xi(\theta_j)^n\right| e^{im\theta_j}. \tag{12.53}$$

Unfortunately, the simple argument in Lemma 12.13 cannot be applied directly to the individual terms in this expression, since θ_j does not tend to zero when $h\to 0$ for modes corresponding to high wavenumbers $j \sim 1/h$. A possible way of overcoming this obstacle is split the set of wavenumbers into "low" wavenumbers (where $|\zeta(\theta_j) - \xi(\theta_j)| = \mathcal{O}(h^{p+1})$ is small) and "high" wavenumbers (where $|A_j|$ is small and stability implies that $|\zeta(\theta_j)^n - \xi(\theta_j)^n|$ is bounded by an $\mathcal{O}(1)$ constant). For full details see Strikwerda [21, Chap. 10].

   The moral of this section is that showing that a finite difference scheme is a convergent approximation to a hyperbolic PDE will almost certainly involve a combination of technical effort and physical intuition. Low wavenumber terms have to be treated accurately and high wavenumber terms have to approximated in a stable fashion. This is too demanding a programme to pursue at this stage.

12.4 Advanced Topics and Extensions

The material in the following sections is included for the benefit of readers who are seeking specialist knowledge of practical aspects of the numerical solution of wave propagation problems. The advanced material is supported by the inclusion of a set of challenging exercises at the end of the chapter.

12.4.1 Dissipation and Dispersion

This material extends the discussion of travelling wave solutions in the preceding section. Recall that the Fourier mode solution (12.45) to the one-way wave equation is special because $|u(x, t)| = 1$—the amplitude is constant for all time and for waves of all wavenumbers. By way of contrast, the solution of the difference equation (12.48) satisfies $|U_m^n| = |\xi(\theta)|^n$, so that the amplitude varies with both n and $\theta = \kappa h$. From (12.50), we see that the error in propagating a single Fourier mode over one time step is given by $e^{-ic\theta} - \xi(\theta)$ which, being a complex quantity, can be described by its modulus and argument. These components of the error are known as *dissipation* and *dispersion*. They will turn out to be useful concepts for understanding the qualitative behaviour of practical schemes.

Definition 12.14 (*Dissipative approximation method*) A stable finite difference scheme is said to be *dissipative* if $|\xi(\theta)| < 1$ for $\theta \neq 0$. It is dissipative of order $2s$ if there is a positive constant K such that

$$|\xi(\theta)|^2 \leq 1 - K\theta^{2s}.$$

A method is *nondissipative* if $|\xi(\theta)| = 1$ for all $\theta \in [-\pi, \pi]$.

The effect of dissipation is greatest on the modes associated with the highest wave numbers (when $|\theta|$ is not close to zero) and has a smoothing effect on solutions— in accordance with the observation that stability, and not accuracy, is the overriding concern for high wavenumbers. The damping of modes associated with low wavenumbers is relatively small ($\theta \approx 0$) and it diminishes as the order of dissipation increases—this is consistent with the maxim that high accuracy is particularly important for low wavenumbers.

Example 12.15 Determine the order of dissipation of the FTBS scheme (12.20).

The amplification factor of the FTBS scheme satisfies (12.23), that is,

$$|\xi|^2 - 1 = -4c(1 - c)\sin^2 \tfrac{1}{2}\theta.$$

Thus, since $|\sin \tfrac{1}{2}\theta| \leq \tfrac{1}{2}|\theta|$, we deduce that the FTBS method is dissipative of order 2 for $0 < c < 1$. Graphs of the FTBS amplification factor are shown in Fig. 12.13 (left) for $c = 0.45$ (solid line) and $c = 0.75$ (dashed line).   ◇

Using results given in Exercises 12.7 and 12.13 and then repeating the argument in Example 12.15, shows that the Lax–Wendroff scheme (12.28) and the third-order upwind scheme (12.33) are both dissipative of order four. These amplification factors are also plotted in Fig. 12.13. It can be observed that the behaviour of $|\xi(\theta)|$ is very similar for large wavenumbers for all three methods but, at low wavenumbers, the greater damping of the FTBS method is evident. The solid curves corresponding to $c = 0.45$ on the left and right are close to the boundaries of their respective shaded regions because these methods experience maximum damping at $c = 0.5$. The Lax–Wendroff method experiences maximum damping at $c = 1/\sqrt{2}$, which is why the dashed curve, corresponding to $c = 0.75$, is closer to the boundary of its shaded region.

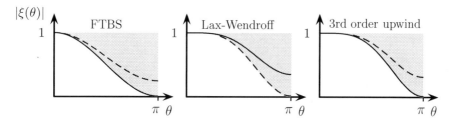

Fig. 12.13 The moduli of the amplification factors $|\xi(\theta)|$ for the FTBS, Lax–Wendroff and third-order upwind methods as functions of $\theta = \kappa h$ for $c = 0.45$ (*solid line*), $c = 0.75$ (*dashed line*). As c varies in the interval $(0, 1]$ the functions $|\xi(\theta)|$ all lie in the shaded areas

It can be shown, more generally, that an explicit method of order p has order of dissipation $2s = p + 1$, when p is odd, and $2s = p + 2$, when p is even (see Exercise 12.21). Both the leapfrog scheme (12.31) and box scheme (12.38) are nondissipative—there is no damping of high wavenumbers—and this is the reason why the numerical solutions in Figs. 12.8 and 12.12 are so noisy.[10]

When evaluated at a grid point (mh, nk), the exact solution of (12.1) with periodic boundary conditions is given by

$$u_m^n = \sum_{|j|\leq\lfloor M/2\rfloor}^{*} A_j \zeta(\theta_j)^n e^{im\theta_j}, \qquad \zeta(\theta) = \exp(-ic\theta), \qquad (12.54)$$

where the coefficients $\{A_j\}$ are the same as those in (12.52) whenever u and its discrete counterpart U satisfy the same initial condition. As noted in Sect. 12.3, the component Fourier modes are all transmitted with the same phase speed ω/κ because the dispersion relation $\omega = a\kappa$ takes such a simple form. Thus, the terms in (12.54) are synchronised and each represents a wave travelling with speed a. Unfortunately, the same is not true for the numerical solution (12.52) because each term (generally) travels with a different speed and the solution breaks up, or disperses. This is known as *dispersion* error.

To determine the speed at which discrete Fourier modes (12.48) evolve, we need to look at the argument of the complex-valued amplification factor ξ and rearrange it into polar form so that

$$\xi(\theta) = |\xi(\theta)|e^{-i\Omega k}, \quad \text{with } \Omega = -\frac{1}{k}\tan^{-1}\left(\frac{\Im(\xi)}{\Re(\xi)}\right). \qquad (12.55)$$

The quantity $\Omega(\kappa)$ represents the dispersion relation of the numerical scheme. Using (12.55) the discrete Fourier mode can be expressed as

$$U_m^n = \xi(\theta)^n e^{im\theta} = |\xi(\theta)|^n e^{i(\kappa mh - \Omega nk)} \qquad (12.56)$$

---

[10]The presence of the square wave in the initial condition further excites the high wavenumbers, by which we mean that the coefficients $\{A_j\}$ in (12.52) are significantly larger for large $|j|$.

which is a discrete analogue of the continuous Fourier mode,

$$u(x,t) = e^{i(\kappa x - \omega t)}. \tag{12.57}$$

Thus, in addition to the error in amplitude when $|\xi(\theta)| < 1$, there will generally be an error in the phase speed: the continuous mode travels with speed $\omega/\kappa = a$, whereas the discrete mode moves at a (generally different) speed $a_h = \Omega/\kappa$.

Example 12.16 Determine the dispersion relation and the phase speed of numerical solutions computed using the FTBS scheme (12.20).

From (12.22) we have $\xi = 1 - c + ce^{-i\theta}$ so that

$$\xi = 1 - c + c\cos\theta - ic\sin\theta = \left(1 - 2c\sin^2\tfrac{1}{2}\theta\right) - ic\sin\theta$$

from which we obtain the numerical speed

$$a_h = \frac{\Omega}{\kappa} = \frac{a}{c\theta}k\Omega = \frac{a}{c\theta}\tan^{-1}\left(\frac{c\sin\theta}{1 - 2c\sin^2\tfrac{1}{2}\theta}\right).$$

When $c = 1$ this reduces to $a_h = a$, the exact phase speed, in accordance with the unit CFL property alluded to in Example 12.7. This expression for the numerical speed in not so informative in its present form but since accuracy depends on the behaviour of the low wavenumber modes (when θ is small), it is appropriate to construct the MacLaurin series expansion of a_h which (using a computer algebra package) gives

$$a_h/a = 1 - \tfrac{1}{6}(1 - 2c)(1 - c)\theta^2 + \mathcal{O}(\theta^4). \tag{12.58}$$

This expansion shows that, in general, $a_h = a + \mathcal{O}(\theta^2)$. We can also see that $a_h = a + \mathcal{O}(\theta^4)$ when $c = 0.5$. This explains why the numerical solution for $c = 0.45$ in Fig. 12.5, though heavily damped, predicts the location of the maxima in the solution remarkably well. The expansion also suggests that $a_h < a$ when $0 < c < 1/2$ (numerical waves travel too slowly) and that $a_h > a$ when $1/2 < c < 1$ (they travel too quickly).   ◊

A graphical illustration of the FTBS dispersion error is given in Fig. 12.14 (left) for $c = 0.45$ and $c = 0.75$, together with corresponding results for the Lax–Wendroff and third-order upwind methods. For the Lax–Wendroff scheme (12.28) the expansion of the numerical phase speed a_h takes the specific form

$$a_h/a = 1 - \tfrac{1}{6}(1 + 2c^2)\theta^2 + \mathcal{O}(\theta^4), \tag{12.59}$$

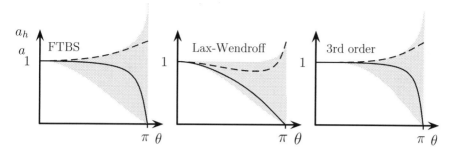

Fig. 12.14 The relative wave speeds $a_h(\theta)/a$ of the FTBS, Lax–Wendroff and third-order upwind methods as functions of $\theta = \kappa h$ for $c = 0.45$ (*solid line*), $c = 0.75$ (*dashed line*). As c varies in the interval $(0, 1]$ the functions $a_h(\theta)/a$ all lie in the *shaded areas*

which also implies second-order phase-speed accuracy. The leading term in the expansion (12.59) suggests that numerical waves computed using Lax–Wendroff are likely to move too slowly ($a_h < a$, for all values of c) compared to the exact solution. This tendency is clearly visible for the high-frequency components of the numerical solutions shown in Fig. 12.7. The corresponding expansion for the third-order method (12.33) is given by

$$a_h/a = 1 - \tfrac{1}{60}(1 - c^2)(1 - 2c)(2 - c)\theta^4 + \mathcal{O}(\theta^6). \qquad (12.60)$$

This shows that the phase-speed error of this scheme is really small—the accuracy is of order six when $c = 0.5$ and when $c = 1$. This explains why the graph of $a_h(\theta)/a$ in Fig. 12.14 (right) remains so close to 1 for much of the interval $0 \le \theta \le \pi$ when $c = 0.45$.

While there are many nondissipative schemes, it is generally not possible to construct methods with zero dispersion error (except for special values of c, for example, where the unit CFL property holds). For convergent methods, dispersion errors are small for small values of θ and normally increase as θ increases. This means that, ideally, methods should have a high order of dissipation (so that $|\xi|$ is very close to 1 for small values of θ) and be such that $|\xi|$ is small for θ close to π. This will mean that the high wavenumber modes—which have the largest dispersion errors—will be rapidly damped and their effect will not be noticeable in the numerical solution (dissipation effects will actually smooth the solution).

The amplification factor can also shed light on other issues relating to the approximation of hyperbolic PDEs. One such issue is group velocity—where groups of waves with similar wavenumbers can travel at a speed that is quite different to that of the individual modes. We refer to Strikwerda [21] or Trefethen [25] for further details.

12.4.2 *Nonperiodic Boundary Conditions*

An important lesson to be taken from Sect. 4.1 is that boundary conditions for the advection equation need to be carefully specified if the resulting initial-boundary value problem is to be well-posed. Further issues emerge when hyperbolic boundary value problems are approximated by finite difference methods. Ways of dealing with some of these issues will be discussed in this section.

To fix ideas, suppose that we want to solve the advection equation with unit wave speed ($a = 1$) in the quarter plane $x \geq 0$, $t \geq 0$ with initial/boundary conditions given by

$$u(x, 0) = g_0(x), \quad x \geq 0; \quad u(0, t) = g_1(t), \quad t \geq 0.$$

This problem is well posed and its exact solution is discussed in Example 4.3. To define a computable finite difference solution the domain must be truncated by choosing two positive numbers L, and τ, and then defining a grid within the rectangle $0 \leq x \leq L$, $0 \leq t \leq \tau$. Clearly L and τ must be sufficiently large that the computational rectangle encloses all interesting phenomena. An obvious observation is that while the numerical solution at a given point (x_m, t_n) will never be influenced by events at later times $t > t_n$, it may (depending on the geometry of the difference stencil) be affected by events for $x > x_m$ (even though the characteristics only transmit data from left to right when $a > 0$). In this case the resulting numerical solution could be (very) sensitive to the specific choice of L. To avoid this possibility a reasonable strategy would be to conduct numerical experiments with at least two different values of L.

With a grid size set by $h = L/M$ and a specific Courant number c the corresponding time step is given by $k = ch$. Computing a solution at all grid points in the rectangle with a stencil that resembles that of the FTBS scheme (Fig. 12.4, left) poses no problem. However, for a spatially centered scheme such as Lax–Wendroff (Fig. 12.6, left), there are difficulties at the outflow $x = L$. For example, when $n = 0$, the update formula (12.28b) can only be applied for $m = 1, 2, \ldots, M - 1$ to provide values of $U_1^1, U_2^1, \ldots, U_{M-1}^1$. The boundary condition at $x = 0$ provides one of the "missing" values: $U_0^1 = g_1(k)$, but there is no corresponding boundary value available at $x = L$ to specify U_M^1—indeed, if there were, the problem would be over-specified.

A standard way of dealing with this issue is to use the FTBS scheme (12.20) at $m = M$, that is

$$U_M^1 = cU_{M-1}^0 + (1 - c)U_M^0, \tag{12.61}$$

which completes the solution at time level $t = t_1$. The process may then be repeated for $n = 1, 2, \ldots$. A different interpretation of the right-side boundary condition is given in Exercise 12.24. In general, only those points marked o in Fig. 12.15 (left) can be computed by the "vanilla" Lax–Wendroff method. The use of the FTBS method as a boundary condition affects the solution at all the grid points indicated in Fig. 12.15 (right), that is, those in the domain of influence of Q. The next example provides

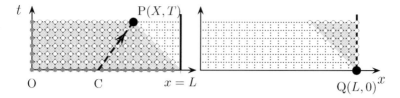

Fig. 12.15 (*Left*) the grid values that can be calculated up to $t = T$ with a method (such as the Lax–Wendroff scheme) with a $(1, 1)$-stencil for a given initial condition on $t = 0$ and a boundary condition along $x = 0$. (*Right*) the domain of influence of the point $(L, 0)$

insight into the extent to which this first-order accurate boundary condition "pollutes" the overall solution.

Example 12.17 Use the Lax–Wendroff method together with the FTBS right-side boundary condition, to solve the advection equation with $a = 1$ on the computational rectangle $L = 1$, $T = 0.45$, with an initial condition $u(x, 0) = x^2$ and a boundary condition $u(0, t) = t^2$, so that the exact solution is $u(x, t) = (x - t)^2$. Compute the solution with $c = 0.45$ using grid size $h = 1/16$ and compare the solution behaviour with that obtained using the leapfrog method with the same boundary conditions.

The exact solution to this problem is quadratic in x and t so the local truncation error of the Lax–Wendroff method (12.30) is identically zero. As a result, the numerical solution must be exact at all grid points in the shaded area in Fig. 12.15 (left). The local truncation error of the FTBS scheme is given by (12.21) and for the specific solution u, is a constant value at each grid point,

$$\mathcal{R}^n_m = (\tfrac{1}{2}ku_{tt} - \tfrac{1}{2}hu_{xx})\big|^n_m = k - h = -h(1 - c). \tag{12.62}$$

The actual global error is plotted in Fig. 12.16 (left) for $c = 0.45$ and $h = 1/16$. The figure suggests that $E^n_m \to$ constant as $n \to \infty$, with a different constant A_m, say, for each m. With a little effort (see Exercise 12.25) it may be shown that

$$A_m = \tfrac{1}{2}(1 - c)^2 h^2 \left(-\frac{1 + c}{1 - c}\right)^{m+1-M}. \tag{12.63}$$

This expression has two important ingredients. First, the global error is $\mathcal{O}(h^2)$, even though the local truncation error of the right-side boundary condition is only first-order accurate. This is good news. The second item of good news is that the error due to the boundary condition decays geometrically as we move away from the outflow boundary. That is, $E^n_m \propto \rho^{M-m}$ as $n \to \infty$ with geometric ratio $\rho := -(1 + c)/(1 - c)$. Clearly $\rho < -1$ for $0 < c < 1$ and the error changes sign and reduces by a factor $|\rho|$ per grid point as we move away from the boundary. The pollution effect increases, both in amplitude and extent, as $c \to 0$ while the effects decrease as $c \to 1$.

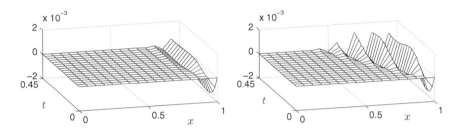

Fig. 12.16 (*Left*) the global error for Example 12.17 with $M = 16$ and $c = 0.45$ using the Lax–Wendroff method with an outflow boundary condition provided by the FTBS scheme. (*Right*) the corresponding global error when using the leapfrog scheme

The global error for the leapfrog method is shown on the right of Fig. 12.16. Its behaviour is similar to that of the Lax–Wendroff method in that it oscillates in the spatial direction while approaching a limit in the t-direction. There is, however, no damping as we move away from the boundary (the counterparts of the limiting values satisfy $A_{m+1} = A_{m-1}$), so the inaccuracy from the boundary condition pollutes the entire domain of influence of the boundary points. The difference in behaviour can be attributed to the fact that the amplification factor of the Lax–Wendroff method satisfies $|\xi(\kappa h)| < 1$ for $\kappa \neq 0$, whereas $|\xi(\kappa h)| = 1$ for the leapfrog scheme.   ◊

When using methods of higher order such as the third-order scheme used in Example 12.10, a more accurate right-side boundary condition is going to be needed. One possibility is to use the (second-order) Warming–Beam scheme

$$U_M^{n+1} = -\tfrac{1}{2}c(1 - c)U_{M-2}^n + c(2 - c)U_{M-1}^n + \tfrac{1}{2}(1 - c)(2 - c)U_M^n. \quad (12.64)$$

Since the third-order scheme method uses two upwind grid values it cannot be used at $x = x_1$ and a simple remedy (which does not affect the order of accuracy of the global error) is to use the Lax–Wendroff stencil at grid points contiguous with the inflow boundary $x = 0$. This remedy does, however, compromise the accuracy of the third-order scheme at grid points in the domain of influence of grid points where the Lax–Wendroff method is applied.

We complete this section by revisiting the boundary value problem for the two-way wave equation described in Sect. 12.2. By following the procedure outlined in Exercise 9.30(f) the two-way wave equation (12.40) may be written as a pair of coupled first-order PDEs $u_t = av_x$ and $v_t = au_x$ and then as a first-order system

$$\boldsymbol{u}_t + A\boldsymbol{u}_x = \boldsymbol{0}, \quad A = \begin{bmatrix} 0 & -a \\ -a & 0 \end{bmatrix}, \quad (12.65)$$

where $\boldsymbol{u} = [u, v]^\mathsf{T}$. This suggests an alternative approach to generating finite difference approximations that is explored in the following example.

Example 12.18 Show how the Lax–Wendroff scheme may be applied to solve the first-order system (12.65) on the strip $0 < x < 1$, $t > 0$ with boundary conditions $u(0, t) = u(1, t) = 0$ and initial conditions $u(x, 0) = g_0(x)$, $u_t(x, 0) = g_1(x)$.

The key is to use the eigenvalues and eigenvectors of A to uncouple the PDEs—see Sect. 9.1.

The eigenvalues are $\lambda_\pm = \pm a$ with corresponding eigenvectors $v_\pm = [1, \mp 1]^T$. Multiplying both sides of (12.65) by v_+^T and v_-^T, in turn, leads to two scalar advection equations

$$(u - v)_t + a(u - v)_x = 0 \tag{12.66a}$$

$$(u + v)_t - a(u + v)_x = 0 \tag{12.66b}$$

having dependent variables $(u - v)$ and $(u + v)$ and wave speeds $+a$ and $-a$, respectively. When these equations are approximated by the Lax–Wendroff method we obtain

$$\begin{aligned} (U_m^{n+1} + V_m^{n+1}) &= (U_m^n + V_m^n) + c\triangle_x(U_m^n + V_m^n) + \tfrac{1}{2}c^2\delta_x^2(U_m^n + V_m^n) \\ (U_m^{n+1} - V_m^{n+1}) &= (U_m^n - V_m^n) - c\triangle_x(U_m^n - V_m^n) + \tfrac{1}{2}c^2\delta_x^2(U_m^n - V_m^n), \end{aligned} \tag{12.67}$$

which involve the two Courant numbers $\mp c$, where $c = ak/h$. By adding and subtracting these we find

$$\begin{aligned} U_m^{n+1} &= U_m^n + c\triangle_x V_m^n + \tfrac{1}{2}c^2\delta_x^2 U_m^n, \\ V_m^{n+1} &= V_m^n + c\triangle_x U_m^n + \tfrac{1}{2}c^2\delta_x^2 V_m^n. \end{aligned} \tag{12.68}$$

These equations may also be written in the matrix-vector form

$$U_m^{n+1} - [I - C\triangle_x + \tfrac{1}{2}C^2\delta_x^2]U_m^n, \tag{12.69}$$

where $U_m^n = [U_m^n, V_m^n]^T$, I is the 2×2 identity matrix and $C = (k/h)A$. Thus a scalar method may be applied to a vector system simply by replacing occurrences of the Courant number c by the Courant matrix $C = (k/h)A$.

The pair (12.67) will be ℓ_2-stable provided that their Courant numbers satisfy the requirements of the von Neumann test which, for the Lax–Wendroff method are $-1 \le c \le 1$. That is, $ak \le h$ for stability (assuming a to be positive). For other matrices A the eigenvalues of C should satisfy the requirements of the von Neumann test. Thus, when A has (real) eigenvalues $\lambda_\pm$ the time step should be restricted by $-h \le k\lambda_\pm \le h$. This illustrates the importance of having finite difference schemes available that are stable for both positive and negative Courant numbers.

The problem is posed with two initial conditions ($u(x, 0) = g_0(x)$ and $u_t(x, 0) = g_1(x)$) for u, but none for v. This is easily remedied since

$$v_x(x, 0) = \frac{1}{a} u_t(x, 0) = \frac{1}{a} g_1(x)$$

may be integrated with respect to x to give $v(x, 0)$.

The internal values U_m^1 and V_m^1 for $m = 1, 2, \ldots, M - 1$ can now be calculated from the initial data using the finite difference equations (12.68). The given boundary conditions $U_0^1 = U_M^1 = 0$ complete the solution for U at $t = t_1$ but no equivalent conditions are provided for V_0^1 or V_M^1. Supplementary conditions, known as numerical boundary conditions, are therefore required for V. Just as it was emphasised in Example 4.3 that boundary conditions for hyperbolic PDEs should be imposed on *incoming* characteristics in order for problems to be well-posed, numerical boundary conditions should be imposed on quantities carried by *outgoing* characteristics so as not to compromise the well-posedness of the problem. For example, specifying a value for V at a boundary point would over-specify the problem.

The outward characteristic at $x = 1$ is governed by (12.66a) (when $a > 0$). The FTBS method (12.20) (with dependent variable $V - U$) is a possible approximation of this equation since it is stable for Courant numbers in the range $0 < c \leq 1$. This leads to

$$V_M^1 = (1 - c)V_M^0 + c(V_{M-1}^0 - U_{M-1}^0), \tag{12.70}$$

where we have used $U_M^0 = U_M^1 = 0$ from the given boundary condition. The determination of a suitable boundary condition for V_0^1 at $x = 0$ is left to Exercise 12.28. The numerical solution at $t = t_1$ is now complete and the whole process is then repeated to determine the solution at times $t = t_2, t_3, \ldots$.   ◊

Treating the two-way wave equation as a system of PDEs is undoubtedly more involved that the direct approach taken in Sect. 12.2 but this may be offset by having greater flexibility (allowing the use of dissipative methods, for example).

12.4.3 Nonlinear Approximation Schemes

Theorem 12.6 is a challenging result: positive type finite difference schemes of the form (12.5) are first-order convergent at best. The aim of this concluding section is to show that this order barrier can be lifted by allowing the coefficients $\{\alpha_j\}$ to depend on the solution at time level n; that is, by devising a *nonlinear* approximation scheme. Our treatment of this issue is heavily based on the expository paper of Sweby [24] in which the earlier works of Harten, van Leer, Roe, Chakravarthy–Osher and others are drawn into a common framework.

Our starting point is the *linear* conservation law (see Sect. 3.2),

$$u_t + f_x(u) = 0, \tag{12.71}$$

with the flux function $f(u) = au$, where $a > 0$. The original methods were designed for use on nonlinear conservation laws (like that in Extension 3.2.4). The approximation we want to study is the explicit method

$$U_m^{n+1} = U_m^n - \beta_m^n \triangle_x^- U_m^n \tag{12.72}$$

which is reminiscent of the FTBS scheme (12.20) except that the coefficient β_m^n is not specified at the moment, but is allowed to vary with m and n (it will, in fact, be chosen to depend on the solution vector U^n at time level n). A sufficient condition for the explicit scheme (12.72) to be of positive type is that

$$0 \le \beta_m^n \le 1. \tag{12.73}$$

A good starting point is the first-order upwind scheme

$$U_m^{n+1} = U_m^n - c\triangle_x^- U_m^n, \tag{12.74}$$

which is of positive type when c is subject to the CFL condition $0 < c \le 1$. Next, note that, with the aid of the identities

$$\triangle_x U_m^n = \triangle_x^- U_m^n - \tfrac{1}{2}\delta_x^2 U_m^n, \qquad \delta_x^2 U_m^n = \triangle_x^- \triangle_x^+ U_m^n, \tag{12.75}$$

(see Exercise 6.1) the Lax–Wendroff scheme (12.28) can be expressed as a correction of the FTBS scheme,

$$U_m^{n+1} = U_m^n - c\triangle_x^- U_m^n - \triangle_x^- \big(\tfrac{1}{2}c(1-c)\triangle_x^+ U_m^n\big), \tag{12.76}$$

which raises its order of consistency from one to two. The additional term $\tfrac{1}{2}c(1 - c)\triangle_x^+ U_m^n$ is known as the *anti-diffusive flux*: its inclusion counters the excessive diffusion in the FTBS method that is evident in Fig. 12.5.

The visual evidence (the oscillatory Lax–Wendroff solutions in Fig. 12.7) indicates that the amount of anti-diffusive flux that is added in (12.76) is excessive. As a compromise, one could define a modified scheme

$$U_m^{n+1} = U_m^n - c\triangle_x^- U_m^n - \tfrac{1}{2}c(1-c)\triangle_x^- \big(\varphi_m^n \triangle_x^+ U_m^n\big), \tag{12.77}$$

which involves a special grid function φ known as a *flux limiter*. Noting that the oscillations in Fig. 12.7 occur where there is a sharp change in the gradient of the solution, it is natural to choose the limiter to be a function of the ratio of successive gradients. That is, in a discrete setting, to choose $\varphi_m^n = \varphi(\rho_m^n)$ where

$$\rho_m^n = \frac{\triangle_x^- U_m^n}{\triangle_x^+ U_m^n}.$$

Note that the grid function ρ_m^n is unbounded if $\triangle_x^+ U_m^n = 0$ (with $\triangle_x^- U_m^n \neq 0$) that is, whenever the numerical solution is about to become flat. The remaining challenge is to construct a limiter function $\varphi(\rho)$ so as to maximize the anti-diffusive flux of (12.77) subject to the modified scheme being of positive type and second-order consistent.

We first observe that $\rho_m^n < 0$ whenever the point x_m is a local maximum (or a local minimum) of U_m^n. To retain the FTBS method in these cases, we will insist that $\varphi(\rho) = 0$ whenever $\rho \leq 0$. Next, we express (12.77) in the form (12.72):

$$\beta_m^n = c + \tfrac{1}{2} c (1-c) \frac{\triangle_x^- (\varphi(\rho_m^n) \triangle_x^+ U_m^n)}{\triangle_x^- U_m^n},$$

and then rearrange it (see Exercise 12.29) to give the explicit expression

$$\beta_m^n = c \left(1 + \tfrac{1}{2}(1-c) \left(\frac{\varphi(\rho_m^n)}{\rho_m^n} - \varphi(\rho_{m-1}^n) \right) \right). \qquad (12.78)$$

It can than be checked that the flux-limited scheme will be of positive type for $0 \leq c \leq 1$ (that is (12.73) holds) when

$$\left| \frac{\varphi(\rho_m^n)}{\rho_m^n} - \varphi(\rho_{m-1}^n) \right| \leq 2. \qquad (12.79)$$

Moreover the condition (12.79) will be satisfied if $\varphi(\rho)$ is chosen so that it lies in the region that is highlighted in Fig. 12.17, that is,

$$0 \leq \varphi(\rho) \leq \min\{2, 2\rho\}. \qquad (12.80)$$

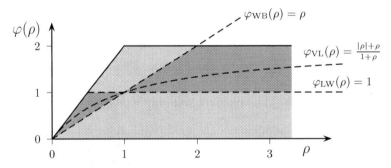

Fig. 12.17 The region where the flux-limited method (12.77) is of positive type (*light shading*) and where it is also consistent of second order (*dark shading*). *Dashed lines* show the Lax–Wendroff, Warming–Beam and van Leer limiters

Next, to ensure second-order consistency of the flux-limited scheme (12.77), we make the observation that any second-order scheme that is defined by the four values $\{U_{m-2}^n, U_{m-1}^n, U_m^n, U_{m+1}^n\}$, will also have to be a linear combination of the Lax–Wendroff (12.28) and Warming–Beam (12.64) methods. Moreover, both methods can be interpreted as a flux-limited scheme (12.77) by choosing the limiter function appropriately: $\varphi_{LW}(\rho) = 1$ and $\varphi_{WB}(\rho) = \rho$, respectively (see Exercise 12.30). As a result, any method having the limiter

$$\varphi(\rho) = (1 - \theta(\rho))\, \varphi_{LW}(\rho) + \theta(\rho)\, \varphi_{WB}(\rho)$$
$$= 1 + \theta(\rho)(\rho - 1), \quad 0 \le \theta(\rho) \le 1,$$

will also be consistent of second order. The flux limiters $\varphi(\rho)$ of interest are thus given by: $\rho \le \varphi(\rho) \le 1$ for $0 < \rho \le 1$ and $1 \le \varphi(\rho) \le \rho$ for $\rho \ge 1$. These two regions are also highlighted (by dark shading) in Fig. 12.17. Many limiters have been proposed in the literature and some of these are listed in Table 12.1. Further details can be found in the excellent book by Leveque [14].

Example 12.19 Solve the advection equation using the scheme (12.77) with the van Leer flux limiter on the interval $0 \le x \le 3$ with $a = 1$, together with the initial condition (12.2) and periodic boundary conditions. Compute the numerical solution with $c = 0.45$ for one complete period ($\tau = 3$) using two spatial grids—a coarse grid ($h = 1/60$, 400 time steps) and a fine grid ($h = 1/240$, 1600 time steps).

Table 12.1 Possible choices for the flux limiter function

Van Leer	$\varphi_{VL}(\rho) = (|\rho| + \rho)/(|\rho| + 1)$
Minmod (Roe)	$\varphi(\rho) = \max\{0, \min(\rho, 1)\}$
Superbee (Roe)	$\varphi(\rho) = \max\{0, \min(2\rho, 1), \min(\rho, 2)\}$
Chakravarthy–Osher	$\varphi(\rho) = \max\{0, \min(\rho, \psi)\}, 1 < \psi \le 2,$

The Van Leer limiter is shown in Fig. 12.17

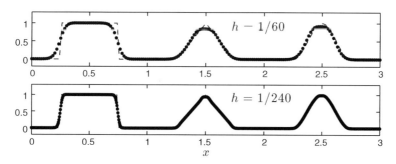

Fig. 12.18 The numerical solution (*dots*) to the problem in Example 12.19 and the exact solution (*dashed line*) using the van Leer flux limiter

The numerical results are presented in Fig. 12.18. Although there is clearly some smoothing of sharp edges, the results show a marked improvement over the schemes used earlier in this chapter. This improvement is achieved with a negligible increase in computational cost.   ◊

Exercises

12.1 ☆ Show that the leading term in the local truncation error (12.21) of the FTBS scheme may be written as $-\frac{1}{2}ah(1-c)\,u_{xx}|_m^n$.

12.2 ☆ Show that the FTFS method (12.25) is stable in both the ℓ_∞ and the ℓ_2 sense for $-1 \le c < 0$.

12.3 ☆ Show that the FTCS method (12.26) does not satisfy the requirements of a positive type operator for a fixed $c \ne 0$. Show also that its amplification factor satisfies $|\xi| = 1 + c^2$ for $\kappa h = \pi/2$ demonstrating that the FTCS method is unconditionally unstable when $h, k \to 0$ with c fixed.

12.4 ☆ Show that the scheme

$$U_m^{n+1} = U_m^n - c\triangle_x U_m^n + \tfrac{1}{2}|c|\,\delta_x^2 U_m^n$$

reduces to either the FTBS method (12.20) or the FTFS method (12.25) depending on whether $c > 0$ or $c < 0$.

12.5 Consider the Lax–Friedrichs scheme

$$U_m^{n+1} = U_m^n - c\triangle_x U_m^n + \tfrac{1}{2}\delta_x^2 U_m^n$$

for solving the advection equation.

(a) Sketch the stencil of the method.
(b) Determine the principal part of the local truncation error and confirm that the scheme is first-order consistent.
(c) Are there values of c for which the scheme is of positive type?
(d) Find the range of values of c for which the scheme is ℓ_2 stable.
(e) What benefit would there be in using this method in preference to the FTBS scheme in (12.20)?
(f) Show that the scheme is a second-order consistent approximation of a modified equation of the form $u_t + au_x = \varepsilon\partial_x^\ell u$. Determine ℓ and the relationship between ε and then parameters c, h and k. How would you expect the scheme to perform when solving the problem in Example 12.1?

12.6 ☆ Calculate the Lax–Wendroff scheme solution U_m^1 for $m = 1, 2, 3, 4$ and $c = 0.5$ of the advection equation starting from the discontinuous initial condition $U_m^0 = g(mh)$, where $g(x) = 1$ for $x < 2.5h$ and $g(x) = 0$ for $x > 2.5h$ Sketch the solution and compare it to the exact solution after one time step.

12.7 Show that the amplification factor for the Lax–Wendroff scheme (12.28) is given by

$$\xi = 1 - ic \sin \kappa h - 2c^2 \sin^4 \tfrac{1}{2}\kappa h$$

and that it satisfies

$$|\xi|^2 - 1 = -4c^2(1 - c^2) \sin^2 \tfrac{1}{2}\kappa h.$$

Deduce that the method is ℓ_2 stable for $-1 \le c \le 1$.

12.8 Verify that the coefficients of the FTBS method (12.20), the Lax–Wendroff method (12.28) and the third-order method (12.33) can all be determined by substituting suitable values of μ and ν into the formula (12.17).

12.9 Use the technique outlined in Sect. 12.1.1 to construct a (μ, ν)-method of the form (12.5) with $\mu = 2$, $\nu = 0$ that is a second-order consistent approximation of the advection equation. (Hint: you are following in the footsteps of Warming and Beam.)

12.10 The Newton backward difference formula that is given by

$$\Phi(x) = U_m^n + \sum_{j=1}^{p} \binom{s+j-1}{j} (\triangle_x^-)^j U_m^n$$

with binomial coefficient $\binom{s+j-1}{j} = \tfrac{1}{j!}(s+j-1)(s+j-2)\cdots s$ and with $s = (x - x_m)/h$, interpolates the finite difference solution values at the points x_{m+j} ($j = -p, -p+1, \ldots, 0$). By using this formula in (12.15) with $p=1$ and $p=2$ (rather than the Lagrange interpolant) derive the FTBS scheme (12.20) and the Warming–Beam scheme (12.64), respectively.

12.11 Leith's method for solving the advection–diffusion equation

$$u_t + au_x = \varepsilon u_{xx}$$

in situations where $0 < \varepsilon \ll |a|$ is based on a Lax–Wendroff approximation of the terms $u_t + au_x$ to which is added the standard second-order approximation of the diffusion term. The resulting scheme is

$$U_m^{n+1} = U_m^n - c\triangle_x U_m^n + (r + \tfrac{1}{2}c^2)\delta_x^2 U_m^n,$$

where $c = ak/h$ and $r = \varepsilon k/h^2$.

(a) What is the order of consistency of this approximation scheme?
(b) Why do the leading terms in the local truncation error suggest that this scheme is particularly appropriate for solving advection–dominated problems?
(c) Determine the range of parameters r and c for which Leith's scheme is ℓ_2 stable. Show that the stability conditions are independent of the parameters a and ε

when expressed in terms of the scaled grid sizes $\widehat{h} = ah/\varepsilon$ and $\widehat{k} = a^2k/\varepsilon$. Hence sketch the region in the h-k plane which leads to a stable method.

12.12 Determine the leading term in the local truncation error of the leapfrog scheme (12.31) for approximating the advection equation and show that it is second order in both space and time.

Show that the scheme has solutions of the form $U_m^n = \xi^n e^{i\kappa mh}$ provided that ξ satisfies a certain quadratic equation. Deduce that both roots satisfy $|\xi| = 1$ for all κ if, and only if, $-1 \le c \le 1$. (This establishes the conditional ℓ_2 stability of the leapfrog scheme.)

12.13 *For the third-order scheme (12.33), show that

(a) the scheme reduces to $U_m^{n+1} = U_{m-c}^n$ when $c = -1, 0, 1, 2$ and the exact solution of the PDE $u_t + au_x = 0$ is reproduced.
(b) the local truncation error is $\mathcal{O}(h^3) + \mathcal{O}(k^3)$.
(c) the CFL condition is $-1 \le c \le 2$.
(d) the von Neumann amplification factor satisfies

$$|\xi|^2 - 1 = -\tfrac{4}{9}c(1 - c^2)(2 - c)\sin^4 \tfrac{1}{2}\theta\big[4c(1 - c)\sin^2 \tfrac{1}{2}\theta + 3\big],$$

where $\theta = \kappa h$.
(e) $|\xi|^2 \le 1$ for all θ if, and only if, $0 \le c \le 1$.

12.14 Show that the leading term in the local truncation error of the BTBS method (12.34) is given by (12.37).

12.15 Suppose that C is the $M \times M$ circulant matrix (12.35) that arises in the study of the BTBS approximation of the advection equation and that $\boldsymbol{v}_j \in \mathbb{C}^M$ ($j = 1, 2, \ldots, M$) is the vector whose mth component is $e^{2\pi i m \, j/M}$ ($m = 1, 2, \ldots, M$). Show that $\boldsymbol{v}_j$ is an eigenvector of C corresponding to the eigenvalue $1/\xi(2\pi jh)$.

12.16 The BTFS scheme

$$(1 - c)U_m^{n+1} + c\,U_{m+1}^{n+1} = U_m^n$$

may be obtained by changing the direction of the spatial difference operator in (12.34). Show that the leading term in the local truncation error is $-\tfrac{1}{2}ah(1 - c)\,u_{xx}|_m^{n+1}$. Examine the ℓ_2 stability of the scheme and compare it with the CFL stability condition.

12.17 Suppose that $\xi_F(\theta)$ (see (12.22)) and $\xi_B(\theta)$ (see Exercise 12.16) denote the amplification factors for the FTBS and BTFS methods, respectively, where $\theta = \kappa h$. Show that $\xi_F(\theta)\,\xi_B(\theta) = e^{-i\theta}$ and hence deduce the ℓ_2 stability restrictions of the BTFS method from those of the FTBS method.

12.18★ Show that the leading term in the local truncation error of the box scheme (12.38) is given by[11] $-\frac{1}{6}ah^2(1-c^2)u_{xxx}$.

12.19☆ Show that the amplification factor (12.39) of the box scheme may be written as $\xi = e^{i\kappa h}\eta/\eta^*$, where $\eta = (1-c)+(1+c)e^{-i\kappa h}$ and η^* is its complex conjugate. Deduce that $|\xi| = 1$ for all wave numbers, so that the box scheme is unconditionally ℓ_2 stable.

12.20 Deduce from the quadratic polynomial (12.44) that (i) the roots form a complex conjugate pair with modulus equal to 1 for all wavenumbers when $c^2 \leq 1$ and (ii) there are wavenumbers κ for which the roots are real when $c^2 > 1$. Explain why one of the roots must have magnitude greater than one when $\kappa \neq 0$ and the roots are real.

12.21★ Suppose that a (μ, ν)-method of the form (12.5) is used to solve the advection equation with initial condition $u(x, 0) = \exp(i\kappa x)$. Show that the local truncation error can be expressed in the form (12.50). Use this result together with the error expansion

$$kR_m^n = h^{p+1}C_{p+1} \left.\partial_x^{p+1}u\right|_m^n + h^{p+2}C_{p+2} \left.\partial_x^{p+2}u\right|_m^n + \mathcal{O}(h^{p+3})$$

to show that the amplification factor of a pth-order method satisfies

$$\xi(\theta) = e^{-ic\theta} - (i\theta)^{p+1}C_{p+1} - (i\theta)^{p+2}C_{p+2} + \mathcal{O}(h^{p+3}),$$

where the constants C_{p+1} and C_{p+2} are defined by (12.11). By considering the cases where p is even and p is odd separately, deduce that

$$|\xi(\theta)|^2 - \begin{cases} 1 - 2(-1)^{(p+1)/2}\theta^{p+1}C_{p+1} + \cdots & p \text{ is odd}, \\ 1 - 2(-1)^{p/2+1}\theta^{p+2}(cC_{p+1} + C_{p+2}) + \cdots & p \text{ is even}. \end{cases}$$

What is the order of dissipation of these methods?

12.22★ Use a computer algebra package to verify that the phase speeds $a_h = \Omega/\kappa$ of the Lax–Wendroff scheme and the third-order scheme are given by (12.59) and (12.60), respectively.

12.23 Suppose that the BTBS scheme (12.34) is used to solve the advection equation in the semi-infinite strip $0 \leq x \leq 1$ with the initial condition $u(x, 0) = g(x)$. Use the CFL condition to show that the method will not converge if (i) $c > 0$ unless the boundary condition $u(0, t) = g_0(t)$ is placed at $x - 0$ or, (ii) if $c \leq -1$ unless the boundary condition $u(1, t) = g_1(t)$ is placed at $x = 1$.

---

[11] Note that this is zero when $c = \pm 1$, as are all subsequent terms. This is consistent with the fact that the method is exact for these Courant numbers—a generalisation of the unit CFL property.

12.24 Suppose that the Lax–Wendroff scheme is used to solve the advection equation in the computational rectangle $0 \leq x \leq L, 0 \leq t \leq T$ with boundary/initial conditions specified along $t = 0$ and $x = 0$. Suppose also that an additional column of grid points is added at $x = L + h$ and an artificial boundary condition $\delta_x^2 U_M^n = 0$ imposed. Show that (i) this implies that the points (x_m, U_m^n) $(m = M - 1, M, M + 1)$ are collinear and (ii) that this boundary condition is equivalent to using the FTBS scheme (12.61).

12.25 * Show that the global error for the problem in Example 12.17 satisfies

$$E_m^{n+1} = \tfrac{1}{2}c(1 + c)E_{m-1}^n + (1 - c^2)E_m^n + \tfrac{1}{2}c(c - 1)E_{m+1}^n,$$
$$M - n \leq m < M, \; n = 2, 3, \ldots$$
$$E_M^{n+1} = cE_{M-1}^n + (1 - c)E_M^n - c(1 - c)h^2, \quad n = 0, 1, 2, \ldots$$

together with $E_m^n = 0$ for $0 \leq m \leq M - n, n = 0, 1, \ldots$.
Next, by assuming that $E_m^n \to A_m$ as $n \to \infty$, show that

$$(1 + c)A_{m-1} - 2cA_m + (c - 1)A_{m+1} = 0,$$

together with boundary conditions $A_M - A_{M-1} = -(1 - c)h^2$ and $A_m \to 0$ as $m \to -\infty$. Verify that a solution of these equations is given by (12.63).

12.26 Determine the characteristic speeds of the system

$$u_t + 2u_x + v_x = 0$$
$$v_t + u_x + 2v_x = 0$$

and explain why the FTBS approach is an appropriate method for their solution? What is the largest grid ratio k/h for which FTBS gives a stable approximation? Calculate the numerical solution at the point $x = 5h, t = k$ when $h = 0.1, k = 0.025$ and the initial conditions are $u(x, 0) = x, v(x, 0) = x^2$.

12.27 * Can you explain why neither the FTCS nor the FTBS methods can be used to give a stable numerical solution of the first-order system (12.65)?

(a) Determine an updating formula for U_m^{n+1} in terms of U_m^n by applying the FTCS method to (12.66a) and the BTCS method to (12.66b).
(b) The scheme described in Exercise 12.4 combines the FTCS and FTBS methods. A vector version applied to $\mathbf{u}_t + A\mathbf{u}_x = 0$ is given by

$$\mathbf{U}_m^{n+1} = \mathbf{U}_m^n - (k/h)A\triangle_x \mathbf{U}_m^1 + \tfrac{1}{2}(k/h)|A|\,\delta_x^2 \mathbf{U}_m^n.$$

In order to evaluate $|A|$, we suppose that $A = V\Lambda V^{-1}$, where V is the matrix of eigenvectors of A and $\Lambda = \mathrm{diag}(\lambda_1, \lambda_2, \ldots)$ is the corresponding matrix of eigenvalues. Then $|A| = V|\Lambda|V^{-1}$, where $|\Lambda| = \mathrm{diag}(|\lambda_1|, |\lambda_2|, \ldots)$. Derive an expression for U_m^{n+1} when A is the matrix appearing in (12.65) and show that it is the same as that obtained in part (a).

(c) Show how the Lax–Friedrichs scheme (12.5) can be applied to solve the system (12.65). Under what condition is the expression obtained for U_m^{n+1} the same as that obtained in part (b)?

12.28 Devise a numerical boundary condition at $x = 0$ for Example 12.18 based on the FTFS scheme and following a process similar to that described at $x = 1$.

12.29 Verify that the flux-limited method (12.77) can be written in the form (12.74) with β_m^n given by (12.78).

12.30 Verify that the flux-limited scheme (12.77) reduces to the Lax–Wendroff scheme (12.28) and the Warming–Beam scheme (12.64) when the flux limiters are set to $\varphi(\rho) = 1$ and $\varphi(\rho) = \rho$, respectively.

12.31 Sketch the graphs of the limiters listed in Table 12.1.

12.32 ⋆ Suppose that $u(x, t)$ satisfies the conservation law $u_t + f(u)_x = 0$, where the flux function $f(u)$ is assumed to be a smooth function. Deduce that $u_{tt} = (f'(u) f(u)_x)_x$, where $f'(u) = df(u)/du$.

The MacCormack method for numerically solving this conservation law is a two stage process. The first stage determines the intermediate quantities $\overline{U}_m^n$ for all m from the formula

$$\overline{U}_m^n = U_m^n - (k/h)\Delta_x^+ f(U_m^n)$$

and these are used to determine the solution at the next time level via

$$U_m^{n+1} = \tfrac{1}{2}(\overline{U}_m^n + U_m^n) - \tfrac{1}{2}(k/h)\Delta_x^- f(\overline{U}_m^n).$$

Show that

(a) the method is identical to the Lax-Wendroff method method when $f(u) = au$ and a is a constant

(b) the method is consistent of second order with the nonlinear conservation law.

Chapter 13
Projects

Abstract The final chapter identifies thirteen projects, involving both theory and computation, that are intended to extend and test understanding of the material in earlier chapters.

In this final chapter we present a baker's dozen of projects designed to be tackled by individuals or small groups of students. The first project is a warm-up exercise in analysis. Several of the projects have been set as open-book examination questions (with a 24 h deadline) for postgraduate students. The majority of projects are based on papers from scientific journals—the references are not given for obvious reasons—citations can be found in the solutions.

Project 13.1 (*Parabolic smoothing*)
Suppose that $u(x, t)$ satisfies the heat equation $u_t = u_{xx}$ on the interval $0 < x < 1$ for $t > 0$ with boundary conditions $u_x(0, t) = u_x(1, t) = 0$ and a given initial condition at $t = 0$. Use energy methods to establish the identities

$$\|u(\,, t)\|^2 - \|u(\cdot, 0)\|^2 = -\int_0^t \|u_x(\cdot, t)\|^2 \, ds \qquad (13.1a)$$

and

$$\frac{d}{dt} \|u_x(\cdot, t)\|^2 = -\|u_t(\cdot, t)\|^2, \qquad (13.1b)$$

where $\|u(\cdot, t)\|^2 := \int_0^1 (u(x, t))^2 \, dx$ denotes the standard L_2 norm. By integrating both sides of the second identity over the interval $s < t < \tau$, and then integrating the result with respect to s over the interval $0 < s < \tau$, show that

$$\|u_x(\cdot, \tau)\| \leq \frac{1}{\sqrt{\tau}} \|u(\cdot, 0)\|.$$

This provides a quantitative measure of how the derivative u_x decays with time.

© Springer International Publishing Switzerland 2015
D.F. Griffiths et al., *Essential Partial Differential Equations*, Springer
Undergraduate Mathematics Series, DOI 10.1007/978-3-319-22569-2_13

Project 13.2 (*Convergence of a fourth-order approximation method*)
This project concerns the high-order finite difference scheme in Exercise 6.32 for approximating a second-order ODE with Dirichlet end conditions. One would like to establish that the scheme converges at a fourth-order rate even when the ODE is approximated by the standard second-order approximation at grid points adjacent to the boundary.

Accordingly, consider the boundary value problem $-u''(x) = f(x)$ on the interval $0 < x < 1$ with the boundary conditions $u(0) = u(1) = 0$. Taking a standard grid of points $x_m = mh$ ($h = 1/M$, $m = 0, 1, 2, \ldots, M$), suppose that $\mathcal{L}_h^{(5)}$ denotes the 5-point approximation of the second derivative given by

$$
\mathcal{L}_h^{(5)} U_m = \begin{cases} -h^{-2}\delta^2 U_m, & \text{for } m = 1, M-1 \\ -h^{-2}\big(\delta^2 - \tfrac{1}{12}\delta^4\big) U_m, & \text{for } m = 2, 3, \ldots, M-2. \end{cases}
$$

and that the grid function U satisfies the boundary conditions $U_0 = U_M = 0$.

A standard analysis of this difference scheme shows that if u and its first six derivatives are bounded on $(0, 1)$ then the local truncation error $\mathcal{R}_m$ is $\mathcal{O}(h^4)$ at $x = x_m$ ($m = 2, 3, \ldots, M-2$), but is only $\mathcal{O}(h^2)$ at $x = x_1, x_{M-1}$. Although the operator $\mathcal{L}_h^{(5)}$ is *not* of positive type, show that it can be factored as the product $\mathcal{M}_h \mathcal{L}_h$ of two positive type operators, where $L_h := -h^{-2}\delta^2$ and

$$
\mathcal{M}_h U_m = \begin{cases} U_m, & \text{for } m = 1, M-1 \\ U_m - \tfrac{1}{12}\delta^2 U_m, & \text{for } m = 2, 3, \ldots, M-2. \end{cases}
$$

Show also that the global error $E := u - U$ satisfies the coupled equations

$$
\mathcal{M}_h V = \mathcal{R}_h, \qquad \mathcal{L}_h E = V.
$$

Next, suppose that $\mathcal{R}_h$ is written as $\mathcal{R}_h = \mathcal{R}_h^* + \mathcal{R}_h^\bullet$, where

$$
\mathcal{R}_h^* \big|_m = \begin{cases} R_m \\ 0 \end{cases} \quad \text{and} \quad \mathcal{R}_h^\bullet \big|_m = \begin{cases} 0, & \text{for } m = 1, M-1 \\ R_m, & \text{for } m = 2, 3, \ldots, M-2, \end{cases}
$$

and show that $V = V^* + V^\bullet$ and $E = E^* + E^\bullet$ satisfy the equations

$$
\mathcal{M}_h V^\bullet = \mathcal{R}_h^\bullet, \quad \mathcal{L}_h E^\bullet = V^\bullet,
$$
$$
\mathcal{M}_h V^* = \mathcal{R}_h^*, \quad \mathcal{L}_h E^* = V^*.
$$

Show that $\mathcal{M}_h$ and $\mathcal{L}_h$ are inverse monotone when supplemented by Dirichlet boundary conditions and hence deduce that both $|V^\bullet|$ and $|E^\bullet|$ are $\mathcal{O}(h^4)$. The behaviour of E^* still needs to be examined carefully because $\mathcal{R}_h^* \big|_m = \mathcal{O}(h^2)$ for $m = 1$ and $m = M-1$.

Can it be shown that $E^* = \mathcal{O}(h^4)$ in the case $V^* = \mathcal{O}(h^2)$? One idea is to suppose that there is negligible interaction between the effects of R_1 and R_{M-1}. The effect of R_1 can then be determined by solving the difference equations $\mathcal{M}_h V_m^* = 0$ (for $m = 2, 3, \ldots$) with end conditions $V_1^* = R_1$ and $V_m^* \to 0$ as $m \to \infty$. Verify that these equations are satisfied by $V_m^* = R_1 z^{m-1}$, where z is the smaller of the roots of the quadratic $z^2 - 14z + 1 = 0$. Next, determine the constant C so that the general solution of the form of the equations $\mathcal{L}_h E_m^* = V_m^*$ ($m = 2, 3, \ldots$) has the form $E_m^* = A + Bm + Cz^m$ for arbitrary constants A and B. Finally, by enforcing the end conditions $2E_1^* - E_2^* = h^2 R_1$ and $E_m^* \to 0$ as $m \to \infty$, show that $E^* = \mathcal{O}(h^4)$.

What else is striking about the behaviour of E^*?

Project 13.3 (*The Allen-Southwell-Il'in scheme*)

This project concerns a specialised finite difference scheme for solving the one-dimensional advection–diffusion problem in Example 6.14. To construct the scheme, suppose that $v(x)$ and $w(x)$ satisfy the independent boundary value problems

$$\left.\begin{array}{l} -\varepsilon v'' - av' = 0, \quad -h < x < 0 \\ v(-h) = 0, \quad v(0) = 1, \end{array}\right\} \qquad \left.\begin{array}{l} -\varepsilon w'' - aw' = 0, \quad 0 < x < h \\ w(0) = 1, \quad w(h) = 0, \end{array}\right\}$$

where ε, a and h are positive constants, and let $g(x)$ be defined so that

$$g(x) := \begin{cases} v(x), & -h \le x < 0 \\ w(x), & 0 \le x \le h \end{cases}.$$

Calculate $\int_{-h}^{h} g(x) \, dx$. Sketch a graph of g when $0 < \varepsilon \ll 1$.

Next, suppose that a grid of points $x = x_m := mh$ is defined on the interval $[0, 1]$, where $h = 1/M$, and that u satisfies the advection–diffusion equation $-\varepsilon u''(x) + au'(x) = f(x)$ in $0 < x < 1$ together with Dirichlet end conditions.

Use integration by parts to show that the identity

$$\int_{x_{m-1}}^{x_m} v(x - x_m)\left(-\varepsilon u'' + au' - f\right) dx + \int_{x_m}^{x_{m+1}} w(x - x_m)\left(-\varepsilon u'' + au' - f\right) dx = 0$$

can be simplified to give

$$-\varepsilon v'(-h)u_{m-1} + \varepsilon\left(v'(0) - w'(0)\right)u_m + \varepsilon w'(h)u_{m-1} = \int_{x_{m-1}}^{x_{m+1}} gf \, dx.$$

By determining the quantities $v'(-h)$, $v'(0)$, $w'(0)$ and $w'(h)$, show that this generates an exact[1] finite difference replacement of the advection–diffusion equation

---

[1]This wonderful property (the finite difference solution is exact at every grid point) *does not* generalise to advection–diffusion problems in two or more space dimensions.

$$- \hat{\varepsilon} h^{-2} \delta^2 u_m + a h^{-1} \triangle u_m = h^{-1} \int_{x_{m-1}}^{x_{m+1}} gf \, \mathrm{d}x, \qquad (13.3a)$$

in which $\hat{\varepsilon} = \varepsilon \mathrm{Pe}_h \coth \mathrm{Pe}_h$, and $\mathrm{Pe}_h = ah/(2\varepsilon)$ is the mesh Peclet number.

The scheme was discovered independently by Allen and Southwell in 1955 and Il'in in 1969. It bears a strong resemblance to the central difference scheme in Example 6.14 (in the case $a = 2$) except that the diffusion coefficient and the right hand side need to be suitably modified. Show that

$$\lim_{h \to 0} \hat{\varepsilon} = \varepsilon \text{ and } \lim_{\varepsilon \to 0} \hat{\varepsilon} = \tfrac{1}{2} ah.$$

What are the corresponding limits of the integral on the right hand side of (13.3a)?

Project 13.4 (*Saul'Yev schemes for advection–diffusion equations*)
The use of implicit methods such as BTCS to solve second-order parabolic PDEs typically involves the solution of systems of linear equations at each time level. An LR factorisation of the coefficient matrix A, say, is performed at the start of the simulation and systems of the form $L\boldsymbol{v} = \boldsymbol{f}$, $R\boldsymbol{u} = \boldsymbol{v}$ are solved by forward and backward substitution in order to determine the solution $\boldsymbol{u}$ at each new time level. In one dimension, when A is tridiagonal, the matrices L and R are lower and upper bidiagonal (having at most two nonzero entries on each row). The idea behind this project is that some of the computational effort can be avoided by designing a "pre-factorised" pair of methods.

As a starting point, consider the following pair of finite difference schemes for solving the advection–diffusion equation $u_t + a u_x = \varepsilon u_{xx}$,

$$-(r + \tfrac{1}{2}c)U_{m-1}^{n+1} + (1 + r + \tfrac{1}{2}c)U_m^{n+1} = (1 - r + \tfrac{1}{2}c)U_m^n + (r - \tfrac{1}{2}c)U_{m+1}^n \tag{13.4a}$$

$$(1 + r - \tfrac{1}{2}c)U_m^{n+1} - (r - \tfrac{1}{2}c)U_{m+1}^{n+1} = (1 - r - \tfrac{1}{2}c)U_m^n + (r + \tfrac{1}{2}c)U_{m-1}^n, \tag{13.4b}$$

in which $r = \varepsilon k/h^2$ and $c = ak/h$. These are generalisations of schemes originally designed for the heat equation ($a = 0$) by Saul'Yev (1967) (see Richtmyer and Morton [18, Sect. 8.2]). Write down the associated stencils and use a domain of dependence argument to show that neither scheme, used on its own, can provide convergent approximations unless $k/h \to 0$ as $h, k \to 0$.

As a consequence of this the two schemes (13.4a) and (13.4b) will be combined. Thus, to approximate the PDE on the interval $0 < x < 1$ together with Dirichlet end conditions, the scheme (13.4a) is used on even-numbered time levels $n = 0, 2, 4, \ldots$. The value U_0^{n+1} is provided by the boundary condition at $x = 0$ and then (13.4a) is used with $m = 0, 1, 2, \ldots$ to determine the numerical solution on the entire time level. When n is odd, U_M^{n+1} is obtained from the boundary condition at $x = 1$, then (13.4b) is used with $m = M - 1, M - 2, \ldots 1$.

By making the substitution $U_m^n = \xi^n e^{i\kappa mh}$, determine the amplification factors ξ_1 and ξ_2, say, of the individual methods (13.4a) and (13.4b). Next, show that when the formulae are used alternately, $U_m^{n+2} = \xi_1\xi_2 U_m^n$ so that their combined amplification factor is $\xi = \xi_1\xi_2$, that is, the product of the individual factors.

It may be shown that

$$|\xi|^2 - 1 = -16rs\frac{1 + (2r - c)(2r + c)s}{(1 + (2 + 2r - c)(2r - c)s)(1 + (2 + 2r + c)(2r + c)s)},$$

where $s = \sin^2(\frac{1}{2}\kappa h)$. Deduce that the combined scheme (a) is stable in the sense that $|\xi|^2 \le 1$ provided that $c^2 \le 1 + 4r^2$ when $r > 0$, (b) is unconditionally stable in the case of pure diffusion ($c = 0$), and (c) is nondissipative in the case of pure advection ($r = 0$).

In the case $c = 0$, write the methods in difference operator notation and show that, when implemented in the odd-even manner described above,

$$(1 - r\triangle_x^+)(1 + r\triangle_x^-)U_m^{n+2} = (1 - r\triangle_x^-)(1 + r\triangle_x^+)U_m^n. \tag{13.4c}$$

Show that the local truncation error for the resulting scheme may be written as $\mathcal{R}_h = R_1 + R_2$, where R_1 is the expression for the local truncation error of the Crank–Nicolson scheme with a time step $2k$ (so that it is $\mathcal{O}(h^2) + \mathcal{O}(k^2)$), and where $R_2 = -\frac{r^2}{2k}\delta_x^2(u_m^{n+2} - u^n)$. Show that requiring that $R_2 \to 0$ as $h, k \to 0$ requires that k/h must also tend to zero. The time step is therefore limited by accuracy rather than stability. Why is this a potentially perilous situation?

Is there an analogous result for the case of pure advection?

Project 13.5 (*Approximation of coupled diffusion equations*)
This project concerns the coupled PDEs

$$u_t = u_{xx} + v, \qquad v_t = \varepsilon v_{yy} - u,$$

where $0 < \varepsilon < 1$, defined on the domain $\Omega = \{(x, y) : 0 \le x \le 1, 0 \le y \le 1\}$ for $t > 0$ and subject to the boundary conditions $u = 0$ on the vertical boundaries $x = 0, 1$ with $v = 0$ on the horizontal boundaries $y = 0, 1$, together with initial conditions $u(x, y, 0) = f(x, y)$, $v(x, y, 0) = g(x, y)$. Use energy arguments to show that

$$\int_\Omega \left(u(x, y, t)^2 + v(x, y, t)^2\right) d\Omega$$

is a strictly decreasing function of t so that $\|u(\cdot, \cdot, t)\|_2 \to 0$ and $\|v(\cdot, \cdot, t)\|_2 \to 0$ as $t \to \infty$.

An explicit FTCS approximation of these equations on a square grid with time step k and grid spacing $h \times h$ in the (x, y) plane is given by

$$\left.\begin{aligned} U_{\ell,m}^{n+1} &= (1 + r\delta_x^2)U_{\ell,m}^n + kV_{\ell,m}^n \\ V_{\ell,m}^{n+1} &= (1 + \varepsilon r\delta_y^2)V_{\ell,m}^n - kU_{\ell,m}^n \end{aligned}\right\} \tag{13.5a}$$

where $r = k/h^2$. The von Neumann stability of this scheme may be examined by making the substitution

$$\begin{bmatrix} U_m^n \\ V_m^n \end{bmatrix} = \xi^n e^{i(\kappa_1 \ell + \kappa_2 m)h} c$$

where $\kappa_1, \kappa_2 \in \mathbb{R}$ are the wavenumbers and c is a constant 2×1 vector. Show that this leads to a homogeneous system $A(\xi)c = 0$ (in which $A(\xi)$ is a 2×2 matrix) that will admit nontrivial solutions if, and only if, $\det(A(\xi)) = 0$. Deduce that there are choices of κ, κ_2 such that $|\xi| > 1$ for any $k > 0$ so that the FTCS method is unconditionally unstable.[2]

Compare and contrast the two alternative semi-implicit solution methods

$$\left.\begin{aligned} U_{\ell,m}^{n+1} &= (1 + r\delta_x^2)U_{\ell,m}^n + kV_{\ell,m}^{n+1} \\ V_{\ell,m}^{n+1} &= (1 + \varepsilon r\delta_y^2)V_{\ell,m}^n - kU_{\ell,m}^n \end{aligned}\right\} \tag{13.5b}$$

and

$$\left.\begin{aligned} U_{\ell,m}^{n+1} &= U_{\ell,m}^n + r\delta_x^2 U_{\ell,m}^{n+1} + kV_{\ell,m}^{n+1} \\ V_{\ell,m}^{n+1} &= (1 + \varepsilon r\delta_y^2)V_{\ell,m}^n - kU_{\ell,m}^n \end{aligned}\right\} \tag{13.5c}$$

with regard to local truncation error and ℓ_2 stability.

Project 13.6 (*Fisher's equation and travelling waves*)
Consider the nonlinear parabolic PDE $u_t = u_{xx} + u(1 - u)$, known as Fisher's equation, on the interval $-\infty < x < \infty$ for $t > 0$. For spatially constant solutions $u(x, t) = v(t)$ show that $u(x, t) \to 1$ at $t \to \infty$ when $v(0) > 0$. When $v(0) < 0$, show that there is a time $t^* > 0$ such that $u(x, t) \to -\infty$ as $t \to t^*$. This illustrates the importance of non-negative initial data.

The focus will be on travelling wave solutions of Fisher's equation such that $u(x, t) \to 1$ as $x \to -\infty$ and $u(x, t) \to 0$ as $x \to \infty$. If such solutions take the form $u(x, t) = \varphi(z)$, where $z = x - at$ is the *travelling coordinate*, and a (the wave speed) is constant, show that φ satisfies the ODE

$$\varphi''(z) + a\varphi'(z) + \varphi(z) - \varphi^2(z) = 0, \quad -\infty < z < \infty. \tag{13.6a}$$

---

[2]For real numbers a and b the roots of the quadratic polynomial $p(\xi) = \xi^2 + a\xi + b$ lie strictly within the unit circle, i.e., $|\xi| < 1$, if, and only if, $p(1) > 0$, $p(-1) > 0$ and $p(0) < 1$. These are known as the *Jury conditions* —for a proof see Griffiths and Higham [7, Lemma 6.10].

(i) If $\varphi(z) \approx Ae^{-\lambda z}$ ($\lambda > 0$) as $z \to \infty$, then $\varphi^2(z)$ is negligible compared to $\varphi(z)$ and so φ satisfies a linear ODE under these conditions. Determine the relationship between a and λ and deduce that the associated wave speed must satisfy $a \geq 2$.

(ii) Analogously, if $\varphi(z) \approx 1 - Be^{\mu z}$ ($\mu > 0$) as $z \to -\infty$, use a similar argument to show that $a \in (-\infty, \infty)$. What is the relationship between λ and μ if the wave speeds are to be the same speed as $x \to \pm\infty$?

(iii) Suppose that

$$s(t) = \int_L^\infty \varphi(x - at)\, dx,$$

where L is a constant. Show that $s'(t) \to a$ as $t \to \infty$ if the conditions in parts (i) and (ii) hold.

Finally, verify that

$$\varphi(z) = \frac{1}{(1 + C \exp(bz))^2}, \quad a = 5/\sqrt{6}, \quad b = 1/\sqrt{6}, \tag{13.6b}$$

is a solution of (13.6a) for any constant C. Why should C be a positive constant? Sketch the graph of $\varphi(z)$. Show that there are values of λ and μ such that $\varphi(z)$ satisfies the properties in (i) and (ii). How do the predicted wave speeds in (i) and (ii) compare with that of the solution in (13.6b)?

Project 13.7 (*Approximation of a quasi-linear PDE*)
Determine the most general solution of the PDE

$$u_t = uu_{xx} + u(1 - u) \tag{13.7a}$$

taking the form $u(x, t) = X(x)/T(t)$. Show that there are constants a (dependent on h), b (dependent on k) and C such that

$$a^2\delta_x^2 X(x_m) = X(x_m) - C, \quad T(t_n) - (1 + b)T(t_{n+1}) = bC,$$

where $x_m = mh$ and $t_n = nk$. Use these results to show that the finite difference scheme

$$U_m^{n+1} = U_m^n + b\left[a^2 U_m^{n+1}\delta_x^2 U_m^n + (1 - U_m^{n+1})U_m^n\right] \tag{13.7b}$$

also has exact solutions of the form $U_m^n = X(x_m)/T(t_n)$ for the same functions X and T. Show that

$$a^2 = h^{-2} - \tfrac{1}{12} + \mathcal{O}(h^2), \quad b = k + \tfrac{1}{2}k^2 + \mathcal{O}(k^3),$$

and hence verify that (13.7b) is consistent with (13.7a).

In order to gain some insight into the sensitivity of the scheme to perturbations in the initial data, it is used to compute an approximation to the solution $u(x, t) = 1$ on the domain $0 < x < 1, t > 0$ with a spatial grid size $h = 1/M$. The boundary conditions are $U_0^n = U_M^n = 1$ and the initial condition is taken to be $U_m^0 = 1 + \varepsilon g_m$, where $\varepsilon \ll 1$ and $|g_m| \leq 1$. By writing $U_m^n = 1 + \varepsilon V_m^n$ show that V is the solution of a linear finite difference scheme with constant coefficients when terms in ε^2 are neglected. Deduce, using von Neumann's method, that this is stable in ℓ_2 when $b \leq 2 \tanh^2 \frac{1}{2} h$.

Project 13.8 (*High-order schemes for advection–diffusion equations*)
The solution of the advection–diffusion equation $u_t + a u_x = \varepsilon u_{xx}$, where ($\varepsilon > 0$) and a are constants, can be constructed explicitly for $k > 0$ (see Example 4.7),

$$u(x, nk + k) = \frac{1}{\sqrt{\pi}} \int_{-\infty}^{\infty} e^{-z^2} u(x - ak + 2z\sqrt{\varepsilon k}, nk) \, dz. \tag{13.8a}$$

This formula provides a means of determining the exact solution at the $(n+1)$st time level from that at the nth time level.

A numerical solution may be generated by approximating u at time $t = nk$ by a polynomial $\Phi(x)$, say, leading to the scheme

$$U_m^{n+1} = \frac{1}{\sqrt{\pi}} \int_{-\infty}^{\infty} e^{-z^2} \Phi(x_m - ak + 2z\sqrt{\varepsilon k}) \, dz. \tag{13.8b}$$

Show that this leads to Leith's method (see Exercise 12.11)

$$U_m^{n+1} = U_m^n - c\triangle_x U_m^n + (r + \tfrac{1}{2}c^2)\delta_x^2 U_m^n,$$

where $r = \varepsilon k/h^2$ and $c = ak/h$, when $\Phi(x)$ is the quadratic polynomial that satisfies the interpolation conditions $\Phi(jh) = U_j^n$ for $j = m - 1, m, m + 1$ (see Example 12.4). Show that increasing the degree of Φ to be a cubic polynomial and imposing the additional interpolating condition $\Phi(mh - 2h) = U_{m-2}^n$ leads to the alternative scheme

$$U_m^{n+1} = \left[1 - c\triangle_x + (r + \tfrac{1}{2}c^2)\delta_x^2 + \tfrac{1}{6}c(1 - 6r - c^2)\triangle_x\delta_x^2 \right] U_m^n. \tag{13.8c}$$

This generalises the third-order scheme in Example 12.10 to advection-diffusion problems. Use a suitable change of variable in the integrand to show that (13.8b) may be written in the alternative form

$$U_m^{n+1} = \int_{-\infty}^{\infty} K(s) \Phi(x_m - sh) \, ds,$$

and determine the exact form of K. Graph the function $K(s)$ for $-5 \leq s \leq 5$ for $r = 0.125, 0.25, 0.5, 0.75, 1$ when $c = 0$. What insight do these graphs provide for judging the range of mesh ratios r for which these methods can be expected to provide accurate solutions of (13.8b)? [Hint: $\int_{-\infty}^{\infty} z^j e^{-z^2} dz$ has the values $\sqrt{\pi}, 0, \frac{1}{2}\sqrt{\pi}, 0$ for $j = 0, 1, 2, 3$, respectively.]

Project 13.9 (*Stability with an uneven spatial grid*)
This project concerns the stability of boundary conditions when approximating the heat equation $u_t = u_{xx}$ on the interval $0 < x < 1$ for $t > 0$ with Dirichlet end conditions. Suppose that we use an FTCS finite difference method based on a grid of size k in time and a grid in space made up of the points $\{x_m = mh : m = 0, 1, \ldots, M, x_{M+1} = 1\}$ where $h = 1/(M + \alpha)$ and $0 < \alpha < 1$. That is, the grid is uniform except in the final subinterval where the distance between grid points is αh. Using the approximation (10.26) with $h_- = h$ and $h_+ = \alpha h$ to approximate u_{xx} at $x = x_M$ leads to

$$U_m^{n+1} = (1 + r\delta_x^2)U_m^n, \quad m = 1, 2, \ldots, M - 1, \tag{13.9a}$$

$$U_M^{n+1} = \frac{2r}{1+\alpha}U_{M-1}^n + (1 - \frac{2r}{\alpha})U_M^n + \frac{2r}{\alpha(1+\alpha)}U_{M+1}^n. \tag{13.9b}$$

We would like to asses the stability of this approximation. To this end, suppose that $U_0^n = U_{M+1}^n = 0$. Show that $|U_m^{n+1}| \leq \|U_\cdot^n\|_{h,\infty}$ whenever

$$0 < r \leq \min \left\{ \frac{1}{2}, \frac{\alpha(1+\alpha)}{1+2\alpha} \right\}.$$

A necessary condition for ℓ_2 stability requires that (13.9a) should be stable in the von Neumann sense (with $|\xi| \leq 1$), but a separate test will be needed to assess (13.9b). The basis of this test (see Strikwerda [21, Sect. 11.3]) is that a solution of the total set of difference equations is sought in the form $U_m^n = \xi^n z^m$. When this is substituted into (13.9a) and (13.9b) (with $U_{M+1}^n = 0$), a quadratic equation and a linear equation are obtained in z, respectively, whose coefficients depend (linearly) on ξ. This sets up a one–to–one correspondence between possible values of z and ξ. The condition for ℓ_2-stability then requires that $|\xi| \leq 1$ in all cases where z has solutions with $|z| > 1$. Show that the z roots are both real, and that the ℓ_2-stability test leads to the condition

$$r \leq \begin{cases} \alpha\sqrt{1 - \alpha^2}, & 0 < \alpha \leq 1/\sqrt{2}, \\ \frac{1}{2}, & 1/\sqrt{2} \leq \alpha \leq 1. \end{cases}$$

Finally, show that the scheme retains its explicit nature if the finite difference approximation at $x = x_M$ is replaced by the BTCS-type scheme

$$-\frac{2r}{1+\alpha}U_{M-1}^{n+1} + (1 + \frac{2r}{\alpha})U_M^{n+1} - \frac{2r}{\alpha(1+\alpha)}U_{M+1}^{n+1} = U_M^n, \tag{13.9c}$$

and that the overall process will be stable with respect to the maximum norm whenever $0 < r \leq 1/2$.

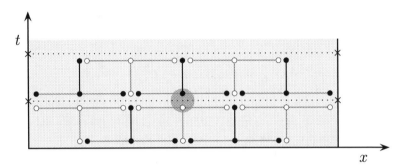

Fig. 13.1 The disposition of FTCS (•) and BTCS (○) methods at the first two time levels for the hopscotch scheme. The stencils are slightly offset in the interests of clarity. Eliminating the grid value at the highlighted point leads to the DuFort–Frankel scheme

Project 13.10 (*Convergence of the hopscotch scheme*)

This project concerns the *hopscotch* scheme devised by Gordon in 1965. It is an ingenius way of combining the FTCS and BTCS schemes (11.8b) and (11.22b) for approximating the heat equation $u_t = u_{xx}$ with Dirichlet end conditions ,

$$U_m^{n+1} = rU_{m-1}^n + (1 - 2r)U_m^n + rU_{m+1}^n, \tag{13.10a}$$

$$-rU_{m-1}^{n+1} + (1 + 2r)U_m^{n+1} - rU_{m+1}^{n+1} = U_m^n. \tag{13.10b}$$

We suppose that the FTCS method is applied at even-numbered grid points to provide values for $U_2^1, U_4^1, \ldots$ at the first time level (see Fig. 13.1). The solution $U_1^1, U_3^1, \ldots$ at the odd numbered grid points can then be computed *without* the need to solve algebraic equations—the method is explicit. At the second time step, the values $U_1^2, U_3^2, \ldots$ are computed by the FTCS method and $U_2^2, U_4^2, \ldots$ by BTCS, again in an explicit fashion. This odd-even pattern is then repeated over subsequent time steps to give a method that has the same computational cost as the FTCS method alone, but having the potential for greater stability because it also incorporates the BTCS method.

The formulae for the individual methods have to be combined into one for the purposes of analysis. To do this we select a point (x_m, t_n) where $m + n$ is even (the grid point highlighted in Fig. 13.1, for example). The value of U_m^n may then be eliminated from (13.10a) (with n replaced by $n - 1$) and (13.10b) to give

$$(1 + 2r)U_m^{n+1} = 2r(U_{m-1}^n + U_{m+1}^n) + (1 - 2r)U_m^{n-1}. \tag{13.10c}$$

This scheme is also known as the Du Fort–Frankel (1953) method. Show that the solution at points where $m + n$ is even can be found independently of the points where $m + n$ is odd.

(a) Determine what restrictions the CFL places on the hopscotch scheme.

(b) Determine (following appropriate scaling) the local truncation error of the method and deduce that it does not tend to zero as $h, k \to 0$ unless $k/h \to 0$.

(c) Show that $|U_m^{n+1}| \leq \|U_\cdot^n\|_{h,\infty}$ for $m = 1, 2, \ldots, M - 1$ for $r \leq \frac{1}{2}$.

(d) Show that the amplification factor of the Du Fort–Frankel scheme satisfies a quadratic equation and deduce that the method is unconditionally stable.

Describe how the FTCS and BTCS approximations (11.61b) and (11.64b) of the two-dimensional heat equation $u_t = u_{xx} + u_{yy}$ may be deployed to give an analogous method when the domain is a square in the x-y plane and a Dirichlet boundary condition is applied on its boundary.

Project 13.11 (*Unsteady advection–diffusion in* $\mathbb{R}^2$)
Suppose that $u(\vec{x}, t)$, with $\vec{x} = (x_1, x_2)$ satisfies the constant coefficient advection-diffusion equation[3]

$$u_t + \sum_{j=1,2} a_j \partial_{x_j} u = \sum_{j=1,2} \varepsilon_j \partial_{x_j}^2 u,$$

where $\varepsilon_j > 0$. A finite difference grid with grid size h_j in the jth coordinate directions is constructed from the lines $x_{1,\ell} = \ell h_1$, $x_{2,m} = m h_2$ and on which the PDE is approximated by the FTCS scheme

$$U_{\ell,m}^{n+1} = \left(1 - \sum_{j=1,2} c_j \triangle_j + \frac{1}{2} \sum_{j=1,2} r_j \delta_j^2\right) U_{\ell,m}^n.$$

with Courant numbers $c_j = a_j k/h_j$, mesh ratios $r_j = 2k\varepsilon_j/h_j^2$ and $\triangle_j$, δ_j^2 are finite difference operators in the jth coordinate direction. Find an expression for the amplification factor for this scheme by substituting $U_{\ell,m}^{n+1} = \xi^n e^{i(\kappa_1 \ell h + \kappa_2 m h)})$ and express the result in terms of the scaled wavenumbers $\theta_j = \kappa_j h_j$.

The purpose of this project is to establish ℓ_2-stability in the sense that $|\xi(\boldsymbol{\theta})| \leq 1$ for all $\boldsymbol{\theta} = (\theta_1, \theta_2) \in [-\pi, \pi] \times [-\pi, \pi]$ if, and only if,

$$\text{(i)} \ r_j \leq 1 \ (j = 1, 2), \quad \text{(ii)} \ \sum_{j=1,2} r_j \leq 1 \text{ and (iii)} \ \sum_{j=1,2} \frac{c_j^2}{r_j} \leq 1$$

which generalise the results in Example 11.24.

---

[3]A slightly different notation is adopted that uses x_1, x_2 in lieu of x, y so as to facilitate the presentation of the scheme and its analysis and also to allow ready generalisation to higher space dimensions. Note also that a factor 2 has been introduced into the definition of r_j so as to avoid fractions occurring at a later stage.

Necessity: Deduce that conditions (i) and (ii) are necessary by examining $\xi(\pi, 0)$, $\xi(0, \pi)$ and $\xi(\pi, \pi)$. Show that the Maclaurin expansion of $|\xi|^2$, for small values of its arguments, may be written as

$$|\xi(\boldsymbol{\theta})|^2 = 1 - \tfrac{1}{4}\boldsymbol{\theta}^{\mathsf{T}}(R - \boldsymbol{c}\boldsymbol{c}^{\mathsf{T}})\boldsymbol{\theta} + \cdots ,$$

where $R = \mathrm{diag}(r_1, r_2)$ and $\boldsymbol{c} = [c_1, c_2]^{\mathsf{T}}$ (see Exercise 11.24). Deduce, by finding the eigenvalues of the 2×2 matrix $(R - \boldsymbol{c}\boldsymbol{c}^{\mathsf{T}})$, that condition (iii) is necessary for stability.

Sufficiency: Show, by writing $c_j \sin \theta_j = (c_j / \sqrt{r_j})(\sqrt{r_j} \sin \theta_j)$ and using the Cauchy–Schwatrz inequality, that

$$\Big(\sum_{j=1,2} c_j \sin \theta_j \Big)^2 \leq \sum_{j=1,2} r_j \sin^2 \theta_j$$

if condition (iii) holds. Hence show that $|\xi(\boldsymbol{\theta})|^2 \leq 1$.

Project 13.12 (*MacCormack's method*)
MacCormack's method can be used to approximate a first-order hyperbolic equation with a nonlinear source term

$$u_t + a u_x = f(u), \quad a > 0. \tag{13.12a}$$

The method can be written as

$$U_m^{n+1} = \frac{1}{2}[\underline{U}_m^n + \overline{U}_m^n], \tag{13.12b}$$

and involves intermediate quantities $\underline{U}_m^n$ and $\overline{U}_m^n$, computed via

$$\overline{U}_m^n = U_m^n - c\Delta_x^+ U_m^n + kf(U_m^n), \quad \underline{U}_m^n = U_m^n - c\Delta_x^- \overline{U}_m^n + kf(\overline{U}_m^n),$$

where k is the time step and $c = ak/h$ the Courant number.

(i) In the case $f(u) = 0$, show that the method reduces to the Lax–Wendroff scheme.

(ii) When $f(u) = -\alpha u$ (where $\alpha \geq 0$ is a constant), show that the amplification factor satisfies

$$|\xi|^2 - 1 = 4(c^2 - (1 - \sigma)^2)c^2 s^2 - 2\sigma(2 - \sigma)c^2 s - \tfrac{1}{4}\sigma(2 - \sigma)((1 - \sigma)^2 + 3),$$

where $\sigma = \alpha k$, and $s = \sin^2 \tfrac{1}{2}\kappa h$. Deduce that $|\xi| \leq 1$ if $0 \leq \sigma \leq 2$ and $c^2 \leq \tfrac{1}{4}((1 - \sigma)^2 + 3)$. Describe the behaviour that you would expect from a sequence of numerical experiments when $\alpha \gg 1$ in which the step sizes k tend to zero with the Courant number remaining constant. The sequence begins with

$k = 3/\alpha$. Would there be any significant difference between the cases when c is small and c is close to 1?

(iii) Suppose that $f(u)$ is a continuously differentiable function of u and that (13.12b) has a constant solution $U_m^n = U^*$ for all n, m with $f'(U^*) \neq 0$. Show that U^* must satisfy $f(U^*) + f(\overline{U}^*) = 0$, where $\overline{U}^* = U^* + kf(U^*)$. In the particular case when $f(u) = \alpha u(1 - u)$, $\alpha > 0$. Show that the scheme has four possible constant solutions corresponding to $U^* = 0$, $U^* = 1$ and two further values that are functions of σ and are given by the roots of a certain quadratic equation. Under what conditions are these roots real?

Project 13.13 (*A conjecture on isospectral matrices*)
A separation of variables solution $u(x, t) = e^{-i\lambda t} v(x)$ of the advection equation $u_t + u_x = 0$ with periodic boundary conditions leads to the eigenvalue problem

$$iv'(x) = \lambda v(x), \quad 0 \leq x \leq 1, \tag{13.13a}$$

with $v(x + 1) = v(x)$ for all $x \in \mathbb{R}$. Determine all the eigenvalues of this problem along with the corresponding eigenfunctions.

The eigenvalue problem will be approximated by finite differences on a *nonuniform* grid in which the interval $[0, 1]$ is divided into M subintervals by the grid points $x_0 = 0 < x_1 < x_2 < \cdots < x_M = 1$. Consider the two alternative approximations to $\mathcal{L}v(x) := iv'(x)$ given by

$$\mathcal{L}_h V_m = i\frac{\Delta V_m}{\Delta x_m}, \tag{13.13b}$$

$$\mathcal{M}_h V_m = i\frac{1}{\Delta x_m}\left(\Delta^+ x_m \frac{\Delta^- V_m}{\Delta^- x_m} + \Delta^- x_m \frac{\Delta^+ V_m}{\Delta^+ x_m}\right). \tag{13.13c}$$

Note that both reduce to the standard second-order approximation $v'(x_m) = h^{-1}\Delta v$ $(x_m) + \mathcal{O}(h^2)$ when the grid points are equally spaced and $h = 1/M$. Periodic boundary conditions imply that $V_0 = V_M$ so the approximations are only needed for $m = 1, 2, \ldots, M$. Furthermore, $V_{M+1} = V_1$ and $\Delta^+ x_M = \Delta^- x_1$.

Show that these approximations of (13.13a) lead to eigenvalue problems for the matrices $A_1 = iD^{-1}(L - L^T)$ and $A_2 = iD^{-1}(RL - R^{-1}L^T)$ respectively, where $h_j = \Delta^- x_j$ $(j = 1, 2, \ldots, M)$, D and R are the $M \times M$ diagonal matrices

$$D = \begin{bmatrix} h_1+h_2 & & & \\ & h_2+h_3 & & \\ & & \ddots & \\ & & & h_M+h_1 \end{bmatrix}, \quad R = \begin{bmatrix} h_2/h_1 & & & \\ & h_3/h_2 & & \\ & & \ddots & \\ & & & h_1/h_M \end{bmatrix},$$

and L is the bidiagonal circulant matrix

$$L = \begin{bmatrix} 1 & & & & -1 \\ -1 & 1 & & & \\ & & \ddots & \ddots & \\ & & & -1 & 1 \end{bmatrix}.$$

Confirm that A_1 is singular having rank $M - 1$ when M is odd, and rank $M - 2$ when M is even (Hint: permute the rows of $L - L^T$ so that the first row becomes the last.) Identify a set of linearly independent vectors that span the nullspace of A_1 in each case.

Next, show that the rank of A_2 is at most $M - 1$ and identify a nonzero vector in its nullspace. When M is even, suppose that the grid is chosen such that $h_m = \alpha h$ for $m = 1, 2, \ldots, M - 1$ and that $h_M = \beta h$ ($\alpha, \beta > 0$ and $h = 1/(\beta + (M - 1)\alpha)$). Show that the rank of A_2 is exactly $M - 1$ when $M = 6$, unless $\alpha = \beta$.

We conjecture that the matrices A_1 and A_2 are isospectral—they have the same eigenvalues.[4]

Prove that the eigenvalues of A_1 are real and that, for every eigenvalue λ with eigenvector v, there is an eigenvalue $-\lambda$ with eigenvector $\bar{v}$ (the complex conjugate of v).

Use an appropriate software package to compute the eigenvalues λ_m of A_1 and μ_m of A_2 for a range of values of M when the grid points $x_1 < x_2 < \cdots < x_{M-1}$ are located randomly in the interval $(0, 1)$. Calculate, in each case, the maximum relative difference, that is

$$\max_m \left| \frac{\lambda_m - \mu_m}{\lambda_m} \right|,$$

where the maximum is taken over all nonzero eigenvalues. Comment on the nature of the eigenvectors corresponding to zero eigenvalues particularly when M is even.

A more robust test of the conjecture is to use a computer algebra package to show that the characteristic polynomials of the two matrices are the same or, more appropriately, that the determinant of $(i(L - L^T) - \lambda D)$ is identical to that of $(i(RL - R^{-1}L^T - \mu D)$. Carry out these computations for $M = 3, 4, 5, 6$. Confirm that the polynomials have a root of multiplicity 1 when M is odd and multiplicity 2 when M is even. If this conjecture is true then the earlier results on rank would imply that A_2 is a defective matrix—it has fewer than M linearly independent eigenvectors.

---

[4]This is a mundane exercise compared with the celebrated article "Can one hear the shape of a drum?" by Mark Kac [10], where it transpires that the answer is no because the Laplacian operator in two dimensions can have the same eigenvalues on two different domains (drum shapes).

Appendix A
Glossary and Notation

The study of PDEs and, particularly their numerical solution has a rapacious appetite for variable names, constants, parameters, indices, and so on. This means that many symbols become overloaded—they are required to take on different meanings in different contexts. It is hoped that the list provided here will help in avoiding confusion.

$\mathcal{B}$:   an operator used to represent boundary conditions

$\mathcal{B}_h$:   a finite difference approximation of $\mathcal{B}$

BC: boundary condition

BVP: boundary value problem—that is a PDE together with boundary and/or initial conditions

c: the wave speed in the wave equation; the Courant number $c = ak/h$ when approximating the advection equation

∂_x, ∂_t: shorthand notation for partial derivatives with respect to x and t; so $u_x = \partial_x u$, $u_{xx} = \partial_x^2 u$, $u_{xt} = \partial_x \partial_t u$

$\Delta^+, \Delta^-, \Delta, \delta$:   forward, backward and alternative central difference operators for functions of one variable; $\Delta_x^+, \Delta_x^-, \Delta_x, \delta_x$ apply to the x variable of a function of several variables

$\partial\Omega$:   the boundary of a typical domain Ω

$\mathcal{F}, \mathcal{F}_h$:   source term in a differential equation $\mathcal{L}u = \mathcal{F}$ or in a finite difference equation $\mathcal{L}_h U = \mathcal{F}_h$,

$\mathscr{F}, \mathscr{F}_h$:   source data in a boundary value problem $\mathscr{L}u = \mathscr{F}$ or its finite difference approximation $\mathscr{L}_h U = \mathscr{F}_h$,

IC: initial condition

IVP: initial value problem

IBVP: initial-boundary value problem

κ: the conductivity coefficient in the heat equation; the wave number in a von Neumann analysis of stability

$\mathcal{L}, \mathcal{M}$:   differential operators involving derivatives with respect to space variables only, such as $\mathcal{L} = -\partial_x^2$

$\mathcal{L}_h, \mathcal{M}_h$:   finite difference approximations of differential operators $\mathcal{L}, \mathcal{M}$

$\mathscr{L}, \mathscr{M}$:   differential operators together with associated boundary conditions

© Springer International Publishing Switzerland 2015

D.F. Griffiths et al., *Essential Partial Differential Equations*, Springer Undergraduate Mathematics Series, DOI 10.1007/978-3-319-22569-2

$\mathscr{L}_h, \mathscr{M}_h$:   finite difference approximations of differential operators $\mathscr{L}$, $\mathscr{M}$

$\vec{n}$:   the outward pointing normal direction; $\vec{n}(\vec{x})$ is the outward normal vector at a point $\vec{x}$ on the boundary

$\vec{\nabla}$:   the gradient vector; $\vec{\nabla}u = [u_x, u_y]$ for a function $u(x, y)$ of two variables

∇^2:   Laplacian operator; $\nabla^2 u = u_{xx} + u_{yy}$ for a function $u(x, y)$ of two variables

Ω, Ω_h: typical domain of a PDE or its approximation by a grid of points,

ODE: ordinary differential equation

PDE: partial differential equation

p: coefficient of first derivative in a differential equation; order of consistency of a finite difference approximation, as in $\mathcal{O}(h^p)$

φ, Φ: comparison functions for continuous ($\varphi \geq 0$ and $\mathscr{L}\varphi \geq 1$) and discrete problems ($\Phi \geq 0$ and $\mathscr{L}_h\Phi \geq 1$), respectively

u, U: typical solution of a boundary value problem $\mathscr{L}u = \mathscr{F}$ or its finite difference approximation $\mathscr{L}_h U = \mathscr{F}_h$,

r: the mesh ratio; $r = k/h^2$ when approximating the heat equation; the radial coordinate in polar coordinates

$\vec{r}$:   position vector; $\vec{r} = (x, y)$ in two dimensions

$\mathcal{R}_h, \mathscr{R}_h$:   local truncation error—$\mathcal{R}_h := \mathcal{L}_h u - \mathcal{F}_h$ and $\mathscr{R}_h := \mathscr{L}_h u - \mathscr{F}_h$, where u solves $\mathcal{L}u = \mathcal{F}$ and $\mathscr{L}u = \mathscr{F}$.

x^*:   Hermitian (or complex conjugate) transpose of a d-dimensional vector x

ξ: the amplification factor in a von Neumann analysis of stability

$\| \cdot \|$:   norm operator; maps d-dimensional vectors x (or real-valued functions) onto non-negative real numbers

$\| \cdot \|_h$:   discrete norm operator; maps grid functions U onto non-negative real numbers; we use two flavours, the maximum (or ℓ_∞-) norm $\| \cdot \|_{h,\infty}$ and the ℓ_2-norm $\| \cdot \|_{h,2}$

$\langle \cdot, \cdot \rangle$:   inner product; maps two d-dimensional vectors (or pairs of functions) onto real numbers

$\langle \cdot, \cdot \rangle_w$:   inner product associated with a positive weight function w; maps pairs of functions onto real numbers.

Appendix B
Some Linear Algebra

B.1 Vector and Matrix Norms

Norms provide a convenient way of measuring the length of vectors and the magnifying ability of matrices.

Definition B.1 A norm on a vector x is denoted by $\|x\|$ and is required to have the properties

(a) positivity: $\|x\| > 0$ for $x \neq 0$,
(b) uniqueness: $\|x\| = 0$ if and only if $x = 0$,
(c) scaling: $\|ax\| = |a| \, \|x\|$ for any complex scalar a,
(d) triangle inequality: $\|x + y\| \leq \|x\| + \|y\|$ for any vector y having the same dimension as x.

When $x = [x_1, x_2, \ldots, x_d]^\mathsf{T}$ a popular family are the so-called ℓ_p norms, and are given by

$$\|x\|_p = \left(\sum_{j=1}^{d} |x_i|^p \right)^{1/p} \tag{B.1}$$

with $1 \leq p < \infty$. Typical choices are $p = 1, 2, \infty$, where the case $p = \infty$ is interpreted as

$$\|x\|_\infty = \max_{1 \leq i \leq d} |x_i|$$

and is known as the *maximum norm*. A particularly useful result is given by *Hölder's inequality* which states that, for any two complex vectors x, y of the same dimension,

$$|x^* y| \leq \|x\|_p \|y\|_q, \quad 1/p + 1/q = 1, \tag{B.2}$$

© Springer International Publishing Switzerland 2015
D.F. Griffiths et al., *Essential Partial Differential Equations*, Springer
Undergraduate Mathematics Series, DOI 10.1007/978-3-319-22569-2

where x^* denotes the Hermitian (or complex conjugate) transpose of x. When $p = q = 2$ this is also known as the Cauchy–Schwarz inequality

$$|x^* y| \leq \|x\|_2 \|y\|_2. \tag{B.3}$$

Definition B.2 A norm on a matrix A is denoted by $\|A\|$ and is required to have the properties

(a) $\|A\| > 0$ for $A \neq 0$,
(b) $\|A\| = 0$ if and only if A is the zero matrix,
(c) $\|aA\| = |a| \|A\|$ for any complex scalar a,
(d) $\|A + B\| \leq \|A\| + \|B\|$ for any matrix B of the same dimension as A,
(e) $\|AB\| \leq \|A\|\|B\|$ for any matrix B for which the product AB is defined.

The requirement that (e) holds is unconventional. (A standard definition would only stipulate (a)–(d).) To find examples of matrix norms where (e) is satisfied, we note that analyses involving norms usually involve both vector and matrix norms, and it is standard practice in numerical analysis to use norms that are *compatible* in the sense that

$$\|Ax\| \leq \|A\| \, \|x\|. \tag{B.4}$$

One way in which this can be achieved is to first define a vector norm $\|x\|$ and then use

$$\|A\| = \max_{x \neq 0} \frac{\|Ax\|}{\|x\|}$$

to *induce* a matrix norm. When the vector norm is a p-norm, this approach leads to the matrix p-norms defined, for a $d \times d$ matrix A having a_{ij} in the ith row and jth column, by

$$\|A\|_1 = \max_{1 \leq j \leq d} \sum_{i=1}^{d} |a_{ij}|, \quad \|A\|_\infty = \|A^\mathsf{T}\|_1, \tag{B.5}$$

while $\|A\|_2 = \sqrt{\lambda}$, where λ is the largest eigenvalue of $A^* A$. These matrix p-norms are examples for which property (e) automatically holds. We refer to Trefethen and Bau [26] for further discussion.

B.2 Symmetry of Matrices

A real $n \times n$ matrix A is said to be *symmetric* if it remains unchanged when its rows and columns are interchanged,[1] so that $A^{\mathsf{T}} = A$. If $A = (a_{ij})$, with a_{ij} denoting the entry in the ith row and jth column, then symmetry requires $a_{ij} = a_{ji}$. To exploit symmetry we note that $A^{\mathsf{T}} = A$ implies that, for vectors $x, y \in \mathbb{C}^n$,

$$y^*(Ax) = y^*Ax = (A^{\mathsf{T}}y)^*x = (Ay)^*x. \tag{B.6}$$

Thus $y^*(Ax) = (Ay)^*x$ so the (complex) scalar product of y and Ax is the same as the scalar product of Ay and x.

To take this idea further we need to introduce a special notation for scalar product—this is generally referred to as an *inner product*. Specifically, we define

$$\langle x, y \rangle \equiv y^*x \tag{B.7}$$

then, from (B.6),

$$\langle x, Ay \rangle = \langle Ax, y \rangle \tag{B.8}$$

for a symmetric matrix A.

B.3 Tridiagonal Matrices

Definition B.3 (*Irreducible tridiagonal matrix*) The $n \times n$ tridiagonal matrix A given by

$$A = \begin{pmatrix} b_1 & c_1 & 0 & \cdots & & 0 \\ a_2 & b_2 & c_2 & & & 0 \\ & & \ddots & \ddots & \ddots & \\ & & & a_{n-1} & b_{n-1} & c_{n-1} \\ 0 & \cdots & 0 & & a_n & b_n \end{pmatrix} \tag{B.9}$$

is said to be *irreducible* if the off-diagonal entries a_j and c_j are all nonzero.

Suppose that $D = \text{diag}(d_1, d_2, \ldots, d_n)$ is an $n \times n$ diagonal matrix and let A be an irreducible tridiagonal matrix. The product matrix

---

[1] We shall restrict ourselves to real matrices, the analogous property for complex matrices is that $A^* = A$, where A^* is the Hermitian (or complex conjugate) transpose.

$$DA = \begin{pmatrix} d_1b_1 \ d_1c_1 & 0 & \cdots & & 0 \\ d_2a_2 \ d_2b_2 & d_2c_2 & & & 0 \\ & \ddots & \ddots & \ddots & \\ & & d_{n-1}a_{n-1} & d_{n-1}b_{n-1} & d_{n-1}c_{n-1} \\ 0 & \cdots & 0 & d_na_n & d_nb_n \end{pmatrix}$$

will be a symmetric matrix if the entries of D are chosen so that

$$d_ja_j = d_{j-1}c_{j-1}, \quad j = 2, 3, \ldots, n.$$

Setting $d_1 = 1$, the fact that $a_j \neq 0$ means that $d_2, d_3, \ldots, d_n$ can be successively computed. The fact that $c_{j-1} \neq 0$ implies that $d_j \neq 0$, $j = 2, \ldots, n$, which ensures that the diagonal matrix D is *nonsingular*. It follows from this result that A will have real eigenvalues when $a_j/c_{j-1} > 0$, for $j = 2, 3, \ldots, n$ (see Exercise B.4). This is the linear algebra analogue of the clever change of variables for ODEs in Sect. 5.3.2 combined with (part of) Theorem 5.11.

The $n \times n$ tridiagonal matrix

$$T = \begin{bmatrix} b & c & 0 & \cdots & 0 \\ a & b & c & & \\ & \ddots & \ddots & \ddots & \\ & & a & b & c \\ 0 & & & a & b \end{bmatrix} \tag{B.10}$$

with constant diagonals, has eigenvalues

$$\lambda_j = b + 2\sqrt{ac} \cos \frac{j\pi}{n+1}, \quad j = 1, 2, \ldots, n. \tag{B.11}$$

with corresponding eigenvectors

$$v_j = [\sin(\pi j x_1), \sin(\pi j x_2), \ldots, \sin(\pi j x_m), \ldots, \sin(\pi j x_n)]^{\mathsf{T}},$$

where $x_m = m/(n+1)$ (see Exercise B.5). Generalizations of this result can be found in Fletcher and Griffiths [3].

B.4 Quadratic Forms

Definition B.4 If $x \in \mathbb{R}^n$ and A is a real $n \times n$ symmetric matrix, then the function

$$Q(x) = x^{\mathsf{T}} A x$$

is called a *quadratic form*.

The form $Q(x)$ is a homogeneous quadratic function of the independent variables $x_1, x_2, \ldots, x_n$.[2] In the simplest (two-dimensional) case, we have $x \in \mathbb{R}^2$, and

$$Q(x) = \begin{bmatrix} x & y \end{bmatrix} \begin{bmatrix} a & b \\ b & c \end{bmatrix} \begin{bmatrix} x \\ y \end{bmatrix} = ax^2 + 2bxy + cy^2. \tag{B.12}$$

We shall be concerned with the *level curves* of the quadratic form, that is points where $Q(x) = $ constant. If we make a change of variables: $x = Vs$, where V is a nonsingular 2×2 matrix, then the quadratic form becomes

$$Q(Vs) = s^T (V^T A V) s$$

which is a quadratic form having coefficient matrix $V^T A V$. The idea is to choose V in such a way that $V^T A V$ is a diagonal matrix,

$$V^T A V = \begin{bmatrix} \alpha & 0 \\ 0 & \beta \end{bmatrix},$$

so that, with $s = [s, t]^T$ we have

$$Q(Vs) = \alpha s^2 + \beta t^2. \tag{B.13}$$

We can picture the form in (B.13) geometrically. The points $Q = $ constant in the s-t plane are (a) ellipses if $\alpha\beta > 0$ (i.e., α and β have the same sign), (b) hyperbolae if $\alpha\beta < 0$ (i.e., they have opposite signs) and (c) straight lines[3] if one of α or β is zero (i.e., $\alpha\beta = 0$).

Two basic questions need to be answered at this point:

Q1. How do we construct the matrix V?
Q2. Different matrices V will lead to different values for α and β but will their signs remain the same?

To answer the first question: a natural candidate for V is the matrix of eigenvectors of A. To confirm this choice, suppose that A has eigenvalues λ_1, λ_2 with corresponding eigenvectors v_1 and v_2:

$$Av_j = \lambda_j v_j, \quad j = 1, 2.$$

Since A is symmetric, the eigenvalues are real and the two eigenvectors are orthogonal: $v_1^T v_2 = 0 = v_2^T v_1$. Setting $V = [v_1, v_2]$, we get

---

[2] Homogeneous means that $Q(cx) = c^2 Q(x)$ for $c \in \mathbb{R}$; this is the case because there are no linear terms and no constant term.

[3] When $Q(x)$ also contains linear terms in x then $Q(Vs)$ will, in general, contain linear terms in s. In such cases the level curves will be parabolae.

$$V^{\mathsf{T}} A V = \begin{bmatrix} \lambda_1 v_1^{\mathsf{T}} v_1 & \lambda_2 v_1^{\mathsf{T}} v_2 \\ \lambda_1 v_2^{\mathsf{T}} v_1 & \lambda_2 v_2^{\mathsf{T}} v_2 \end{bmatrix} = \begin{bmatrix} \lambda_1 v_1^{\mathsf{T}} v_1 & 0 \\ 0 & \lambda_2 v_2^{\mathsf{T}} v_2 \end{bmatrix}$$

so that we have (B.13) with $\alpha = \lambda_1 v_1^{\mathsf{T}} v_1$ and $\beta = \lambda_2 v_2^{\mathsf{T}} v_2$. If the eigenvectors are normalised to have unit length then $\alpha = \lambda_1$ and $\beta = \lambda_2$. In both cases the sign of the product $\alpha\beta$ is the same as that of $\lambda_1\lambda_2$ so the level curves will be ellipses if $\lambda_1\lambda_2 > 0$ and hyperbolae if $\lambda_1\lambda_2 < 0$. Moreover, since the characteristic polynomial of the matrix A is given by

$$\det(A - \lambda I) = \lambda^2 - (a + c)\lambda + (ac - b^2),$$

then the product of the eigenvalues is given by $\lambda_1\lambda_2 = ac - b^2 = \det(A)$, so we see that it is not necessary to compute the eigenvalues in order to determine the nature of the level curves. This is formally stated in the following result.

Theorem B.5 *The level curves of the quadratic form $Q(x, y) = ax^2 + 2bxy + cy^2$ are hyperbolae if $b^2 - ac > 0$ (when $Q(x, y)$ has two distinct real factors) and ellipses if $b^2 - ac < 0$ (when $Q(x, y)$ has no real factors). In the intermediate case, $b^2 - ac = 0$ (when $Q(x, y)$ is a perfect square) the level curves are straight lines.*

If the eigenvectors are normalised then $V^{\mathsf{T}} V = I$ and $x = V s$ may be inverted to give

$$s = \begin{bmatrix} s \\ t \end{bmatrix} = V^{\mathsf{T}} x = \begin{bmatrix} v_1^{\mathsf{T}} \\ v_2^{\mathsf{T}} \end{bmatrix} x$$

so that $s = v_1^{\mathsf{T}} x$ is the component of x in the direction of v_1 and $t = v_2^{\mathsf{T}} x$ is the component of x in the direction of v_2. In the nondenegerate case ($b^2 \neq ac$), a further rescaling of the components of s to new variables $\boldsymbol{\xi} = [\xi, \eta]^{\mathsf{T}}$ defined by $\xi = s\sqrt{|\lambda_1|}$ and $\eta = t\sqrt{|\lambda_2|}$ can be applied. This gives $s = D\boldsymbol{\xi}$ where

$$D = \begin{bmatrix} 1/\sqrt{|\lambda_1|} & 0 \\ 0 & 1/\sqrt{|\lambda_2|} \end{bmatrix}$$

is a nonsingular diagonal matrix with $Q(V D\boldsymbol{\xi}) = \pm 1(\xi^2 + \eta^2)$ in the elliptic case and $Q(V D\boldsymbol{\xi}) = \pm 1(\xi^2 - \eta^2)$ in the hyperbolic case. Thus, with this special rescaling, the contours are either circles or rectangular hyperbolae in the ξ-η plane. The process is illustrated by the following example.

Example B.6 Determine the level curves of the quadratic form associated with the matrix

$$A = \begin{bmatrix} 13 & -4 \\ -4 & 7 \end{bmatrix}.$$

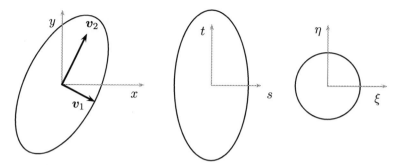

Fig. B.1 Level curves of $Q(\boldsymbol{x})$ (*left*), $Q(V\boldsymbol{s})$ (*centre*) and $Q(VD\boldsymbol{\xi})$ (*right*) for Example B.6. The mapping from s-t to x-y corresponds to a simple rotation of the coordinate system

Computing the eigenvalues and eigenvectors of A gives

$$\lambda_1 = 15, \quad \boldsymbol{v}_1 = \frac{1}{\sqrt{5}}\begin{bmatrix} 2 \\ -1 \end{bmatrix}, \quad \lambda_2 = 5, \quad \boldsymbol{v}_2 = \frac{1}{\sqrt{5}}\begin{bmatrix} 1 \\ 2 \end{bmatrix}.$$

Setting $V = [\boldsymbol{v}_1, \boldsymbol{v}_2]$ and changing variables gives $s = (2x - y)/\sqrt{5}$, $t = (x + 2y)/\sqrt{5}$, $\xi = (2x - y)\sqrt{3}$ and $\eta = x + 2y$ with associated quadratic forms

$$Q(V\boldsymbol{s}) = 15s^2 + 5t^2, \quad Q(VD\boldsymbol{\xi}) = \xi^2 + \eta^2.$$

We note that both of these correspond to writing

$$Q(x, y) = 3(2x - y)^2 + (x + 2y)^2.$$

The level curves $Q(\boldsymbol{x}) = 25$ in the x-y plane, $Q(V\boldsymbol{s}) = 25$ in the s-t plane and $Q(VD\boldsymbol{\xi}) = 1$ in the ξ-η plane are shown in Fig. B.1. The ellipse in the s-t plane is a rotation of the ellipse in the x-y plane anticlockwise through the angle $\tan^{-1}(1/2)$. The eigenvectors $\boldsymbol{v}_1$ and $\boldsymbol{v}_2$ can be seen to be directed along the minor and major axes of the ellipse associated with $Q(\boldsymbol{x})$. $\Diamond$

Returning to the second question above, we will see that suitable transformation matrices V can be defined without knowledge of the eigenvalues or eigenvectors of A. The key element in the construction of such matrices is Sylvester's law of inertia, stated below. Two definitions will be needed beforehand.

Definition B.7 (*Congruence*) Assuming that V is a nonsingular matrix, the matrix (triple-) product $V^\mathsf{T} A V$ is called a *congruence transformation* of A.

The two matrices $V^\mathsf{T} A V$ and A are said to be congruent. The set of eigenvalues of two congruent matrices will generally be different (unless V is an orthogonal matrix). A congruence transformation does, however, retain just enough information about the eigenvalues to be useful in the current context. This information is called the *inertia* of a matrix.

Definition B.8 (*Inertia*) The *inertia* of a symmetric matrix is a triple of integers (p, z, n) giving the number of positive, zero and negative eigenvalues, respectively.

Theorem B.9 (Sylvester's law of inertia) *The inertia of a symmetric matrix is invariant under a congruence transformation.*

Sylvester's law of inertia guarantees that the qualitative nature of the level curves are invariant when the matrix is subject to congruence transformations. Thus in two dimensions the level curves of a matrix having inertia $(2, 0, 0)$ or $(0, 0, 2)$ will be ellipses while an inertia of $(1, 0, 1)$ will lead to hyperbolae.

One natural possibility for a congruence transformation is to use Gaussian elimination. When $a \neq 0$ we subtract b/a times the 1st row of A from the second row in order to create a zero in the $(2, 1)$ position. The elimination process can be represented by defining V^T so that

$$V^\mathsf{T} = \begin{bmatrix} 1 & 0 \\ -b/a & 1 \end{bmatrix}, \quad V^\mathsf{T} A = \begin{bmatrix} a & b \\ 0 & (ac - b^2)/a \end{bmatrix}, \quad V^\mathsf{T} A V = \begin{bmatrix} a & 0 \\ 0 & (ac - b^2)/a \end{bmatrix}$$

so we generate (B.13) with $\alpha = a$ and $\beta = (ac - b^2)/a$. The congruence transformation explicitly highlights the role of the discriminant $(ac - b^2)$ of the underlying quadratic form. Next, computing the inverse transformation matrix gives

$$V^{-1} = \begin{bmatrix} 1 & b/a \\ 0 & 1 \end{bmatrix},$$

and setting $s = V^{-1} x$, we find that $s = x + by/a, t = y$. Thus, from (B.13),

$$Q(x, y) = a \left(x + \frac{by}{a} \right)^2 + \left(c - \frac{b^2}{a} \right) y^2$$

which can be seen to be equivalent to "completing the square".

Example B.10 (Example B.6 revisited) Completing the square in the quadratic form associated with the matrix

$$A = \begin{bmatrix} 13 & -4 \\ -4 & 7 \end{bmatrix}$$

we find that

$$Q(x, y) = 13x^2 - 8xy + 7y^2 = 13 \left(x - \frac{4}{13} y \right)^2 + \frac{75}{13} y^2.$$

This suggests an alternative mapping from (x, y) to (s, t) via $s = x - (4/13) y$ $(= x + by/a)$, $t = y$. The change of variables leads to

$$Q(Vs) = 13s^2 + \frac{75}{13}t^2,$$

and defines the ellipse in the s-t plane that is associated with the Gaussian elimination congruence transformation $V^T A V$. Making the further scaling $\xi = s/\sqrt{13}$, $\eta = t\sqrt{13/75}$ gives $Q(V D\xi) = \xi^2 + \eta^2$, whose level curves in the ξ-η plane are again circles. ◊

B.5 Inverse Monotonicity for Matrices

A real matrix A or a real vector x which has entries that are all nonnegative numbers is called a *nonnegative* matrix or vector. They can be identified by writing $A \geq 0$ and $x \geq 0$. The matrix interpretation of an inverse monotone discrete operator is called a monotone matrix.

Definition B.11 (*Monotone matrix*) A nonsingular real $n \times n$ matrix A is said to be *monotone* if the inverse matrix A^{-1} is nonnegative: equivalently, $Ax \geq 0$ implies that $x \geq 0$.

The standard way of showing that a matrix is monotone is to show that it is an M-matrix.

Definition B.12 (*M-matrix*) A real $n \times n$ nonsingular matrix A is an M-matrix if

(a) $a_{ij} \leq 0$ for $i \neq j$ (this means that A is a Z-matrix), and
(b) the real part of every eigenvalue of A is positive.

An M-matrix is guaranteed to be monotone. To expand on condition (b): first, a symmetric matrix A has real eigenvalues. They are all positive numbers if and only if the quadratic form $Q(x) = x^T A x$ is positive for every nonzero vector x (such a matrix is said to be *positive definite*). Second, an irreducible nonsymmetric matrix will satisfy (b) if it is also diagonally dominant; that is if $a_{ii} \geq \sum_{j=1, j \neq i}^{n} |a_{ij}|$ with strict inequality for at least one value of i. (This result immediately follows from the Gershgorin circle theorem.[4])

---

[4]This states that: let $r_i = \sum_{i \neq j} |a_{i,j}|$ denote the sum of the moduli of the off-diagonal entries of an $n \times n$ matrix A, then every eigenvalue of A lies in at least one of the disks of radius r_i centered on a_{ii}.

Exercises

B.1 Prove that an $n \times n$ matrix A is symmetric if and only if $\langle x, Ay \rangle = \langle Ax, y \rangle$ for all vectors $x, y \in \mathbb{R}^n$. (Hint: Choose $x = e_i$ and $y = e_j$, where e_k is the kth unit vector in $\mathbb{R}^n$, that is, the vectors whose only nonzero entry is one in the kth position.)

B.2 Suppose that A is an irreducible $n \times n$ tridiagonal matrix. Construct the nonsingular diagonal matrix $D = \text{diag}(1, d_2, \ldots, d_n)$ that makes the product matrix AD symmetric.

B.3 Suppose that A is a tridiagonal matrix and let D be a nonsingular diagonal matrix that makes the matrix DA symmetric. Prove that AD^{-1} is a symmetric matrix.

B.4 Suppose that the matrix A in (B.9) is irreducible and that its elements are real with $a_j/c_{j-1} > 0$ for $j = 2, 3, \ldots, n$. Show that its eigenvalues are real.
   [Hint: $Av = \lambda v$ for an eigenvector $v \in \mathbb{C}^n$ and corresponding eigenvalue λ. Now consider the inner product $\langle v, ADv \rangle$.]

B.5 Suppose that $U_m = \sin(\pi jm/(n+1))$ so that $U_0 = 0$ and $U_{n+1} = 0$ when j is an integer. Show, using the trigonometric identity for $\sin(A \pm B)$, that

$$aU_{m-1} + bU_m + cU_{m+1} = \lambda_j U_m,$$

where λ_j is the jth eigenvalue of the tridiagonal matrix (B.10).

B.6 Sketch the level curves of the quadratic form $Q(x)$ when

$$A = \begin{bmatrix} -10 & 10 \\ 10 & 5 \end{bmatrix}.$$

B.7 Consider a quadratic form $Q(x)$ of the form (B.12) with $a \geq c > 0$ and $b^2 < ac$. Suppose that $\hat{a} = \frac{1}{2}(a+c)$ and that r, θ are defined by $a = \hat{a}(1 + r\cos\theta)$, $c = \hat{a}(1 - r\cos\theta)$ and $b = \hat{a}r\sin\theta$ $(0 \leq r < 1, -\pi/2 \leq \theta \leq \pi/2)$. Show that

$$r^2 = 1 - 4\frac{ac - b^2}{(a+c)^2}$$

and that the matrix of the quadratic form has eigenvalues $\lambda_\pm = \hat{a}(1 \pm r)$ with corresponding eigenvectors

$$v_+ = (\cos\tfrac{1}{2}\theta, \sin\tfrac{1}{2}\theta), \quad v_- = (\sin\tfrac{1}{2}\theta, -\cos\tfrac{1}{2}\theta),$$

where $\tan\theta = 2b/(a-c)$.
   If the major axis of the ellipse $Q(x) = \text{constant}$ makes an angle ϕ with the y-axis, show that $-\pi/4 \leq \phi \leq \pi/4$.

B.8 This builds on the previous exercise. Attention is drawn in Sect. 10.2.5 to quadratic forms where the coefficients satisfy $b^2 < ac \le a|b|$. Show that the minimum value of r, subject to these inequalities, is $1/\sqrt{2}$. Hence show that the ratio of the lengths of the two axes of the ellipse $Q = $ constant (that is, $\sqrt{\lambda_+/\lambda_-}$) is greater than, or equal to, $\sqrt{2} + 1$.

B.9 Find the stationary points of the function $\phi(x, y) = 2x + 2ye^x + y^2$.

The leading terms in the Taylor expansion of a function of two variables about a point $x = a$ may be expressed in the matrix-vector form

$$\phi(x) = \phi(a) + (x - a)^\mathsf{T} g + (x - a)^\mathsf{T} H(x - a) + \cdots$$

where $x = (x, y)^\mathsf{T}$, $g = (\phi_x(a), \phi_y(a))^\mathsf{T}$ is the gradient of ϕ at a and

$$H = \begin{bmatrix} \phi_{xx}(a) & \phi_{xy}(a) \\ \phi_{yx}(a) & \phi_{yy}(a) \end{bmatrix}$$

is the matrix of second derivatives (or Hessian matrix) evaluated at a. Use the expansion to determine whether the stationary points of $\phi(x, y)$ are maxima, minima or saddle points.

Appendix C
Integral Theorems

This appendix reviews the most important results in vector calculus: these are generalisations of the *fundamental theorem of integral calculus*, that is

$$\int_a^b f'(x)\,dx = f(b) - f(a).$$

Theorem C.1 *Suppose that* $\Omega \subset \mathbb{R}^3$ *is a closed bounded region with a piecewise smooth boundary* $\partial\Omega$ *and let* $\vec{n}$ *denote the unit outward normal vector to* $\partial\Omega$. *If* ϕ *and* $\vec{F}$ *denote, respectively, scalar and vector fields defined on a region that contains* Ω, *then*

$$(i) \quad \iiint_\Omega \operatorname{grad} \phi \, dV = \iint_{\partial\Omega} \phi\, \vec{n} \, dS$$

$$(ii) \quad \iiint_\Omega \operatorname{div} \vec{F} \, dV = \iint_{\partial\Omega} \vec{F} \cdot \vec{n} \, dS$$

$$(iii) \quad \iiint_\Omega \operatorname{curl} \vec{F} \, dV = \iint_{\partial\Omega} \vec{n} \times \vec{F} \, dS.$$

These three identities are associated with the famous names of George Green and Carl Friedrich Gauss. The result (ii) is perhaps of greatest importance in applications and is often referred to as the *divergence theorem*. What might not be obvious is that if any one of the identities in Theorem C.1 is true then the other parts follow. This is established below. It will be assumed that, in Cartesian coordinates, $\vec{F} = [F_x, F_y, F_z]$ with a similar notation for $\vec{n}$.

© Springer International Publishing Switzerland 2015
D.F. Griffiths et al., *Essential Partial Differential Equations*, Springer
Undergraduate Mathematics Series, DOI 10.1007/978-3-319-22569-2

- (i)⇒(ii) If we take $\phi = F_x$ and write down the first (or the x-) component of the vector-valued identity (i), we get

$$\iiint_\Omega \frac{\partial F_x}{\partial x} \, dV = \iint_{\partial\Omega} F_x n_x \, dS$$

where n_x is the first component of $\vec{n}$. Moreover, taking $\phi = F_y$ and $\phi = F_z$ and writing down the second and third components of $\vec{n}$ gives

$$\iiint_\Omega \frac{\partial F_y}{\partial y} \, dV = \iint_{\partial\Omega} F_y n_y \, dS, \quad \iiint_\Omega \frac{\partial F_z}{\partial z} \, dV = \iint_{\partial\Omega} F_z n_z \, dS.$$

Adding these results gives identity (ii) written in Cartesian coordinates.
- (ii)⇒(i) If we take $\vec{F} = [\phi, 0, 0]$ then (ii) gives

$$\iiint_V \frac{\partial \phi}{\partial x} \, dV = \iint_S \phi n_x \, dS,$$

which is the first component of result (i). Taking $\vec{F} = [0, \phi, 0]$ and $\vec{F} = [0, 0, \phi]$ give the second and third components. Hence, (i) follows from (ii).
- (i)⇒(iii) The first component of (iii) is

$$\iiint_V \left(\frac{\partial F_z}{\partial y} - \frac{\partial F_y}{\partial z} \right) dV = \iint_S \left(F_z n_y - F_y n_z \right) dS.$$

To establish this result, the second component of (i) with $\phi = F_z$ must be subtracted from the third component of (i) with $\phi = F_y$, that is

$$\iiint_V \frac{\partial F_z}{\partial y} \, dV = \iint_S F_z n_y \, dS \quad \text{and} \quad \iiint_V \frac{\partial F_y}{\partial z} \, dV = \iint_S F_y n_z \, dS.$$

The other two components of (iii) can be established in exactly the same way.
- (iii)⇒(i) Suppose that $\vec{v}$ is a constant vector. Taking $\vec{F} = \phi\vec{v}$ in identity (iii) gives

$$\iiint_V \mathrm{curl}(\phi\vec{v}) \, dV = \iint_S \vec{n} \times (\vec{v}\phi) \, dS.$$

Next, using the vector identity

$$\mathrm{curl}(\phi\vec{v}) = \phi \, \mathrm{curl}\, \vec{v} - \vec{v} \times \mathrm{grad}\, \phi$$

the left hand side simplifies (since $\vec{v}$ is a constant vector, its curl is zero) to give

$$-\iiint\limits_V \vec{v} \times \text{grad}\,\phi \,dV = \iint\limits_S \vec{n} \times (\vec{v}\phi)\,dS = -\iint\limits_S \phi\vec{v} \times \vec{n}\,dS$$

where we have used $\vec{a} \times \vec{b} = -\vec{b} \times \vec{a}$ on the right hand side. Since $\vec{v}$ is a constant vector, it can be taken outside the integrals to give

$$-\vec{v} \times \iiint\limits_V \text{grad}\,\phi \,dV = -\vec{v} \times \iint\limits_S \phi\vec{n}\,dS.$$

Finally, since the above result is valid *for all* constant vectors $\vec{v}$, identity (i) must always hold. $\qquad\square$

Appendix D
Bessel Functions

The ODE

$$x^2 u'' + xu' + (x^2 - \nu^2)u = 0 \qquad \text{(D.1)}$$

defined on the semi-infinite real line $0 < x < \infty$ and involving a real parameter ν is known as Bessel's equation. This is a linear second-order ODE and therefore has two linearly independent solutions. These are usually denoted by $J_\nu(x)$ and $Y_\nu(x)$ and are referred to as Bessel functions of the first and second kind, respectively, of order ν.

The equation cannot be solved in terms of standard elementary functions and so the method of Frobenius is used to construct a series solution. That is, coefficients $\{a_n\}$ are sought such that a solution may be expressed in the form

$$u(x) = \sum_{n=0}^{\infty} a_n x^n \qquad \text{(D.2)}$$

It is sufficient for our purposes to consider the case $\nu = 0$. Thus, on substituting (D.2) into (D.1) and collecting terms with like powers of x, we find

$$x^2 u'' + xu' + x^2 u = a_1 x + \sum_{n=2}^{\infty} \left(n^2 a_n + a_{n-2} \right) x^n$$

and the right hand side is zero when $a_1 = 0$ and

$$a_n = -\frac{1}{n^2} a_{n-2}, \quad n = 2, 3, \ldots. \qquad \text{(D.3)}$$

© Springer International Publishing Switzerland 2015
D.F. Griffiths et al., *Essential Partial Differential Equations*, Springer
Undergraduate Mathematics Series, DOI 10.1007/978-3-319-22569-2

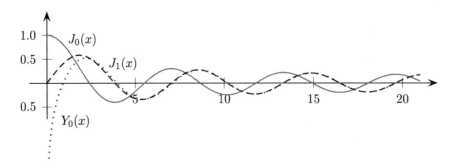

Fig. D.1 The Bessel functions $J_0(x)$ (*solid*), $J_1(x)$ (*dashed*) and $Y_0(x)$ (*dotted*) for $0 \le x \le 21$

When $n = 3, 5, \ldots$ we find that all odd-numbered terms in (D.2) vanish while, writing $n = 2m$, the even-numbered coefficients are

$$a_{2m} = (-1)^m \frac{1}{2^{2m}(m!)^2} a_0, \quad m = 0, 1, 2, \ldots \tag{D.4}$$

in which a_0 is an arbitrary constant The remaining coefficient a_0 is used to normalise the solution and, by choosing $a_0 = 1$, we find that $u(x) = J_0(x)$, where

$$J_0(x) = \sum_{n=0}^{\infty} (-1)^n \frac{1}{(n!)^2} \left(\frac{x}{2}\right)^{2n} \tag{D.5}$$

is the Bessel function of the first kind of order zero. Note that $J_0(0) = 1$ and $J_0'(0) = 0$.

Bessel functions of the second kind have the property that $Y_\nu(x) \to -\infty$ as $x \to 0^+$ and, since we make no direct use of them in this book, they will not be discussed further. Graphs of $J_0(x)$, $J_1(x)$ and $Y_0(x)$ are shown in Fig. D.1. The graphs of $J_1(x)$ and $Y_0(x)$ are indistinguishable for $x > 3$. Their oscillatory behaviour is evident and the zeros play an important role when using separation of variables, as illustrated in Example 8.5. A selection of the zeros of $J_0(x)$ are given in Table D.1. It can be shown that, for large values of x,

Table D.1 Selected zeros ξ_k of $J_0(x)$ compared with $(k - \frac{1}{4})\pi$ from the approximation (D.6) with $\nu = 0$

k	1	2	3	5	10	20
ξ_k	2.4048	5.5201	8.6537	14.9309	30.6346	62.0485
$(k - \frac{1}{4})\pi$	2.3562	5.4978	8.6394	14.9226	30.6305	62.0465
	0.0486	0.0223	0.0143	0.0084	0.0041	0.0020

The bottom row gives the difference between the preceding two rows

$$J_\nu(x) \approx \sqrt{\frac{2}{\pi x}} \cos\left(x - (\nu + \tfrac{1}{2})\frac{\pi}{2}\right), \tag{D.6}$$

which suggests that the zeros of $J_0(x)$ approach $(k - \tfrac{1}{4})\pi$ as $k \to \infty$. The table confirms that this is a reasonable approximation even for small values of k.

Bessel functions have played an important role in applied mathematics for more than a century. One reason for this prominent position is that those equations that can be solved in Cartesian coordinates by the use of sines and cosines require Bessel functions when the equations are expressed in polar coordinates. Bessel functions have been studied extensively over the years. A comprehensive review can be found in the celebrated book by G.N. Watson [27] (originally published in 1922), whereas Kreyszig [12] provides an accessible introduction to their properties.

Exercises

D.1 Use the result of Exercise 5.4 to show that solutions of (D.1) are oscillatory.

D.2 By differentiating (D.1) when $\nu = 0$, show that $u = cJ_0'(x)$ satisfies Bessel's equation with $\nu = 1$ and the initial condition $u(0) = 0$ for any constant c.

D.3 Follow the process leading to (D.5) to show that (D.1) with $\nu = 1$ has a solution $u(x) = J_1(x)$, where

$$J_1(x) = \sum_{n=0}^{\infty} (-1)^n \frac{1}{n!(n+1)!} \left(\frac{x}{2}\right)^{2n+1}. \tag{$\star$}$$

Verify, by term by term differentiation, that $J_1(x) = -J_0'(x)$, confirming the result of the previous exercise with $c = -1$.

D.4 Show that equation (D.1) with $\nu = 0$ can be rewritten in the form $(xu')' + xu = 0$. Hence show that

$$\int_0^x s J_0(s)\, ds = x J_1(x).$$

D.5 By multiplying (D.1) with $\nu = 0$ by $2u'$, establish the result

$$\frac{d}{dx}\left(x^2 u(x)'^2 + x^2 u(x)^2\right) = 2xu^2(x).$$

Then, using the fact that $u = J_0(x)$ is a solution of (D.1), show that

$$\int_0^\xi x J_0^2(x)\, dx = \tfrac{1}{2}\xi^2 J_1^2(\xi), \tag{$\ddagger$}$$

where ξ is any zero of J_0.

D.6 By making an appropriate change of variable in ($\ddagger$) establish the relation

$$\int_0^a r J_0^2(r\xi/a)\,\mathrm{d}r = \tfrac{1}{2}a^2 J_1^2(\xi).$$

D.7 Using a search engine, or otherwise, explore the *ratio test* for convergence of an infinite series. Use the ratio test to show that the series for $J_0(x)$ in equation (D.5) and $J_1(x)$ in Exercise D.3 converge for all values of x.

Appendix E
Fourier Series

Consider the case of a real-valued function $u(x)$ that is defined on the real line $-\infty < x < \infty$ and is *periodic*, of period L. Thus $u(x + L) = u(x)$ for all x and knowledge of u on any interval of length L is sufficient to define it on the entire real line. Let us suppose that $u(x)$ is to be determined on the interval $(0, L)$. Such a function $u(x)$ has a well-defined Fourier series expansion into complex exponentials (sines and cosines)

$$u(x) = \sum_{j=-\infty}^{\infty} c_j e^{2\pi i j x/L}, \tag{E.1}$$

with complex-valued coefficients c_j that are constructed to ensure that the series converges at almost every point x. Since we know (from Exercise 5.19) that the functions $\{\exp(2\pi i j x/L)\}$ $(j = 0, \pm 1, \pm 2, \ldots)$ are mutually orthogonal with respect to the complex $L_2(0, L)$ inner product (5.17), we can determine a general Fourier coefficient c_k by multiplying (E.1) by $\exp(-2\pi k x/L)$ and integrating with respect to x over the interval. This gives

$$c_k = \frac{1}{L} \int_0^L e^{-2\pi i k x/L} u(x)\, dx. \tag{E.2}$$

Expanding (E.2) in terms of sines and cosines, and then taking the complex conjugate gives

$$c_k = \frac{1}{L} \int_0^L \{\cos(2\pi i k x/L) - i \sin(2\pi i k x/L)\}\, u(x)\, dx \tag{E.3}$$

$$c_k^* = c_{-k} = \frac{1}{L} \int_0^L \{\cos(2\pi i k x/L) + i \sin(2\pi i k x/L)\}\, u(x)\, dx. \tag{E.4}$$

D.F. Griffiths et al., *Essential Partial Differential Equations*, Springer Undergraduate Mathematics Series, DOI 10.1007/978-3-319-22569-2

Combining these results gives new coefficients a_k, b_k, so that

$$a_k := c_k + c_{-k} = \frac{2}{L} \int_0^L \cos(2\pi kx/L)\, u(x)\, dx \qquad (E.5)$$

$$b_k := i\,(c_k - c_{-k}) = \frac{2}{L} \int_0^L \sin(2\pi kx/L)\, u(x)\, dx. \qquad (E.6)$$

Next, rearranging the expressions in (E.5) and (E.6) gives $c_k = \frac{1}{2}(a_k - ib_k)$ and $c_{-k} = \frac{1}{2}(a_k + ib_k)$, and substituting these expressions into (E.1) and rearranging (see Exercise E.1) generates the *standard form* (no complex numbers!)

$$u(x) = \frac{1}{2}a_0 + \sum_{k=1}^{\infty} \{a_k \cos(2\pi kx/L) + b_k \sin(2\pi kx/L)\}. \qquad (E.7)$$

Another direct consequence of the mutual orthogonality of the complex exponentials is *Parseval's relation* (see Exercise E.2)

$$\sum_{k=-\infty}^{\infty} |c_k|^2 = \frac{1}{L} \int_0^L |u(x)|^2\, dx. \qquad (E.8)$$

This implies that the Fourier coefficients are square summable if, and only if, the function is square integrable. This also shows that the coefficients a_k and b_k will need to decay with increasing k if the Fourier series (E.1) is to converge: coefficients of the form $(1, 1, 1, 1, \ldots)$ will not be allowed, whereas $(1, \frac{1}{2}, \frac{1}{3}, \frac{1}{4}, \ldots)$ will be just fine (because $1 + \frac{1}{4} + \frac{1}{9} + \frac{1}{16} + \cdots < \infty$).

Example E.1 Construct a piecewise continuous function (see Definition 8.3) that has Fourier coefficients that decay harmonically, that is $|c_k| = \frac{a}{|k|}$ for $k \neq 0$, where a is a constant (so that the series (E.1) converges).

Integrating the right hand side of (E.2) by parts we get

$$c_k = -\frac{1}{2\pi ik}\, u(x)\, e^{-2\pi ikx/L}\Big|_0^L + \frac{1}{2\pi ik} \int_0^L e^{-2\pi ikx/L} u'(x)\, dx. \qquad (E.9)$$

The factor $1/k$ in the first term of (E.9) suggests the specific choice $u(x) = x$ (the constant derivative means that the second integral is zero). Note that the periodic extension of this function is *discontinuous* at the boundary points: $u(0^+) = 0 \neq u(L^-) = L$. When these two limits are substituted into the first term in (E.9) we get $c_k = -\frac{L}{2\pi ik}$ so that $|c_k| = \frac{L}{2\pi|k|}$, as required.

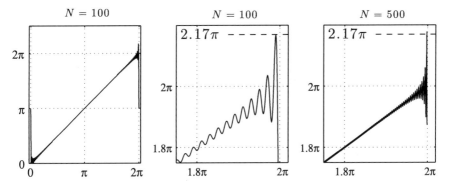

Fig. E.1 The Fourier series of the function $u(x) = x$ for $x \in (0, 2\pi)$ truncated to $N = 100$ real terms (*left*). An expanded view in the *square* around the point $(2\pi, 2\pi)$ is shown with $N = 100$ (*centre*) and $N = 500$ (*right*)

This slow decay of the coefficients has significant implications. The Nth partial sum with $L = 2\pi$ is (since $c_0 = \pi$)

$$S_N(x) = \sum_{k=-N}^{N} c_k e^{ikx} = \pi - \sum_{k=1}^{N} \frac{2}{k} \sin(kx)$$

and is shown on the left of Fig. E.1 with $N = 100$. It is seen to give a coherent approximation of $u(x) = x$ except at points close to the discontinuities at $x = 0, 2\pi$. The expanded view in the central figure reveals a highly oscillatory behaviour. When N is increased to 500, the oscillations in the rightmost figure appear to be confined to a smaller interval but their amplitude is undiminished—a dashed horizontal line is drawn at $y = 2.17\pi$ for reference. This is an illustration of Gibb's phenomenon and is a consequence of using continuous functions in order to approximate a discontinuous one. The "overshoot" in the partial sum is at least 8.5 % (closer, in fact, to 9 %) of the total jump (2π) at $x = 2\pi$. This, and other features, are succinctly explained by Körner [11, pp. 62–66]. ◊

The following theorem relates the smoothness of the function u to the rate of decay of the coefficients c_k as the index k increases.

Theorem E.2 *Let $s \in \{1, 2, \ldots\}$ be a parameter. If the function u and its first $(s - 1)$ derivatives are continuous and its sth derivative is piecewise continuous on $(-\infty, \infty)$, then the Fourier coefficients satisfy*

$$|c_k| \le a/|k|^{s+1} \tag{E.10}$$

for $k \ne 0$, where a is a constant.

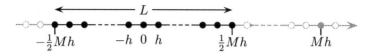

Fig. E.2 Grid points over a full period $-\frac{1}{2}Mh \le x \le \frac{1}{2}Mh$ when M is even

Proof Suppose that $s = 1$. The function u is continuous (and periodic) so $\lim_{x\to 0^+} u(x) = \lim_{x\to L^-} u(x)$. Thus, unlike the previous example, the first term in (E.9) is zero, so that

$$|c_k| = \frac{1}{k} \cdot \left| \underbrace{\frac{1}{2\pi i} \int_0^L e^{-2\pi ikx/L} u'(x)\, dx}_{c_k'} \right|.$$

Note that c_k' is a scalar multiple of the Fourier coefficient of the (piecewise continuous) derivative function, so the coefficients $|c_k'|$ will decay harmonically (at worst). The general result may be obtained by repeating the above argument $s - 1$ times (each integration by parts gives another factor of k in the denominator). $\square$

Next, suppose that our periodic function u is sampled on the grid of points $x_m = mh$, $m = 0, \pm 1, \pm 2, \ldots$ where $h = L/M$ (as illustrated in Fig. E.2), so that

$$u(x_m) = \sum_{k=-\infty}^{\infty} c_k\, e^{2\pi ikm/M}. \tag{E.11}$$

The periodicity is reflected in the fact that $\exp(2\pi ik'm/M) = \exp(2\pi ikm/M)$ for $k' = k + \ell M$ where ℓ is an integer. In this situation we say that the wave numbers $2\pi k$ and $2\pi k'$ are *aliases* of each other. If we now collect all the like coefficients together, then (E.11) can be written as a finite sum of distinct Fourier modes

$$u(x_m) = \sum_{k=0}^{M-1} \tilde{c}_k\, e^{2\pi ikm/M}, \quad \tilde{c}_k = \sum_{\ell=-\infty}^{\infty} c_{k+\ell M}, \tag{E.12}$$

with modified coefficients $\tilde{c}_k$ that are periodic with period M: $\tilde{c}_{k+M} = \tilde{c}_k$ for all k. Note that if M is an even number (as illustrated above) then the function u can be defined over the interval $(-\frac{M}{2}h, \frac{M}{2}h)$ by simply summing over $M+1$ Fourier modes (instead of M modes) and halving the contribution of the first and last terms (since they are equal). This leads to the alternative representation[5]

$$u(x_m) = \sideset{}{^*}\sum_{|k|\le M/2} \tilde{c}_k\, e^{2\pi ikm/M}. \tag{E.13}$$

---

[5]The first and last term adjustment is indicated by the asterisk in the summation.

Note that if M is odd then the range of $|k|$ will be $\lfloor M/2 \rfloor$ and no adjustment of the two end contributions is needed.

The exact representation of the periodic function in (E.12) (or equivalently (E.13)) involves a summation of an infinite number of the coefficients c_k. In a practical computation the exact coefficients $\tilde{c}_k$ will be approximated by (discrete) coefficients C_k and the associated *discrete* Fourier series is constructed

$$U_m = \sum_{k=0}^{M-1} C_k e^{2\pi i k m / M}, \tag{E.14}$$

where $\{U_m\}$ are a set of periodic grid values ($U_{m+M} = U_m$ for all integers m) that approximate the exact grid values $u(x_m)$. The construction of the discrete Fourier coefficients C_k mirrors the construction used in the continuous case. All that is needed is a suitable inner product.

Theorem E.3 *The discrete Fourier modes* $\{e^{2\pi i k m / M}\}_{m=0}^{M-1}$ *associated with distinct wave numbers k are mutually orthogonal with respect to the discrete inner product*[6]

$$\langle U, V \rangle_h = \frac{1}{M} \sum_{m=0}^{M} {}'' U_m V_m^*. \tag{E.15}$$

Proof By construction

$$\left\langle e^{2\pi i k m / M}, e^{2\pi i \ell m / M} \right\rangle_h = \frac{1}{M} \sum_{m=0}^{M-1} e^{2\pi i (k-\ell) m / M} = \frac{1}{M} \sum_{m=0}^{M-1} z^m,$$

where $z = e^{2\pi i (k-\ell)/M}$ is one of the Mth roots of unity, so $z^M = 1$. The formula for the sum of a geometric progression then gives

$$\left\langle e^{2\pi i k m / M}, e^{2\pi i \ell m / M} \right\rangle_h = \begin{cases} M & \text{when } k = \ell \\ \frac{1}{M} \frac{1-z^M}{1-z} = 0 & \text{when } k \neq \ell. \end{cases}$$

$\square$

The general coefficient C_k can thus be determined by multiplying (E.14) by $\exp(-2\pi i k m / M)$ and summing over m. This gives

$$C_k = \frac{1}{M} \sum_{m=0}^{M-1} U_m e^{-2\pi i k m / M}. \tag{E.16}$$

---

[6]The associated ℓ_2 norm $\|U\|_{h,2}$ is also defined in (11.41). The primes on the summation symbol signify that the first and last terms of the sum are halved.

The discrete Fourier coefficients will be periodic: $C_{k+M} = C_k$ for all k, and mirroring the exact coefficients c_k they satisfy $C_{-k} = C_k^*$ and $C_{M-k} = C_k^*$: thus we only need half of the coefficients in order to represent a real-valued grid function U. A classic algorithm that can be used to efficiently compute the coefficients $\{C_k\}$ (in both real- and complex-valued cases) is the *Fast Fourier Transform*. The algorithm is especially effective when M is a power of 2. Further details can be found in the book by Briggs [2].

Following the argument in the lead-up to (E.13) shows that we can write

$$U_m = \sum_{|k| \le \lfloor M/2 \rfloor}^* C_k e^{2\pi i k m / M}, \tag{E.17}$$

in place of (E.14). We can also establish (see Exercise E.5) a discrete version of Parseval's relation,

$$\sum_{k=0}^{M-1} |C_k|^2 = \frac{1}{M} \sum_{m=0}^{M-1} |U_m|^2 = \|U\|_{h,2}^2, \tag{E.18}$$

which implies that the discrete ℓ_2 norm of a periodic function is intrinsically connected to the sum of squares of the discrete Fourier coefficients.

The book by Strang [20, Chap. 4] is recommended for an overview of Fourier analysis and its role in applied mathematics.

Exercises

E.1 Show also that (E.1) can be written as (E.7) when $c_k = \frac{1}{2}(a_k - ib_k)$ and $\{a_k\}$, $\{b_k\}$ are real sequences.

E.2 Show that Parseval's relation (E.8) follows from (E.2).

E.3 Verify that the discrete inner product (E.15) satisfies all the properties of the function inner product (5.17) that are listed in Exercise 5.11.

E.4 Show that the coefficients in (E.16) satisfy $C_{-\ell} = C_\ell^*$ when U is a real sequence.

E.5 Show that the discrete version of Parseval's relation (E.18) follows from (E.16) using an argument analogous to that in Exercise E.2.

References

1. U.M. Ascher, *Numerical Methods for Evolutionary Differential Equations* (SIAM, Philadelphia, 2008)
2. W.L. Briggs, V.E. Henson, *The DFT: An Owners' Manual for the Discrete Fourier Transform* (SIAM, Philadelphia, 1987)
3. R. Fletcher, D.F. Griffiths, The generalized eigenvalue problem for certain unsymmetric band matrices. Linear Algebra Appl. **29**, 139–149 (1980)
4. B. Fornberg, A finite difference method for free boundary problems. J. Comput. Appl. Math. **233**, 2831–2840 (2010)
5. N.D. Fowkes, J.J. Mahon, *An Introduction to Mathematical Modelling* (Wiley, New York, 1996)
6. G.H. Golub, C.F.V. Loan, *Matrix Computations*, 4th edn., Johns Hopkins Studies in the Mathematical Sciences (Johns Hopkins University Press, Maryland, 2012)
7. D.F. Griffiths, D.J. Higham, *Numerical Methods for Ordinary Differential Equations*, Springer Undergraduate Mathematics Series (Springer, London, 2010)
8. Iscrles A, S P. Nørsett, Order Stars (Chapman & Hall, London, 1991)
9. R. Jeltsch, J.H. Smit, Accuracy barriers of difference schemes for hyperbolic equations. SIAM J. Numer. Anal. **24**, 1–11 (1987)
10. M. Kac, Can one hear the shape of a drum? Am. Math. Mon. **73**, 1–23 (1966)
11. T.W. Körner, *Fourier Analysis* (Cambridge University Press, Cambridge, 1989)
12. E. Kreyszig, *Advanced Engineering Mathematics: International Student Version*, 10th edn. (Wiley, Hoboken, 2011)
13. B.P. Leonard, A stable and accurate convection modelling procedure based on quadratic upstream differencing. Comput. Methods Appl. Mech. Eng. **19**, 59–98 (1979)
14. R.J. LeVeque, *Finite difference methods for ordinary and partial differential equations* (SIAM, Philadelphia, 2007)
15. K.W. Morton, *Numerical Solution of Convection-Diffusion Problems* (Chapman & Hall, Philadelphia, 1996)
16. J.D. Pryce, *Numerical Solution of Sturm-Liouville Problems*, Monographs in Numerical Analysis (Oxford University Press, New York, 1993)
17. R. Rannacher, Discretization of the heat equation with singular initial data. ZAMM **62**, 346–348 (1982)
18. R.D. Richtmyer, K.W. Morton, *Difference Methods for Initial Value Problems* (Wiley Interscience, New York, 1967)
19. H.-G. Roos, M. Stynes, L. Tobiska, *Numerical Methods for Singularly Perturbed Differential Equations, Convection Diffusion and Flow Problems*, vol. 24 (Springer Series in Computational Mathematics. Springer, Berlin, 1996)

© Springer International Publishing Switzerland 2015
D.F. Griffiths et al., *Essential Partial Differential Equations*, Springer Undergraduate Mathematics Series, DOI 10.1007/978-3-319-22569-2

20. G. Strang, *Introduction to Applied Mathematics* (Wellesley-Cambridge Press, Wellesley, 1986)
21. J.C. Strikwerda, *Finite Difference Schemes and Partial Differential Equations*, 2nd edn. (SIAM, Philadelphia, 2004)
22. M. Stynes, Steady-State Convection-Diffusion Problems, in *Acta Numerica*, ed. by A. Iserles (Cambridge University Press, Cambridge, 2005), pp. 445–508
23. E. Süli, D. Mayers, *An Introduction to Numerical Analysis* (Cambridge University Press, Cambridge, 2003)
24. P.K. Sweby, High resolution schemes using flux limiters for hyperbolic conservation laws. SIAM J. Numer. Anal. **21**, 995–1011 (1984)
25. L.N. Trefethen, Group velocity in finite difference schemes. SIAM Rev. **24**, 113–136 (1982)
26. L.N. Trefethen, D. Bau III, *Numerical Linear Algebra* (SIAM, Philadelphia, 1997)
27. G.N. Watson, *A Treatise on the Theory of Bessel Functions*, 2nd edn. (Cambridge Mathematical Library. Cambridge University Press, Cambridge, 1995)

Index

© Springer International Publishing Switzerland 2015
D.F. Griffiths et al., *Essential Partial Differential Equations*, Springer
Undergraduate Mathematics Series, DOI 10.1007/978-3-319-22569-2

Printed in the United States
By Bookmasters